Praise for
Elderflora

"Jared Farmer brings both classic and state-of-the-art botany alive. Farmer shows a wonderfully deep understanding of the scientific process. He deftly communicates the research produced by generations of scientists. Moreover, he shows singular insight into *how* we do what we do—and perhaps more importantly—*why* we spend our lives studying trees."

—Hope Jahren, bestselling author of *Lab Girl*

"Farmer writes of ancient trees with a wisdom and eloquence worthy of the sacred values they have long embodied for so many people around the world. *Elderflora* is both a delight and a revelation."

—William Cronon, author of *Nature's Metropolis*

"In *Elderflora*, Farmer has created something that has never existed before. It appears to be a series of interwoven journeys—biological, cultural, scientific—spanning Earth and revealing five thousand years of living trees and our amazement in their presence. But the author's mastery is unprecedented. Beautifully told, full of pathos, this book is itself a force of nature."

—Dan Flores, *New York Times*–bestselling author of *Coyote America*

"Farmer has written a history that is as big and bold as the ancient trees at its center. These trees have stood for ages and endured the unremitting assault of modern society. In them, Farmer finds not only a rich organic archive but also the wisdom of elders—wisdom that surely deserves our heed."

—Jack E. Davis, author of *The Gulf*

"Farmer has given us a stunning, globe-spanning, deep-time history, one that is also moving and intimate: a story of ancient trees in all their beauty and complexity that is also a story of how we imagine the best and worst of ourselves. From the hills of Lebanon, to the trails of Sequoia National Park, to Polish old growth, *Elderflora* transports us, and asks us to seek a better world for trees—and ourselves." —Bathsheba Demuth, author of *Floating Coast*

"Through the engrossing stories of long-lived tree species and long-lived trees, historian Farmer immerses us in 'tree time,' vividly evoking a deeper past and a vital, enduring future."

—Michelle Nijhuis, author of *Beloved Beasts*

"While it is true that the trees have no tongues, that doesn't mean they don't speak to us. Having once cored old-growth trees on the grounds of Thomas Jefferson's Monticello to decipher riddles of our past, I know something about the remarkable stories that elderflora tell us about our environment and our history. Read Farmer's lucid and fascinating book to discover the other mysteries told by elderflora."

—Michael E. Mann, author of *The New Climate War*

ELDERFLORA

Also by Jared Farmer

Glen Canyon Dammed

On Zion's Mount

Trees in Paradise

JARED FARMER

ELDERFLORA

A MODERN HISTORY OF ANCIENT TREES

BASIC BOOKS

New York

Copyright © 2022 by Jared Farmer

Cover design by Chin-Yee Lai
Cover image: Juan Carlos Muñoz / Alamy Stock Photo
Cover copyright © 2022 by Hachette Book Group, Inc.

Hachette Book Group supports the right to free expression and the value of copyright. The purpose of copyright is to encourage writers and artists to produce the creative works that enrich our culture.

The scanning, uploading, and distribution of this book without permission is a theft of the author's intellectual property. If you would like permission to use material from the book (other than for review purposes), please contact permissions@hbgusa.com. Thank you for your support of the author's rights.

Basic Books
Hachette Book Group
1290 Avenue of the Americas, New York, NY 10104
www.basicbooks.com

Printed in the United States of America

First Edition: October 2022

Published by Basic Books, an imprint of Perseus Books, LLC, a subsidiary of Hachette Book Group, Inc. The Basic Books name and logo is a trademark of the Hachette Book Group.

The Hachette Speakers Bureau provides a wide range of authors for speaking events. To find out more, go to www.hachettespeakersbureau.com or call (866) 376-6591.

The publisher is not responsible for websites (or their content) that are not owned by the publisher.

Print book interior design by Amy Quinn.

Library of Congress Cataloging-in-Publication Data

Names: Farmer, Jared, 1974- author.
Title: Elderflora : a modern history of ancient trees / Jared Farmer.
Description: First edition. | New York : Basic Books, 2022. |
    Includes bibliographical references and indexes.
Identifiers: LCCN 2022014665 | ISBN 9780465097845 (hardcover) |
    ISBN 9780465097852 (epub)
Subjects: LCSH: Trees. | Longevity. | Civilization, Modern. |
    Dendrochronology. | Climatic changes. | Time perception. |
    Landscape assessment. | Forests and forestry. | Human ecology.
Classification: LCC QK477 .F37 2022 | DDC 582.1609—dc23/eng/20220608
LC record available at https://lccn.loc.gov/2022014665

ISBNs: 9780465097845 (hardcover), 9780465097852 (ebook)

LSC-C

Printing 1, 2022

To the caretakers,
living and dead,
of Green-Wood Cemetery

There rolls the deep where grew the tree.
	O earth, what changes hast thou seen!
	There where the long street roars, hath been
The stillness of the central sea.

—Alfred Tennyson (1850)

The wrongs done to trees, wrongs of every sort, are done in the darkness of ignorance and unbelief, for when light comes, the heart of the people is always right.

—John Muir (ca. 1900)

I read widely on the present world situation, but am not frightened or discouraged. The life of a species has been calculated to cover about 9,000,000 years, so perhaps man will have gained some necessary wisdom before he has run his course. The world has been a going concern for about three billion years. With such a long-range view, today's conflicts and confusion, however disagreeable to live with, cannot make one feel pessimistic or hopeless.

—Winifred Goldring (1950)

CONTENTS

PROLOGUE: WPN-114

ONE SUMMER DAY IN 1988, WHILE YELLOWSTONE BURNED, MY FATHER and I drove through Utah's "West Desert." Just beyond the Nevada border lay our destination: Great Basin National Park. Leaving US 50, the "Loneliest Road in America," we went up and up the Wheeler Peak Scenic Drive—a route so smooth it felt too easy—and killed the engine in a subalpine parking lot. From there, we hiked the short trail to the cirque beneath the second-highest mountain in the Silver State. We gazed at the "glacier"—not much more than a rock wall chocked with dirty ice. During its retreat to near nothingness, the cirque glacier had exposed layers of moraines. On a soilless field of quartzite blocks stood trees that looked geological as much as botanical. My father, a scientist at Brigham Young University, must have told me this population of pines postdated the Pleistocene. I was a teenager, so maybe I didn't care. I can't remember. I do recall being possessed by the peak. I desired to climb it.

A few years later, after scampering up the ridge—my first Thirteener—I left my name in the notebook in the mailbox on the summit. Two entries stuck with me. A European who had toured the charismatic red-rock country of southern Utah conveyed relief at the grayness of the Great Basin. And a local man from White Pine County, Nevada, shared his heartache at high altitude. I'm so lonely, he wrote. I just want a boyfriend.

Two decades passed, and I revisited the cirque, my attention on pines. Prior to drafting a manuscript on ancient trees, I wanted to pay respects. Also, I admit, I hoped for some kind of revelation within the rocky grove. Habits of magical thinking acquired in my religious upbringing die hard. I knew that somewhere in the shadow of the everlasting peak was a former living being, an almost sacred thing—the oldest tree ever known to science. I longed to be near, though not too close. I resisted the urge to pilgrimage to a stump. Being a professor of history, I historicized the object of my yearning as a cultural fetish.

The very next year, I returned. My heart had changed my mind. I had to see and touch the remains of the Great Basin bristlecone pine known as WPN-114.

INTRODUCTION

WHAT'S THE OLDEST KNOWN LIVING THING, AND HOW DO WE KNOW? Why should we even want to know? The explanation is a history of curiosity and care. This story has science, and religion, too. Above all, it has relationships with trees, long-term relationships, as long as long can be. Coauthored with millennial plants, my narrative about the modern past concerns the not-yet end of the planet.

TREES ARE PLANTS THAT PEOPLE CALL TREES—A TERM OF DIGNITY, NOT botany. Personification is intrinsic to treeness. A tree is a radically non-human thing, a modular organism, that humans exalt through misunderstanding as a person-like being: an individual with torso and limbs. Among personified plants, *megaflora* and *elderflora*—overlapping categories—receive the highest honors in life and the profoundest sorrows in death. People cherish big trees, old trees, and especially big old trees.

Except for when they don't.

As objects of veneration *and* vandalism, elderflora appear in the oldest surviving mythologies and the earliest extant texts. The ancient trees of the ancient past were hallowed by site-specific associations with gods or heroes, ascetics or seers. They grew—and selectively still grow—in sanctuaries. In the event of arboreal death, or arboricide, guardians planted anew and thus extended the life of the sacred site. A consecrated tree is less aged than time-honored. Its age is expressed not numerically but relationally: "as old as" or "older than."

The Promised Land provides a perfect example. Josephus was a Pharisee from Jerusalem who became a citizen of Rome, where he authored chronicles in Greek. In *Jewish Antiquities*, the expatriate rewrote the

Hebrew Bible, including the scene from Genesis in which Abraham, camped at Mamre, near Hebron, entertains three angels beneath an oak. Josephus identified the tree as "Ogyges," a primeval Grecian name that evoked the deluge that ended the Silver Age. In his other major work, a firsthand account of the Great Revolt of Jews against Romans, Josephus reported that among the present-day monuments of Hebron was a famous terebinth that had been growing since Creation. As old as Zeus, older than the Flood: the ancient historian described *two* antediluvians in the vicinity of Mamre.[1]

At some point, these two ideas fused into one organic form: the "Oak of Mamre," aka "Abraham's Oak." In the fourth century CE, the first Christian emperor, Constantine, constructed a basilica to enclose and purify the designated tree, which pagans still adored with idols and sacrifices. Thanks to imperial attention, Mamre's annual Fair of the Oak—part commerce, part religion—attracted visitors from afar.

This expression of the oldest living thing—a tree that *predated* the world—outlived its host organism. Over the centuries, the Oak of Mamre died multiple times, changed locations, occupied multiple locations simultaneously. Jews, Christians, and Muslims made it a pan-Abrahamic pilgrimage site. Crusaders and medieval pilgrims did recurrent damage by stripping organic souvenirs. In the nineteenth century, tourists with paintbrushes and cameras fixed their gaze on a picturesque Palestine oak with a trinity of trunks. When a snowstorm broke one bough, the British consulate in Jerusalem seized the opportunity. Ignoring admonitions of local Arabs, who said, reportedly, that anyone who maimed this tree would lose their firstborn son, the consul's dragoman requisitioned camel loads of oaken relics destined for London. In 1871, the rural land around the disfigured tree was purchased from Muslims by the Russian Orthodox Church for a monastery that has since been enveloped by modern Hebron, a Palestinian city in the West Bank. In the last days of the twentieth century—counting from Jesus, a near-contemporary of Josephus—the fenced-in, propped-up, steel-wrapped oak perished yet again. Its trunk finally collapsed in 2019, prompting church arborists to make soil space for the latest replacement.

Globally speaking, few believers follow the flourishing of this dry tree. The Oak of Mamre no longer has world fame, and not just because pilgrims—mainly Russians—must pass through Israeli checkpoints. There are so many more, and more "believable," ancient trees here and there, anointed by science.

THE OLDEST EXTANT LIST OF OLDEST TREES IN "THE WORLD"— specifically, the Roman world—appears in *Natural History* by Pliny, a direct contemporary of Josephus. This anthology of twenty thousand facts was cut short by the author's death, a consequence of observing too closely the Vesuvian eruption. A proto-scientist, Pliny supplemented relational dating with numerical dating when he could provide known or estimated dates of planting. In his posthumous book, the natural historian mentioned, with detectable skepticism, trees older than Rome, as old as Athens.[2]

Long after Pliny died—for a millennium and a half—post-Roman intellectuals continued to favor relational age. Only in the eighteenth century did science become the primary mechanism by which Western people assigned age value to elderflora. In the first English-language treatise on forestry, John Evelyn—a noted man of letters in early modern Britain—compiled an updated list of senior trees. To the usual consecrated plantings Evelyn added naturally occurring specimens. For his audience, "the world" now included the Americas. As part of the turn toward quantification, scholars of Evelyn's time and place began estimating a tree's life span by enumerating the layers of woody tissue called cambium. Tree age became *cambial age*, or the age of the oldest tree ring. One thousand, a number significant to rationalists and evangelicals alike, became the threshold of arboreal antiquity. Evelyn explained how the years of a woody plant could be "vulgarly reckon'd" by counting the number of "solar revolutions, or circles," inscribed within the trunk.[3] Unfortunately for a dendrophile like Evelyn, calculating from tree rings required a felled or fallen tree.

Illustrated lists of olden trees became common features of encyclopedias in the nineteenth century, as "scientific botany" became professionalized in Europe. (The word *scientist* entered English in 1834.) Whereas geologists wondered whether they could use strata fossils to date the Earth—speculations that led to counts of millions and ultimately billions of years—botanists wondered whether a single living organism could take them beyond 6,000 years, beyond biblical chronologies. Geologists used a kind of relative dating, placing events in chronological order by determining which layers of rock were older or younger. With cambium layers, botanists worked toward a goal of absolute dating. In overlapping ways, scholars of rocks and trees sought to "burst the limits of time."[4]

Using the classification and naming system developed by Linnaeus, modern natural historians studied comparative longevity in plant species, and published papers in transactions of learned societies. Research botanists—many of them employed at a new institution, the botanical garden—scrutinized long-lived European trees, especially yews, and especially churchyard yews in Britain. The nation that claimed to have invented modern science tried, with mixed results, to extract empirical evidence from its iconic conifer.

Simultaneously, field scientists in the style of Alexander von Humboldt, Prussian naturalist and international celebrity, followed routes of empire to prospect for megaflora in Africa, the Americas, and the antipodes. They assumed—incorrectly, it turned out—that bigger must mean older, and that the biggest plants would occur in the subtropical zone, where climate favored growth and where the history of human wood use was less intensive than in Eurasia. As estimated by botanical numerologists, contenders for the designation "6,000 and up" included a baobab in Senegal, a dragon tree in Tenerife, a cypress in Oaxaca—and, fleetingly, a cycad in Queensland.

Scientific travelers cleared the way for nature tourists—quintessentially modern figures. For those with knowledge and means, visiting the latest oldest monumental plant was one more experience to consume. In particular locales, notably the Sierra Nevada, where middlebrow Americans rode stagecoach through tunneled sequoias, the modern novelty of arboreal antiquity became commodified.

Parallel to scientific investigations and touristic encounters, eighteenth- and nineteenth-century Western governments instituted "scientific forestry," which instrumentalized tree age. State-sponsored forestry was the high point of sustainability, an early modern concept that grew alongside the idea of the perpetual state as eternal land manager. Imperial conquests and industrial revolutions relied on timber products: wood-stock long guns for capturing lands and peoples; naval vessels with mighty masts for transporting the enslaved and the harvests of their labor; wood-framed tenements and factories for urban proletarians; and wooden pencils for keeping records of it all on paper ledgers.

Forest engineers sorted trees into generational cohorts so that mature specimens could be harvested at optimal times on a sustained-yield basis. This rational approach to crop rotation was facilitated by a German mechanical invention, the increment borer—a hand-cranked tool that extracted core samples, measuring rods of cambial age, with minimal damage to the plants. By counting cored rings, technicians could closely approximate ages. Even as they recategorized conifers as fungible timber stock, state foresters in central Europe inventoried and reserved "remarkable" trees, especially European oak—the species associated with Zeus, Jupiter, and Thor—where morphology suggested antiquity. A secular reenchantment of individuated trees accompanied the scientific disenchantment of the "high forest."

The longest increment borer couldn't get close to the pith of the giant conifers that once grew around the Pacific Rim, from northern California to southern Chile to New Zealand to Taiwan. Botanists deduced their ages from stumps, of which there were plenty—another legacy of the nineteenth century. On the peripheries of empires, before the arrival of forest engineers from metropoles, settlers dispossessed Indigenous peoples and cleared forests with violent abandon. In the process, Westerners made a mockery of Western ideas of sustainability. The combination of settler colonialism and industrial capitalism desacralized and demolished some of the biggest, oldest trees ever known, anywhere. Like Marlow in *Heart of Darkness*, extractivists imagined they were "traveling back to the earliest beginnings of the world, when vegetation rioted on the earth and the big trees were kings."[5] Forest regicide inspired counteractive forest protection movements, some led by descendants of settlers, others by Indigenous activists.

For exploitation *and* preservation *and* investigation of old-growth trees, no region was more important than the North American West during and after US conquest. Here originated the discipline of dendrochronology, literally the study of *tree time*. The basic insight that trees can record climate signals in their growth layers is at least as old as da Vinci. Many nineteenth-century naturalists, including Thoreau, read tree rings for information. Originally, readers approached each tree as a separate text. Then, A. E. Douglass, an astronomer at the University of Arizona, compiled a cross-referenced library of core samples. He figured out how to date wood of unknown age by matching its climate-sensitive rings to those with known dates from living trees of the same species. Conifers from semiarid Arizona produced ideal *timewood*—legible, undecayed, complete from start to end. After building chronologies of Southwestern tree-ring patterns, Douglass could assign absolute dates to wooden beams in Puebloan ruins. Archaeologists celebrated proof that cambial time was calendrical time. Unlike the relative dating of rock layers, annular growth enabled annual resolution.

Douglass's star student, Edmund Schulman, went a different direction, creating networks of samples from various species and genera. In the 1940s and 1950s, Schulman took summer-long western field trips, driving rough and dusty roads, coring conifers, filling the trunk of his sedan with samples and notebooks. Through statistical analyses of aggregated data, Schulman deduced multimillennial climatic changes on the subcontinental level. In the course of his comprehensive fieldwork, he discovered pines older than sequoias. These weren't massive trees in prime locations, but stunted trees on arid, high-altitude slopes. From his data, Schulman developed an anti-Humboldtian principle: "Longevity under adversity."[6]

The scientific inquiry into cambial age that started with European yew culminated with Great Basin bristlecone pine. Smallish yews are full of ancient meanings, but aren't precisely datable by the extraction and interpretation of timewood because they hollow out. Giant sequoias are solid and datable, though only when toppled; they are far too thick for increment boring. Smallish bristlecones function like yews, inverted: precisely datable and replete with modern meanings.

A new forest of elderflora sprang to life through tree-ring science. Gnarled trees without economic value gained instant scientific utility as producers and recorders of empirical evidence. In the 1960s, researchers used cross-dated timewood from Schulman's field sites to calibrate the radiocarbon (C-14) dating method, an outgrowth of atomic weapons research. By examining dead organic matter for remaining levels of isotopic carbon—residues of past atmospheres—scientists could estimate the ages of many things, not just layered things. Once calibrated, radiocarbon dating became an invaluable tool for archaeologists, and for botanists, too, who could at long last approximate the ages of nonwoody plants, and wood with missing or unreliable rings.

Upon close inspection, the cellular layers of bristlecones comprised a codex. Climatologists found proxy data for temperature and precipitation, using them to reconstruct the regional past, and to model the hemispheric future. In the same cellular tissue, astrophysicists found isotopic signatures of radiation events of cosmogenic origin. As prehistoric living media, Great Basin bristlecone pine, *Pinus longaeva*, represents the ne plus ultra modern ancient tree.

After Schulman, the scientific search for the oldest living thing shifted away from individuality, a concept that never truly fit modular organisms. Each branching module is functionally an organism, capable of independent life (the principle upon which fruit tree propagation, an age-old horticultural practice, is based). In effect, the span of a plant's life is the time between the birth of the first module and the death of the last. Current investigations into extreme longevity thus extend to clones and clonal populations. Defined in such terms, plants called trees may be tens of thousands of years old. With a technique known as the "molecular clock," phylogeneticists can also estimate the evolutionary age of an arborescent species, the time since its speciation, which can date back tens or hundreds of millions of years. Darwin may have been correct "that not one living species will transmit its unaltered likeness to a distant futurity," but certain gymnosperms have transmitted their similitude across an ocean of time.[7] In the 1940s and again in the 1990s, in China, then Australia, botanists announced spectacular disclosures of living "dinosaur trees"—anachronisms that closely resembled fossils from the Mesozoic.

Today, dendrochronologists continue to enumerate elderflora—in swamps, in lava fields, and in human haunts. But to measure is to mortalize. The admonition of Ecclesiastes still applies: "For in much wisdom is much grief: and they who increaseth knowledge increaseth sorrow."[8] As soon as the latest oldest is discovered, it is discerned to be dying, and becomes a site tinged with mournfulness. Age fame attracts vandals as well as tourists who love things to death, or who unwittingly carry pathogens. Unlike a consecrated tree, an age-dated tree cannot be renewed. It is ancient only once. Climate change magnifies the forefeeling of loss. More than any previous generation, we know the geocoded coordinates and exact ages of thousand-year survivors—and predict their demise, and anticipate our sorrow, then feel it even sooner than anticipated.

So far, I have evaded my opening question: What is the oldest known living thing? The short answer—the one currently listed in record books—is a Great Basin bristlecone pine, dated by Edmund Schulman, in the lofty White Mountains of California. People call it Methuselah, after the biblical patriarch who lived 969 years. Its thingness has graduated to beingness. The wardens of Methuselah keep its coordinates secret to protect the pine from harm. Who in the world would dare do wrong to a time-being almost 5,000 years old? Automatically, the notion feels sacrilegious. Or, to use antique words from social anthropology, it feels taboo to desecrate a totem.

Time and again, when I thought I was researching evolutionary relicts and history of science, the sources led me to cultural relics and religious studies. No longer am I surprised to encounter secular people in search of the aura of instrumentally dated trees. Their seeking fits a deep pattern.

Humans have long set apart and named special plants; in turn, these cared-for plant-persons help humans make sense of—and make amends for—their capacity to harm. In the shadow of the sanctified tree lurks an oppositional archetype, the tree violator. He—conventionally a man—is Gilgamesh atop Mount Lebanon, pillaging a sacred cedar for a temple door. He is Aśoka, the Indian emperor who takes the path of Buddha after trying—and failing—to kill the bodhi tree. He is al-Mutawakkil, an Abbasid caliph who, as described in Persian texts, meets an untimely death after felling a monumental cypress, old as heaven, planted by Zoroaster.[9] He is Rātā, a Polynesian hero who levels a regal tree to carve a canoe without asking its permission, which angers the gods, who use an incantation to restore the plant. And he is Erysichthon, the impious prince who orders the cutting of Demeter's grove. As canonized by the Roman poet Ovid, this mythological Greek figure earns comeuppance. Rebuking his servants for their refusal to complete the desecration, Erysichthon grabs an axe and hacks down the biggest, eldest oak—a grove in itself. He cuts short the life of the tree's coeval dryad, though not before she curses the waster with insatiable hunger.[10]

With law as with literature: Sanctions against cutting consecrated trees are among the earliest recorded acts of nature protection, appearing independently in Warring States China and classical Greece. Much of what we know about Greek "sacred groves"—surprisingly little—comes from inscriptions on boundary stones. The number of regulations, prohibitions, and penalties suggests that not all Grecians were sufficiently superstitious about deities and dryads to refrain from poaching trees.

Local histories of tree care, including horticulture, coexist with global legacies of deforestation. A million years ago, hominins joined bark beetles as top predators of woody plants (lignophytes). The original technology for terraforming was fire. In the Late Pleistocene, small groups with firesticks catalyzed widespread extinctions; clearance through burning consumed indiscriminately. In the long aftermath, though, traditional practices of swidden favored many types of trees by reducing competition and lowering the probability of intense fire. Only in recent centuries, in combination with fossil fuel burning, have people abandoned agroforestry and returned to scorched-earth clearance. In the tropics, cash-crop farmers torch biodiverse forests to make room for a few domesticated species. Even with state forestry, and large-scale tree-planting initiatives, the planet's total woody phytomass is lesser now than in the immediate postglacial moment.

As much as people have combusted forests for land, they have plundered them for beams. John Evelyn decried "the universal waste and destruction of timber trees."[11] Resinous wood is the organic equivalent of gold—chemical perfection. After the invention of cutting tools, humans coveted the best timbers for prominent purposes. Utilization and veneration have been reciprocal. In various premodern traditions, people conciliated megaflora through ceremony before striking them down. Cultural groups can revere individuated trees—especially those planted on burial grounds—while deforesting whole landscapes, as seen in East Asian history. Agriculturalists converted the lowlands of the Yellow River drainage into fields and paddies many thousands of years ago; in response to a wood crisis in the eleventh century CE, emperors turned to the mountainous south, the Yangtze River drainage, where their conscripts extracted old-growth lumber for palaces in Beijing. Through tax policies, the state encouraged the creation of replacement plantations of merchantable Chinese fir.

Across the world, the felling of one old tree can be auspicious or scandalous, whereas the harvesting of millions of wood-producing plants is unnoteworthy. In the year of my birth, a paper company advertised this truism: "Most people feel that trees are sacred. But if you think about it, some trees have to be ordinary."[12]

Paradoxically, elderflora benefit from the absence of care as well as caretaking. A foundational work of Chinese philosophy, the *Zhuangzi*, contains the recurring allegory of the twisted old tree that woodworkers could hew, but see no worth in hewing. The philosopher praises nonutility as the secret to longevity. An unchoppable tree can live out the years allotted by heaven and thus become a shrine. People should, the philosopher says, plant and tend such worthless things.[13]

Another paradox: Caring for elderflora does not track with eldercare. Because gnarled trees possess personhood without bodily mortality—and because they embody oldness without elderliness—they elicit wonder and esteem, unlike hunched bodies of old people, objects of pity and contempt. Prejudice against the elderly is the norm rather than the exception in recorded history, despite the universal recognition that age and wisdom have a positive correlation.

One more paradox: Elderflora are objects of cross-cultural respect and subjects of intercultural violation. In periods of conflict, people may uproot the oldest things of their antagonists—even if their own scriptures prohibit such acts. There are further canonical stories of men attacking trees. Lucan, a poet in Ovid's shadow, gave an account, apocryphal though not unbelievable, of Julius Caesar, on campaign in Gaul, sentencing a Celtic grove to death. When his soldiers shrink in aversion, Caesar snatches an axe, makes the first gash in a towering oak, and taunts the Druids with blasphemous words.[14] In post-Roman Europe, Christian missionaries struck down thousands of consecrated trees and sacred pillars in the shapes of trees, all the while worshiping the cross. Their world-felling project took more than a millennium. In the meantime, many pagan trees regenerated as Catholic sites of pilgrimage.

Before science and technology, prostrating an Old One was a panic-inducing as well as time-consuming task. To commit such an act required either profanation or propitiation. In the chapters that follow, I share a few extraordinary instances of extractive loggers who refused to maim old heartwood for "irrational" reasons. For the most part, post-Enlightenment Europeans and their colonists learned to shed their fearful reverence for megaflora and elderflora, and to replace those sentiments with rationalism and management (in the case of state forests), or curiosity and care (in the case of natural monuments and national parks). In total, the forest area managed or preserved, though significant, was vastly smaller than the area cleared.

By the nineteenth century, Europeans had reduced Europe's "primeval forest" to a pocket of Russia-partitioned Poland. In this context, Protestant factualists in Germany and Britain assuredly categorized "tree worship" as a relic of "primitive man." James George Frazer, author of the unlikely best seller *The Golden Bough* (1890), posited an evolutionary theory of human consciousness. He imagined a progressive evolution from animism to polytheism to monotheism to secularism. Or, in simpler terms, from magic to religion to science. At his career's end, the Cambridge don lamented all the "wasted time" that humans had spent on irrational behaviors derived from inherited superstitions.[15]

According to scholars like Frazer, ancient cultists had made a category mistake, believing that the sacred tree *itself* was divine. Demonstrating more "advanced" thinking, Greeks and Romans had sanctified groups of trees as properties of the gods. As interpreted by professors of mythology, Catholicism had been a backward step into idolatry, for the cult of the cross renewed the cult of the tree. Victorians missed something that seems obvious in retrospect. Beyond Europe, traditional sacred trees persisted in modernized settings. In Japan, for example, the Meiji state co-opted Shinto shrine groves. Consecrated megaflora survive today in countries as varied as Turkey (tied off with rags), India (wrapped in cotton strings), and South Korea (cordoned with rice straw ropes).

Frazer was similarly blind to all the *secular sacred trees* recognized through science, governance, and tourism. Nineteenth-century Europeans fetishized ancient monuments—archaeological, architectural, arboreal. Their modernity generated a new relationship with antiquity. In the decades before World War II, industrial states, starting with Germany, inventoried old and otherwise remarkable trees, and facilitated their protection. Conservationists from the great powers conceived of the world as a synchronized political sphere, and imagined the senior organic monuments of each nation—validated by science—as separate, equivalent patrimonies. After World War II, parallel to the invention of "world heritage," NGOs and international conventions codified legal norms for small-scale protected areas. Today, on the redwood coast of California, tourists visit "cathedral groves" in UNESCO-listed parks—sanctuaries of the state. On the opposite, dry side of the Golden State, scientific pilgrims follow the path of a professor, Edmund Schulman, to experience the numen of the "Methuselah Grove," a national reserve.

In short, the sacred tree has survived, both materially and culturally, as latter-day tradition. *The Golden Bough* provided some of the fertilizer. Read out of context, it served as a sourcebook for twentieth-century antimoderns, such as neopagans, who wished to revive nature cults. Frazer's mythopoetic influence registers in Tolkien, who gave life to the Ents, the eldest beings of Middle Earth, a primordial race of giant trees. Fantasy fictions—mythologies for modernity—commonly feature tree worship. Tourists can visit an old-growth cypress stand in Japan that informed the anime *Princess Mononoke*, or visit Walt Disney World Resort to see physical simulacra of the digital trees in the sci-fi action film *Avatar*.

Fallacies and fantasies riddle the sources on elderflora, including authoritative tomes. The most common misbeliefs concern age. Relatively few plants can be dated chronometrically. Famous trees claimed to be extremely old—or *exactly this old*—are predominantly not, and the majority of current record holders lacked any fame until instrumental dating. Likewise, any tree-specific place attachment asserted to be ancient is most likely to be modern, though the impulse to affiliate with hoary plants is as old as history.

There's no need to posit an innate, evolutionary urge for these affiliations. Memes without genes explains enough. In Frazerian terms, the automatic emotions people have about Methuselah, the bristlecone pine, are cultural transmissions more than primitive relics. The survival of the oldest requires continuity in literature and law as well as longevity in plants. It requires stories, passed down, of veneration and violation. Stories, like trees, are among the most enduring things in creation.

Despite this cross-cultural (and somewhat conjectural) context, elaborated in the opening chapter, I make no claims to tell a universal story. Rather, I take a Western narrative—the scientific search for the world's oldest living thing—as far as it goes, to someplace unfamiliar, even weird. The destination is more-than-human timefulness. Although the globalization of Western temporalities—linear time, historicism, millenarianism, biblical and civilizational and geological timescales, the Gregorian calendar, Coordinated Universal Time—informs my story, global history is not my ambition. I am attempting a different genre: *place-based planetary history.*

In my usage, *world* and *globe* and *planet* and *Earth* are each distinct. The Earth is singular, a celestial sphere. Its past contains a multitude of habitable planets, or biospheres. The planet always changes, more often slowly, sometimes incredibly fast. In its remaining eons, many additional planets will transpire on Earth. The biosphere into which *Homo sapiens* came into its own—the Holocene epoch within the Cenozoic era—hosted an incredible diversity of human worlds, or worldviews: cultural systems for understanding humanity's place in the cosmos. In the last half millennium of globalization, imperial and (inter-)national worlds ferociously subsumed local worlds and local temporalities. Historical tensions between the local and the global affect current international efforts to address planetary change.

It's no coincidence that curiosity and care for the "oldest living" tracks with Western colonialism and fossil fuel capitalism, two great destroyers of oldness. Even as traditional knowledge—and the elders who kept it alive—became disposable in modernity, elderflora became more valuable to moderns. Numerical validation of arboreal longevity, and preservation of emblematic specimens, occurred against a backdrop of social and ecological transformation. Various old economies, old languages, old habitats, and old species succumbed to violence. Many other olden things persisted, or changed; we should always be skeptical about modernity's overinsistence on its own novelty, its prideful rejection of the dictum in Ecclesiastes that "there is no new thing under the sun."[16] In the end, however, Western obsessions with discovery, accumulation, and the next new thing did produce something measurably novel: a carbon cycle transformed by human activity.

For the modern era, historians have narrated various "long" centuries, with the "long nineteenth century" generally defined as the span between the French Revolution and World War I. My book concerns the *longest nineteenth century*: the period when planetary age, evolutionary age, and arboreal age pulled consciousness far backward in linear time, and, simultaneously, when the energy transition to fossil fuels hurtled human impacts far into the future.

As an environmental historian, I approach long-term, multiscalar thinking through long-lived trees. They are hyperlocals, ultraterrestrials, and supermortals all at once. To play with Einsteinian language: these plants exist in *placetime*. They allow a story in which time operates at multiple speeds at once—the speeds of geology, evolution, and history. Old species have lived through many planets, one place at a time. They are the long ago here. Current old specimens have recorded in their rings the revolution of anthropogenic climate forcing. These keepers of time allow me to narrate the remote, abstract scale of the carbon cycle alongside the intimate, material scale of the churchyard. Each of my chapters shifts between the micro (hyperlocal plants and their ecosystems), the meso (empires, states, nations, bioregions), and the macro (the biosphere). Across temporalties, I bring plants and peoples together to show global connections and planetary consequences.

By sharing stories of human relationships with alien organisms—for ultraterrestrials experience time and place (time *in* place, place *in* time) in ways so earthly they seem otherworldly—I hope to say something hopeful, or at least anti-hopeless, about linear time, including the future. Instead of the sacred linearity of Christian time or the secular linearity of geological time, both defined by catastrophe, I want to emphasize the potential sacrality of elongating the now, postponing the end, and abiding in uncertainty. I think ahead to the next new world, after fossil fuel capitalism, when gardens must grow in our ruins.

As long as I could, I resisted calling myself a "tree guy." I worried that my appreciation for ancient trees could be construed as an apology for the Western men who have studied them. There's something juvenile about chronicles of Oldest! Biggest! Most Extreme! And something risible (and phallic) about the history of naturalists measuring girths to estimate ages. My dendrophilia grew in spite of my research. I became irritated, then exhausted, by various nineteenth-century types—pedantic antiquarians, genteel tourists, frontier hucksters, empire foresters—who waxed about big old trees. It bothered me that moderns fell back on race, nation, civilization—and mawkish mysticism—to give meaning to elderflora even after the insights of dendrochronology.

Rather than renouncing these troubled literatures, including their personification of plants, I decided to stay with them, and work through them. My historical cast—despite a surplus of dead white men and deplorable characters—contains people who tried their best, in their time and place, to think beyond themselves, beyond our species. Olden boughs—and their placetime—allow contemplations that are truly and maturely mind altering. I found vegetal philosophy amid economic botany and tree-ring science. I found *treehood* within and beyond anthropomorphic treeness. The plants that people call "ancient trees" have lives of their own; they transcend ideologies and methodologies; they outlast us all.

It bears mentioning that I identify with the Great Basin—bristlecone country—a land that makes me feel at home like no other. I grew up in Utah as a descendant of colonizers of English and German derivation. Because sections of my book, which I began outlining in Brooklyn, relate to the global influence of peoples and ideas from England and Germany in the nineteenth century, and because my narrative reaches its climax in the twentieth-century Great Basin with a protagonist from Brooklyn, I suppose what follows may seem personally overdetermined, when in fact it surprised and challenged me at every turn. For compelling me to rethink my own place in time, I give gratitude to gymnosperms great and small.

THIS IS WHERE BIOLOGY COMES IN. ALTHOUGH "OLDEST LIVING THING" is a signifier, it cannot be reduced to semiotics. The idea needs matter— organic matter. History gets the historian only so far. People may construct the meaning of "trees" and assign age value to the vascular plants they call "ancient trees," but people cannot themselves create life that grows in place for centuries. Exclusively, solar-powered organisms enact that miracle. Among plants, there are ephemerals, annuals, biennials, perennials—and, beyond them all, *perdurables*, thousand-year woody life-forms.

As a rule, gymnosperms (flowerless plants with naked seeds) grow slower and live longer than angiosperms (flowering plants with fruits). Gymnosperms include ginkgo (a genus of one), cycads, and every kind of conifer—including yews, pines, firs, spruces, cedars, redwoods, cypresses, podocarps, and araucarias. All these lineages began hundreds of millions of years before the divergence of angiosperms. In effect, the newer, faster competition forced slow growers to retreat to exposed sites and poor soils, adverse niches conducive to oldness. Five thousand years is the approximate limit for nonclonal living under adversity. In plants, the potential for extreme longevity seems to be an evolutionary holdover from the deep past. Only about twenty-five plant species can, without human assistance, produce organisms that live beyond one millennium, and they are mainly conifers of primeval lineage. The cypress family contains the most perdurables, followed by the pine family. Many relict conifers hang on in limited, vulnerable habitats. The ice ages didn't help their cause. In general, neither did humans, with their technologies of fire, domestication, and metalworking. Of some six hundred conifer species, roughly one-third are endangered, with many genera reduced to a single species.

A gymnosperm doesn't so much live long as die longer—or, live longer through dying. The interior dead wood—the heartwood—performs vital functions, mechanically and structurally. In comparison, the thin living outer layer is open to the elements. If damaged by an extrinsic event such as fire or lightning, this periderm doesn't heal or scar like animal skin. Instead, new cambium covers the injury, absorbing it as one more historical record alongside its growth rings. Thus, an ancient conifer is neither timeless nor deathless, but timeful and deathful. A few special conifers such as bristlecone pine can live through sequential, sectorial deaths—compartmentalizing their external afflictions, shutting down, section by section, producing fertile cones for an extra millennium with the sustenance of a solitary strip of bark. The final cambium has vitality like the first. Longevity doesn't suppress fecundity. Unlike animals, plants don't accumulate proteins that lead to degenerative diseases.

The strongest correlation with long life (elongated death) is chemical. Longevous conifers produce copious resins—volatile, aromatic hydrocarbons like terpenes—that inhibit fungal rot and insect predation. Chemically, bristlecone is off the charts. Its high-elevation habitat offers additional protection from enemies, competitors, and fire, given that they tolerate dryness and cold. In habitats with chronic stress, conifers grow slower and stockier. Slow woody growth generates more lignin, another organic polymer with defensive properties. Stress-tolerant plants prioritize stability over size. Their stuntedness is equal parts adaptation and tribulation.

If dwarfism provides one path to longevity, gigantism provides another. Megaconifers such as sequoia in California, alerce in Chile, and kauri in New Zealand are more likely to survive any discrete attack. In cycles of forest regeneration, they tend to be first and last. They grow quickly as seedlings, establishing soil space, then keep on growing vertically, claiming canopy space. Long life is necessary because opportunities to establish new populations—conditions following intense disturbances—occur infrequently. This strategy comes with one big downside: the risk of falling. Gravity can be as fatal as rot. Another trade-off, the burden of transporting water skyward, explains why the tallest conifers occur in humid temperate zones.

Regrowth is another pathway to oldness, an adaptation that appears in both gymnosperms and angiosperms. Certain single-boughed woody species—notably ginkgo, redwood, yew, olive—can recover from catastrophic damage, even the death of the bole. These trees never lose their ability to resprout and regenerate. At the organismal level, they do not senesce, meaning they don't lose vitality with age. In theory, such a plant is internally capable of immortality, though some external force inevitably ends its life. With particular species and cultivars, humans can force rejuvenation through grafting, pollarding, or coppicing. Plants that normally die young may live long under horticultural care.

Lastly, some perdurables, mainly angiosperms, grow as clonal colonies with many trunks, invisibly connected; their longevity is contained in the root mass—and the genome—rather than in the cambium aboveground. The clonal age is estimated from the date of sexual germination of the common ancestor. For example, a colony of quaking aspen may live—or, keep on dying—for tens of thousands of years.

Even nonclonal trees are de facto colonies and superorganisms. Treehood is as multitudinous as treeness is reductive. The underworld plant—physiologically and philosophically, the most generative part—is a networked system of roots that communicate and share with other plants through symbiotic associations with fungi. Meanwhile, aboveground, each branch is effectively its own tree with its own history, as recorded in its unique shape. Modular organisms allow for infinite variety through plasticity. After a sudden disturbance, a tree can alter its growth trajectory. On a longer timescale, a population of woody plants can shift from arborescence to shrubbiness (or vice versa) in response to incremental change.

Relatively rapid microevolution at the population level belies the slowness of macroevolution at the species level. To be a woody plant is to be changeful and changeless at once, in different dimensions. On a twin-boughed bristlecone, the two stems are separated by four millennia of somatic mutations, yet that divergence is subtle compared to the fast-evolving bacteria and fungi that live on those branches, changing over millions of generations, all the while trying and largely failing to eat pinewood that might as well be wood from the geologic past. A half billion years after the development of lignin—one of the key outcomes of evolutionary history—lignophytes retain advantages.

The price of longevity is immobility. At the organismal level, a plant cannot migrate like an animal. Its localism is total. Trees take what comes until something indomitable comes along. Extrinsic mortality may result from a distinct catastrophe, such as fire or gale, or multiple, cumulative stressors. There are limits beyond which even the most deeply rooted organisms can no longer function. Thresholds of water, salinity, and temperature are absolute thresholds.

Climatic stability promotes arboreal longevity. The very recent geologic past was a favorable time for elderflora, big and small. According to computer models informed by tree-ring data, the immediate future will bring increased variability, extremity, and precarity. The famous "hockey stick" graph showing global mean temperature increase is one of many steep curves—population growth, resource use, methane and CO_2

emissions, ocean acidification, and more—cumulatively known as the postwar "Great Acceleration."

The flipside is the *great diminution*: fewer big trees and big animals; fewer old trees and old-growth forests; fewer old species and species overall. In biblical terms, humans multiplied and deplenished the Earth. Even protected areas are porous to pollution, desertification, illegal logging, invasive pests and pathogens, stronger storms, hotter droughts, rising temperatures, rising seas. Old Ones, wild and domesticated, face new vulnerabilities. In the US West, past efforts to suppress wildfire have backfired spectacularly, priming forests for destructive megafires. Today's elders of the land, monuments of lost climates, perish before their times—displaced by our times.

Forest dieback means a decline in *chronodiversity*. This concept relates to biodiversity, or the variety of life on Earth, crudely measured as the number of species, or "species richness." Conservation biologists make a precautionary assertion about that enumeration. They argue that safeguarding the maximum possible amount of genetic information created over millions of years of evolutionary history is wise and moral. The complement of species richness is *temporal richness*. The biosphere has further possibilities if it contains species of various evolutionary ages, species of various life strategies and life spans, and specimens of various ages within species. It is an ecological loss of doubled magnitude when a species-rich, age-rich rain forest becomes row upon row of monocultural, monochronic crops. Over the past two centuries, states and corporations— often working against local users and Indigenous activists—have divided the forested areas of the globe into binary zones: large industrial plantations of "ordinary" young trees and small inviolate preserves of "extraordinary" old trees. Before the awareness of anthropogenic climate change, preservation through segregation had its own logic—the logic of permanence. National groves were supposed to last forever. Now, in a time of fateful dynamism, these outdoor museums of olden trees may be doomed by their fixity.

Does a naturally occurring tree of great age have value in itself? Foresters and forest ecologists have long debated this question. A century ago, technicians used words like "overage," "overmature," and "decadent" to describe standing timber past its prime. Commercial managers saw tree life as individual and rotational, and considered postmerchantable growth to be a biological waste of time. Their business—international markets for wood products—encouraged uniformity in age and size. By contrast, forest ecologists studied the communities in, on, and under each tree—each a world in itself—and saw forest life as processual. The cycle of life required dead and dying trees. Today, foresters meet ecologists halfway: old trees provide nutrient cycling, carbon storage, and other "ecosystem services."

Assuming they don't turn to ashes in the ever-warmer meantime.

PERDURABLES ARE MORE THAN SERVICE PROVIDERS. THEY ARE ETHICAL gift givers. They invite us to be fully human—truly sapient—by engaging our deepest faculties: to venerate, to analyze, to meditate. They expand our moral and temporal imaginations. This end-all argument, at once pragmatic and philosophical, is a matter of time.

Big time is the age of the universe, well over 10 billion years; a cosmic narrative that begins with the Big Bang and ends with endless entropy, or something more mysterious. This timescale is arcane—the Einsteinian realm of cosmology.

Deep time is the age of the Earth, beyond 4 billion years; a linear story, a geohistory, visualized vertically, layers upon layers—a stratigraphy of time. The geological timescale was the great collaborative project of science in the eighteenth and nineteenth centuries.

Long time is more than a human lifetime—even counting the doubling of life expectancies in the recent past. It exceeds the span between deceased grandparents and unborn grandchildren. At the upper limit, it's equivalent to the current geological epoch, the Holocene, the past eleven millennia when ice retreated and humans proliferated. This temporality is predictive as well as historical, covering the planetary future of humans as well as the human past. It's the timescale of intergenerational ethics and Earth system modeling—a key collaborative project of science in the current moment.

Short time is the scale of business cycles and news cycles, the user periods of disposable products, the half-lives of fads and online memes—durations of several years, a few financial quarters, or mere days. Even "long-term shareholder value" runs quite short. The shortening of popular culture is a signal accomplishment of consumer capitalism (and now surveillance capitalism) in tandem with Americanization.

Of these temporalities, only big time seems self-contained anymore. The other three collapse into each other. Or melt. The deep, the long, and the short dissolve into one time frame: the "Anthropocene." The historical becomes the geological, and the present becomes the future, when we fathom the human impact on the carbon cycle. Climate change is time, changed. Short-term decisions made—or postponed—may linger for millennia. Throwaway choices, in the aggregate, may have permanent evolutionary consequences. Over million-year cycles, rocks will record modernity in technofossils, as well as fossil absences that mark extinctions.

With geological power comes epochal responsibility—the duty of long-term thinking. This is the problem of our time, a problem *of* time, even as attention spans get shorter and shorter.

As popularized by behavioral economists, the human brain is capable of complementary "fast" and "slow" thinking, System 1 and System 2. On a collective level, small-scale traditional societies exhibit something analogous to these dual modes, with elders and designated knowledge keepers assisting the slower decision-making. Artificial intelligence can now facilitate long-term planning with instant slow thinking, though AI is only as thoughtful as the people who program and deploy it. In large-scale modern polities, including bicameral democracies, immediate trumps deliberate. Nations mobilize quickly (not always smartly) in response to armed invasions, natural disasters, and epidemics. The climate threat—despite occurring at breakneck geologic speed—has yet to inspire commensurate urgency.

From 2017 through 2021, I struggled to compose this book while the presidency of the United States became a weapon of short attention, while future "climate change" became immediate "climate crisis," and while

Covid-19 lay bare the extent to which ageism, falsity, and partisan animosity had become endemic. With the social contract in tatters, could I still imagine an ecological compact? By turns, I felt motivated and dispirited.

While I advocate for long-term thinking, I recognize it doesn't necessarily encourage more-than-human ethics, or humility, or peace. Dual types of futurist visions—For All Time, the End Times—are similarly violent in their certitude. Empires and states insist on their own immortality and use force to prolong that fiction. Authoritarians exhibit long-term hubris, from the pharaohs and their pyramids to Hitler's "Thousand-year Reich." Evangelical millenarianism, a global persuasion since the nineteenth century, turns the tables: the powerful will be laid low in the Last Days before the thousand-year reign of Jesus Christ. Revolutionary communism, another nineteenth-century belief system, betrays a similar temporality.

Various forms of apocalypticism—religious and secular—have made long time contingent on rupture. The postulated onset of the Anthropocene (or death date of the Holocene) might as well be called *anno hominis*. Tech prophets who predict "the singularity" have more than a little in common with believers in Zoroastrian and Abrahamic apocalypses. With or without religiosity, religious time persists. Secular environmentalists can be pious in their doomism, a Cold War frame of thought that became fossilized. Speculative fiction—a genre that began with Mary Shelley's *The Last Man* (1826)—trends dystopian, in part because of the recent commodification of disaster by the entertainment industry, including the cult of the superhero. For many people, including myself on most days, it's easier to believe in future collapse than to keep faith in the future as a continuation of the past and present.

In personal terms, I was raised in a millenarian church, founded in 1830, with the word "Latter-day" in its name—a new religious movement in which adolescent boys can be "elders." Patriarchal temporalities and biblical dispensations now repel me, yet I'm still drawn to thousand-year linear thinking. I'm searching for a sacred timescale without religion, without prescribed or predicted ending, without the eternity that is the "fullness of time."

For tutelage, I bow to trees—partly out of tradition. In mythical form, trees appear in creation stories, present at time's beginning. In graphical form, they represent seasons, cycles, genealogies, algorithms, and systems of knowledge. An olden bough is a bridge between temporalities we feel and those we can only think. This is why Darwin imagined millions of years of evolutionary history as a wide-spreading Tree of Life. Most profoundly, select living conifers—ancient organisms of ancient ancestry—are incarnations of geohistory. Volcanic eruptions, magnetic field reversals, and solar proton events leave signatures in their wood. Through tree-ring science, we see how woody plants register cyclical time *and* linear time, Chronos (durations) *and* Kairos (moments), climate *and* weather, the cosmogenic *and* the planetary. As multitemporal beings— the short, the long, and the deep together, in living form—perdurables allow us to think about the Anthropocene without anthropocentrism. They grant emotional access to timefulness.

Other timeful life-forms—bacteria, viruses, fungi, lichen—fail to activate positive emotions in humans. People decline to grant personhood to jellyfish and flatworms, much less adore them as immortals, despite their actualization of the mystery of perpetual living. People love their companionate mammals, of course, and regret that they senesce like us, only faster, in "dog years" and "cat years." The longest-lived vertebrates—sharks and tortoises—have life spans double that of human supercentenarians, but their undomesticated lives transpire beyond the normal frame of human viewing.

Because big animals are more relatable than plants, even personified plants, people turn first to fossilized megafauna for deep temporal services. An overemphasis on dinosaurs deforms the geologic past. Since the 1980s, people have imagined a cataclysmic rupture between dinosaurian Earth (the Mesozoic) and mammalian Earth (the Cenozoic). We aren't taught to see the big plants among the big lizards; the leaves are too familiar. The five major extinction events in geohistory count primarily as *animal* extinctions. Iconic gymnosperms, including ginkgoes, have continuity since the Jurassic period. Calling them "dinosaur trees" is zoocentric.

Deep time, including the geologic now, needs a phytocentric adjustment. The plant kingdom accounts for some 80 percent of the total biomass on Earth. In the long-ago Paleozoic era, vascular plants took over the land, remade the atmosphere, altered the chemistry of the oceans, and reset the course of evolution by contributing to marine extinctions. These plants, the original Prometheans, changed the climate and the planet itself by adding oxygen and combustible material. The subsequent Mesozoic era could be reimagined as the *Coniferous* era—a golden age for giant conifers that extended into the early Cenozoic. Although the last 30 million years have been far more challenging for gymnosperms, the endurance of certain genera across deep and long temporalities is amazing.

In the marketplace of stories, however, vegetal persistence doesn't command attention like societal collapse, ecocide, and the Sixth Extinction. On days when hope seems distant, I feel the pull of this plot: In a flash, humans join plants as planet changers by burning carbon stored in plants; just as quickly, a changing planet burns civilization.

In 1980, at one of the earliest conferences on climate change, Margaret Mead declared: "What we need to invent—as responsible scientists— are ways in which farsightedness can become a habit of the citizenry of the diverse people of this planet."[17] Considering the lackluster result of this top-down approach—and the disastrous failure of science-informed governance to curtail greenhouse gas emissions—it's easy to despair that short-termism is a flaw in "human nature," another kind of original sin.

History suggests otherwise. The primal architectural form, the burial ground, is a landscape of affiliation between the living and the long-term dead. Religion, at its best, comprises an intergenerational practice of gratitude. More concretely, a number of decision-making institutions—monasteries, universities, family-owned businesses—have been maintained for one thousand years or longer. Capitalism may be averse to obligations across generations, but capitalism is hardly the sum of humanity. Every day, parents, teachers, and librarians engage in world-sustaining actions, as do repairers of infrastructure, and humble planters of saplings. A well-known Chinese proverb (with analogues in other languages) captures the idea: *One generation plants the trees under whose shade another generation rests.*

Considering all these living examples of longsightedness—which could be multiplied—I would modify Mead's advice: Scientists need to augment and amplify timeful habits and traditions already invented by diverse peoples. And humanists need to tell more science-based stories about long-term relationships with nonhuman things and beings. Together, scientists and humanists can advocate for *geohumanism*. This philosophical position puts Earth first—neither in an ecocentric nor an antihuman sense but in simple recognition that Earth is precedent to humanity. It recognizes humankind as a product of evolution and recognizes our species—alongside land plants—as atmospheric agents, with the humbling acknowledgment that the Earth system reacts to climate forcing in unpredictable and ungovernable ways, on a timescale beyond our lived experience.

My ethical position is explicitly temporal: to become wise stewards of the planet, we must learn to think in the fullness of tree time.

Moderns barely had time to adjust to the modern condition—the nonnegotiable contract that the world must change, again and again, fast, now faster—before the once-local idea "the world" scaled up to the global. To the endangerment of the planet, growth for growth's sake and novelty for novelty's sake became end goals of global capitalism. More and more stuff equaled less and less life. Literally speaking, the Earth will outlast this moment of faster, cheaper, crappier. People cannot destroy a celestial sphere any more than they can save it. However, through avarice or mindlessness, the powers that be can indeed destroy the habitat of continuity between the current planet and whatever follows. This temporal orphaning, this dismaturation of the world, somehow doesn't count on the ledger of loss. As part of the trauma of terminal modernity—better known as business as usual—intergenerational discounting, like labor exploitation, is built into the system. Future generations lack standing. The beings of the next new world don't yet exist.

Except for those that do. The future oldest trees of a slower, fairer, post-carbon economy are here already, alive. These plants are not abstractions.

Although they cannot speak, cannot vote, they have an ethical claim. Attending to their longevity means tending our posterity. Nothing could be more pragmatically sacred. Caretakers of elders-to-be create solicitude between worlds, kinship between species. They are coauthors of survival stories.

Even mass diebacks of big old trees, already in progress, need not entail the extinction of elderflora. After an interregnum, renewed old-making should be possible. Because long-lived plants have been introduced widely, hotspots of longevity could be every here and there. Someday, in changed habitats and modified cities, vegetal life-forms should again achieve the status of ancient beings—if humans in the future still venerate such things. Old Ones will survive as long as people hope for and sorrow over—and give care to—young life that might live long.

Unlike the end of the world, this story does not require an ending. Not yet.

I.

VENERABLE SPECIES

CEDAR — OLIVE — GINKGO — PIPAL — BAOBAB

Plants endowed with oldness have been objects of reverence for millennia. Five species stand out for the density and duration of traditions around them: cedar, olive, ginkgo, pipal, and baobab. In a kind of biocultural symbiosis, these trees provide economic and spiritual services, while humans provide care. To one degree or another, each species has been domesticated. Speculation about the limits of their life spans antedates science. Well into the modern period, arboreal longevity remained a matter of approximation—and exaggeration—rather than precision. In recent decades, scientists have validated what nonscientists have long known: particular iconic specimens have persisted in proximity to people across many centuries. For the longest-living individuals of these five species, endurance is symbiotic—a combined function of evolutionary potential and human assistance. Veneration does not, however, ensure against destruction. Global change and regional conflict can terminate assisted longevity. As can ordinary use. The only thing commensurate with dendrophilia is deforestation.

CEDAR

If history is an archive of stories, the original oldest trees may be the Cedars of Lebanon—world-famous symbols of permanence and loss. They stand tall in the Torah and in a literary precursor, the Epic of Gilgamesh. In ancient texts, "cedar" could function as a poetic catchall for "tree," but when paired with the place-name "Lebanon," it usually meant *Cedrus libani*. For millennia, Mount Lebanon's cedars have been adored and demolished. Today, in its emblematic habitat, the species has nowhere higher to retreat, having been pushed to the limit by humans.

It all starts with Gilgamesh. The title character of the Mesopotamian epic may or may not have been a historical person, a king, from the third millennium BCE. In the story, he's a demigod who becomes human. A poem about his coming of age was told and retold in multiple languages—Sumerian, Akkadian, Hittite, Hurrian—then standardized around 1200 BCE, as written in cuneiform in twelve parts. For another thousand years, the epic appeared on clay tablets marked by artisans. It didn't survive as Greek or Latin translations. After a multimillennial life, the epic was forgotten—buried—for two millennia. In the late nineteenth century, British archaeologists excavated an ancient library in Iraq and recovered fragments of the epic. The nature of the text—a story with no original, no official, and no complete version, a story with the weight of ancient civilization but without the burden of modern civilization— lends itself to contemporary readings.

Gilgamesh is tall in stature, short in wisdom. He tyrannizes his own kingdom. To temper the youthful demigod, the gods create a wild man, Enkidu. After fighting to a draw, Gilgamesh accepts Enkidu as an equal. A bromance for the ages begins.

The king wants to make a monument to himself on Cedar Mountain. He dares his new best friend to trespass with him in the dwelling of the gods. Despite premonitions, the demolitionists trek to mountains beyond mountains, cross a final ravine, and enter the primeval forest. The trees grow thick and tall; the canopy calls out in a chorus of cicadas, birds, and monkeys. Fragrant resin drips from the cedars like sticky rain. The heroes waste no time chopping. Humbaba, the forest's tusk-faced guardian-giant, arrives on the scene. He reprimands Enkidu, a former acquaintance, and issues a challenge. At the close of an earth-shattering battle, the ogre curses the intruders with shortened lives. They murder him anyway. After surveying their work, they marvel to themselves: we have reduced the forest to a wasteland. Hoping to salvage the sacrilege, Enkidu removes the loftiest cedar—a tree whose crown had scraped the cope of heaven—as one solid timber, and guides it down the Euphrates to serve as a temple door. Gilgamesh takes his own trophy, Humbaba's head, in a sack.

For their joint offenses, Enkidu pays the price. The gods cause him to waste away, leaving Gilgamesh alone, dejected, and newly conscious of his eventual demise. After a long, failed quest—first for immortality, then for rejuvenation—Gilgamesh accepts his humanity, his mortality; he comes home to his people in the city. His days of violation behind him, he grows in wisdom. He does what wise kings do: he builds a wall. Deep inside the brickwork, he places a box—perhaps a cedar box—containing the original account of his adventure.

The ancient forest of Mount Lebanon contained junipers, firs, and pines, but only cedars became literary metaphors and economic indicators. The reason is resin. Cedarwood contains organic polymers that resist shrinkage, warpage, and rot, making it ideal for woodworking. Additionally, its resin can be refined into medicines and salves as well as agents for caulking, wood preserving, and embalming. When twentieth-century archaeologists exhumed a ship beside the Great Pyramid of Giza, the 4,500-year-old cedar planks still smelled sweet.

Egypt obtained its everlasting wood from Phoenicia, a group of coastal city-states in present-day Lebanon and Syria. Every major power in the ancient Near East traded with Phoenician timber merchants. According to the Torah, some of the best cedar ended up in Jerusalem, after King Solomon of Israel contracted with King Hiram of Tyre. Solomon finished the First Temple in aromatic cedar, and for himself constructed an opulent residence called the House of the Forest of Lebanon.

Solomon's timbers outlasted his buildings. In a wood-scarce region, conquest led to recycling. No city has been conquered more times than

Jerusalem. Through radiocarbon dating, researchers have demonstrated that Al-Aqsa Mosque, the holy building that has occupied the Temple Mount since the eighth century CE, was built in part with cedar beams reclaimed from Roman temples, which themselves were made with material taken from the monuments of Herod, the Jewish king who erected the Second Temple.

The plunder goes back further. Nebuchadnezzar II sacked Solomon's Temple in the sixth century BCE, conquering both Israel and Phoenicia, and taking captives back to Babylon. By the Euphrates, Nebuchadnezzar raised a cedar-roofed palace and a cedar-jointed ziggurat, a structure that may have served as literary inspiration for the skyscraping Tower of Babel. On his edifices, the king inscribed first-person boasts, which he repeated on monuments placed on roads leading to Lebanon. They are all of a piece: *I did what no other king could do, I cut through mountains, I crushed stone, I cut down cedars with my own pure hands.* Nebuchadnezzar seemingly channeled Gilgamesh—a story he would have known.

After the Neo-Babylonians came the conquering Persians, then Greeks, then Romans. All wanted cedar. By the Common Era, Phoenician cultures ceased to exist. In the second century, Emperor Hadrian placed the equivalent of one hundred "No Trespassing" signs around Mount Lebanon. Abbreviated Latin inscriptions on boulders marked the timber on the other side as imperial property. Today, shrubland surrounds these Roman boundary stones, and people dig around them, looking for buried treasure. The evergreen glory of Lebanon is nowhere to be seen.

To explain what happened, modern commentators fall back on a universal fable of hubris and exploitation—in two words, the "Gilgamesh gene."[1] Evidence suggests a more complicated story. The extraction of Lebanon's wood seems to have peaked in the Bronze Age, and again during the Roman period. But the greatest human impact on conifer ecology occurred *after* the fall of the Roman Empire. Low-status mountain dwellers—refugees, ascetics, shepherds—did more than pharaohs and emperors to curtail the kingdom of cedar.

In the early medieval period, for the first time, large numbers of people moved to the Levantine high country. Mount Lebanon became a refuge for ethnoreligious minorities, notably Maronites (eastern Catholics), who cleared forests and terraced the land for cereal crops. On a continuing basis, locals cut trees for firewood and charcoal. Highlanders also tended goats, which nibbled the understory to the ground each season. Conifers did not evolve with mammals, much less grazers. It takes decades for a *Cedrus libani* to reach sexual maturity and produce its distinctive upright cones.

By the late medieval period, *Cedrus* on Mount Lebanon had been reduced to scattered, high-elevation stands. The most prominent—for centuries believed to be the last—occurred within a cirque, elevation 2,000 meters, below the range's highest peak. Rising from a hummocky expanse of talus, the trees assumed layered, asymmetrical forms, nothing like the tall, straight cedars praised in ancient texts. Below the cirque ran Qadisha Valley, a cave-pocked cleft that has sheltered monastic communities since the early Christian era. Maronites from Bsharri, the valley's upper village, guarded the grove for centuries. At altars beneath the trees, they performed mass for the Feast of Transfiguration.

Starting around 1550, European pilgrim-tourists began journeying to the top of Qadisha Valley to see these incorruptible relics of biblical time. Visitors obsessively enumerated the grove's remaining "Ancient Ones"—specimens coeval with Creation, or the Deluge, or the Prophets, or Solomon. No one could agree on a time frame, or a census system. Should they add up only the largest trees, or only those that *looked* old? Sixteenth-century tallies varied from twenty-three to twenty-eight. The problem became proverbial: *The Cedars of Lebanon cannot be counted.*

By the nineteenth century, the number of "Patriarchs" or "Saints" had fallen as low as five or ten. Scientific botanists began to speculate about the future extinction of the species as well as the current age of the oldest individuals. Extrapolating from the tree rings on a cut branch, Joseph Hooker and Andrew Murray—a pair of botanical authorities from Britain—inferred maximum life spans of 2,500 and 5,000 years, respectively. Drawing from geology, they hypothesized that glaciation, followed by warming, had driven the species to refugia. It may be all wrong, wrote Murray, to ascribe cedar decline to "the maladministration of governments, the wastefulness of man, and the desolation of war." He posited a different theory: "climatal change."[2]

Contemporary tourists took a shorter view. They blamed Arabs and their goats, and also themselves for the disappointing raggedness of the "most renowned natural monuments in the universe."[3] Every trunk had been defaced with penknives and stained by campfires. Limbs had been hacked, bark stripped, cones picked clean. The famed adventurer Richard Francis Burton couldn't hide his disdain for the "Cedar Clump," which he called "essentially unpicturesque."[4] Burton and other Britons noted smugly that *Cedrus libani* grew better in the British Isles, where it had been introduced widely for its biblical aura.

At Bsharri, the first bureaucratic conservation efforts dated to 1873–1883, when Rüstem Pasha served as the Ottoman-appointed Christian governor of Mount Lebanon. Taking personal interest in the sacred grove, Pasha issued regulations, appointed a guardian, and authorized construction of a wall around the trees, approximately 450 in number. He wanted to try reforestation, too, but couldn't find a native source for cedar cuttings and seedlings. He had to place an order with the Royal Botanic Garden in Brussels.

New pressures on Lebanon's remnant forests arrived in the twentieth century. The opening decade witnessed the last major logging in the highlands. Resinous timber no longer served as temple doors and palace roofs but as railroad ties. Two wood-intensive regional projects—the Damascus–Medina railway and the Aleppo–Baghdad railway—exacerbated international tensions. With the fall of the Ottoman Empire after World War I, then the end of the French Mandate after World War II, Lebanon became independent. For its emblem, the new nation adopted *Cedrus libani* and printed graphical cedars everywhere. Initially, the state neglected to safeguard its living emblems, as symbolized by the ski resort adjacent to Bsharri's "Old Grove." Finally, in the 1960s, in parallel with the global environmental movement, Lebanon inventoried its twelve relict groves—comprising fewer than three thousand hectares—and began reforestation projects with UN assistance. The Lebanese Civil War (1975–1990) disrupted all conservation, though the placement of landmines inadvertently protected two groves from damage. A new era began in the 1990s. The government gave blanket protection to the national tree, established new reserves, and authorized new plantings of the species. For its part, UNESCO gave World Heritage designation to Qadisha Valley, including the famous grove, now called Arz el-Rab (Cedars of God) in Arabic.

Creating protected areas does not automatically create protections, especially in a country recovering from war. Monitoring reports from UNESCO have complained about illegal developers, unlicensed concessionaires, and reckless tourists at Arz el-Rab. The grove's legal status is highly complex: owned by the Maronite Patriarchate; managed by an NGO; overseen by various local, national, and international entities. Each of the other eleven has unique complications. Collectively, the Cedars of Lebanon suffer from the same conditions that afflict the Lebanese—political instability and economic precarity.

There are ecological threats, too, and they relate to planetary changes. In the 1990s, in multiple groves, swarms of sawflies began defoliating cedars. With international funding, Lebanese and French experts controlled this infestation, but more outbreaks will come in the future as winters continue to get milder and ski resorts become snowless. Interglacial "climatal change" has become anthropogenic "climate change." Modelers predict that by 2100 only a handful of high-altitude locations on Mount Lebanon will be able to support cedar—assuming that people continue to assist their migration and defense. *Cedrus* is a unifying symbol in a nation divided by sectarianism and stymied by corruption. In fall 2019, Lebanese protesters filled the streets to demand the ouster of the political class; in the same season, forest fires of unusual severity burned across the mountains.

At the genomic level, *Cedrus libani* isn't yet endangered. In the Taurus Mountains, Turkey, cedar has better habitat, with room to move upward. In Lebanon, the future promises further domestication. The Lebanese will surely find a way to keep the species in place, if only for symbolic connections to Phoenicians from millennia ago. Lebanon without cedar would be like Kilimanjaro without snow. Above the Cedars of God, just below the summit, a future sacred grove has been planted on abandoned agricultural terraces.

Botanists have feared the worst for Lebanon's cedars since the mid-nineteenth century, when geological time and evolutionary time made extinction newly thinkable. In that same moment of modernity, a story circulated about the oldest introduced cedar in Europe. Supposedly, the plant had been brought to Paris at the king's request. On a prolonged voyage across the Mediterranean, the royal curator of plants went thirsty by choice; he donated his ration of water to the seedling, which he carried in his hat, packed with soil. The collector managed to get the potted-tree-in-a-hat past a disbelieving customs officer, and added it to the Jardin des Plantes, where it grew to monumental size. In its hundredth year, 1837, this tree—the oldest local specimen of the oldest-growing species—was summarily cut down to make room for France's first railway. The plant couldn't survive a "world of changes."[5]

Although none of it was true—except for a large cedar in the Jardin des Plantes—this allegory spoke to anxiety about the pace of technological disruption. This feeling has only grown with time. It should be some small consolation, then, that the cedar in Paris survives to this day. On Mount Lebanon, the oldest specimen ever documented by scientists had 645 growth rings. The Parisian cedar is almost halfway there.

OLIVE

The same ancient civilizations that felled highland cedars cultivated lowland olives. The northern Levant is a leading contender for the "cradle" of olive horticulture. For millennia, *Olea*, the oil-producing tree, has defined the Mediterranean.

If its domestication roughly seven thousand years ago is one of the greatest unrecorded acts of history, *Olea* deserves much of the credit. Mediterranean peoples gave this plant practically nothing and received boundless gifts in return. Unlike citruses and apples, which can only be propagated by grafting, an olive cultivar can be cloned by slicing off some basal material and sticking it in the ground. Olives are preadapted for cultivation.

In the Neolithic period, olive culture fell somewhere between foraging and gardening. Depending on the situation, people grafted cultivars onto wild olives (oleasters) or uprooted wild olives to serve as garden rootstock. Because oleasters and olives cross-fertilize, the categories "wild," "domesticated," and "feral" barely apply. Cultivars produce bigger, oilier fruits, and no spines. Horticulturists prune them into treelike forms conducive to human labor, whereas oleasters occur in shrublands called maquis. Compared to monocultural crop trees, traditional olive groves in the Mediterranean sustain high levels of biodiversity. Functionally, they are native woodlands.

Even more than cereal or rice agriculture, horticulture requires long-term thinking. It takes three or four decades for an olive to reach peak production. Many proverbs convey this message: *I planted my grapevine, but my grandfather planted my olive.* After a conservative wait, olives become low-maintenance members of the family economy. They grow on rocky slopes unsuitable for other crops. Besides pruning and harvesting, they demand little care and no water or fertilizer. Harvest arrives in late autumn to early winter, an otherwise slow time.

The longevity of *Olea* is impressive among fellow angiosperms (broad-leaved, nonconiferous, fruit-bearing plants). The olive's staying power comes from regeneration. In the original botany textbook, Theophrastus argued: "The longest-lived tree is that which in all ways is able to persist, as does the olive by its trunk, by its power of developing side-growth, and by the fact that its roots are so hard to destroy."[6] Olives grow sectionally, meaning that separate branches connect to separate roots. In hard times, an olive can die in sections without wholly dying; in good times, it can grow new sections.

Endowed with unusual properties, the olive tree occupied a special place in Greek mythology, law, and war. The gift of Athena perfected the gift of Prometheus: Olives gave oil, and oil gave light. To hurt such a tree was an act of aggression. During the Peloponnesian War between Athens and Sparta, hoplites routinely attacked the olives of their rivals. Greeks had a verb, *dendrotomeô*, to describe the hostile cutting of fruit trees. For practical and cultural reasons, they rarely laid waste to fruited landscapes.

Olives aren't easily uprooted. Moreover, Greek law protected stumps, for they were considered living fruiters. Sophocles, in one of his plays, praised the self-renewing, indestructible tree that mocks the enemy's spear.[7] Even in wartime, special trees had untouchable status: Spartans spared the olives surrounding the Academy in Athens. The Persians under Xerxes felt no such compunctions when they invaded Attica. While ransacking the Acropolis, they lit Athena's sacred tree on fire. As related by Herodotus, the charred olive sprouted new growth the very next day.[8]

For the same reasons that they are hard to kill, olives are hard to date. Their sectional growth and their adventitious regrowth—plus their propensity to hollow out—defeat tree-ring scientists. Therefore, the oldest believed tree functions as the oldest known tree. Measured in the number of believers, the oldest is a collection of eight—the sacred specimens that grow within the walls of the Garden of Olives, the Franciscan sanctuary in East Jerusalem. They stand as Christian symbols of eternal life.

The Greek-derived *Christ*, like the Hebrew-derived *messiah*, means "anointed"—marked with blessed oil. The New Testament is rich with olive imagery. However, an olive garden as such does not appear in the four Gospels. Rather, the evangelists mention an undefined place called Gethsemane near the Mount of Olives that Jesus frequented during the last spring of his life. The Semitic word *gat-šemānî* means "olive press." Only the Gospel of John mentions a "garden" nearby.[9] In the original Greek, "garden" (*kēpos*) suggests cropland. In Jesus's time, the Mount of Olives contained an oil-processing facility within a cave. The owner of this underground facility may have rented the space to Jesus in the off-season.

The Mount of Olives became a magnet for Christian pilgrims as early as the fourth century CE. Here they commemorated the Agony, the Arrest, and the Ascension of Christ, not to mention the tomb of the Virgin. Byzantines built churches on the mount, as did later Crusaders. Through cycles of conquest and construction, revolt and razing, local olives survived, century after century. As a discrete pilgrimage site, though, the Garden of Olives is a nineteenth-century creation—a material response to interdenominational and international rivalries. After centuries on the periphery, Palestine became a geopolitical pawn. In this context, tourists flocked to the Holy Land, and the built environment changed to accommodate them. By century's end, Christian sightseers could choose between three gardens called Gethsemane—one Roman Catholic, one Greek Orthodox, one Russian Orthodox.

Franciscans owned the traditional site. It featured gnarled, romantic trees—and little else. Lacking permission from the Ottomans to raise a church, the monks walled off their olives in 1847, and subsequently turned the enclosure into an open-air chapel, with stations of the cross. They sold olive-stone rosaries, wooden relics, and oil to tourists. To meet European expectations of a "garden," the custodians planted ornamental flowers. Later, they added French-style formal plantings. "The stiffest garden I ever saw," complained one American.[10] By the end of Ottoman rule, the site had morphed into an arboretum, with palms and cacti alongside olives.

Under the British Mandate—the first Christian administration of Palestine since the Crusades—Franciscans seized the opportunity to erect a basilica next to the trees. After moving their ritualism indoors, the custodians restored the olive yard to its "original" condition: they uprooted flowerbeds, cypresses, picket fences. Today, the naturalistic garden appeals to born-again Christians. Unlike Victorian Protestants, who scoffed at the credulity of Catholic and Orthodox pilgrims, contemporary evangelicals pray under the same trees where Jesus prayed.

Could that possibly be true? According to Josephus, a Jewish eyewitness of the Great Revolt, the Roman commander Titus destroyed all the gardens and fruit trees adjoining Jerusalem in AD 70, leaving a melancholy scene of desolation.[11] It seems doubtful that legionnaires axed every tree. Besides, olives can resurrect from the stump. Believers can therefore believe.

In 2012, at a Vatican news conference, Italian botanists reported on an unprecedented investigation. They had radiocarbon dated the oldest wood at Gethsemane and separately estimated the age span of the absent wood—the missing years. According to the researchers, the hallowed olives were eight to nine hundred years old, coeval with the Kingdom of Jerusalem. They speculated that Crusaders had planted the trees as a group, given that genetic tests indicated that all eight shared the same parent. Whether that parent had occupied the site previously—or as far back as Jesus—was impossible to say, though the scientists seemed disposed to bolster tradition. "Plants of greater age than our olives are not cited in the scientific literature," claimed one of the researchers.[12]

Skeptics and atheists can still be impressed by the endurance of two things at Gethsemane, one institutional, one vegetal. Francis of Assisi assigned custodians to the Holy Land in 1217; his devotees acquired the olive plot in 1681; and since the mid-nineteenth century, Franciscans have maintained it as a garden. Nonetheless, the historical timeline suggests that these venerable plants have lived roughly half their existence in abandonment—half wild, like the species itself. At the Mount of Olives, these trees outlasted the Crusaders, outlived churches made of stone. They survived to become the second holiest Christian site in Jerusalem. Although they lack eternal life, their dispensation exceeds that of all current governments.

The modern state of Israel has a strange relationship to the iconic tree of the ancient Kingdom of Israel. Through the Jewish National Fund, Zionists environmentally "restored" the Promised Land—or, in aesthetic terms, Europeanized it—by planting evergreens by the millions. For afforestation, they turned to pines, not oaks or olives. Israel ranks as a minor producer and consumer of olive oil, though its plantations are distinctively Israeli: high tech, drip irrigated, machine harvested. Unlike oil exported by ancient Jews, Israeli extra virgin is destined for eating, not burning.

Meanwhile, in the West Bank, by financial necessity and by cultural persuasion, Palestinians use traditional methods on older trees. Across the twentieth century, olives became integral to the economy and the identity of this occupied, stateless population. Their unwatered fruit-bearing trees symbolize what Palestinians call *sumud*: steadfastness in hardship, perseverance in place. As legal markers of land tenure, Arab olives became targets of Israeli uprooting, contrary to Talmudic tradition, which forbids the destruction of fruit-producing trees—and wasting in general—even in time of war. Islamic law says much the same, though the holy book records that the Prophet received divine exemption to burn date palms owned by Jews in Medina.

Since the Second Intifada, one particular fruiter in al-Walaja, a village between Jerusalem and Bethlehem, has become a Palestinian emblem. The people of al-Walaja were uprooted during the 1948 Arab-Israeli War. Unfortunately for them, the postwar border bisected their relocated village. In the 2000s, the Israeli military transformed the Green Line into a treeless zone of concrete walls and electric fences. As a result, the villagers of al-Walaja lost olives and the ability to pass freely between remaining groves.

A short distance from the barrier stands a large olive with a multi-iterated trunk. For years, a local man, Salah Abu Ali, guarded this plant. He gave lessons to the political tourists who pilgrimaged here. "This tree is a witness to the tragedies that have been happening to the people of al-Walaja," said Abu Ali in 2012. "This tree is Palestine."[13] The Palestinian Ministry of Culture has expressed a desire to scientifically date the tree and nominate it for the World Heritage List. Foreign supporters of Palestinian nationhood repeat the assertion that al-Walaja's tree is the deepest-rooted olive in the world—1,000, 4,000, even 5,000 years old.

Other states lay claim to the oldest. The Lebanese Ministry of Tourism advertises the Sisters Olive Trees of Noah, an altitudinous grove that supposedly escaped the Flood and served as the source of the twig that the dove carried back to the ark. In Crete, an olive tree museum has been erected next to a monumental specimen publicized as vintage Bronze Age. In 2004, for the Olympic Games in Athens, Greek organizers requested cuttings. The branches were destined to be special wreaths for the marathon champions—a revival of classical tradition. When a "latecomer copycat" village on the opposite side of Crete claimed to have the *real* oldest olive, organizers stemmed the controversy by using material from both trees.[14]

Although appreciation for the longevity of olives is ancient, monetization of what southern Italians call *ulivi secolari* (centuries-old olives) is recent. During the global real estate boom of the early 2000s, old olives, like tall palms, became mobile commodities. Landscapers scoured the villages of Spain, Italy, and Greece for sculptural specimens and offered struggling farmers tens of thousands of euros in cash for fruiters past their productive prime. By truck, ship, and helicopter, olives left the impoverished countryside for châteaus in Bordeaux and resorts in Dubai.

In response, local conservationists have registered monumental olives and attempted to ban their uprooting, which they compare to ivory poaching and archaeological looting. In the Spanish coastal region between Valencia and Barcelona, along the old Roman road Via Augusta, activists have, with EU support, publicized *olivos milenarios*, encouraged culinary tourism around them, and incentivized a market for their oil. There are Chinese consumers who buy age-defying cosmetics derived from ancient trees, just as there are Californians who buy ultra-premium EVOO for life-extending diets.

Of the hundreds of millions of olive trees in Spain, Italy, and Greece, a small percentage count as perdurables by the thousand-year standard. Radiocarbon dating by Spanish scientists along Via Augusta suggests that the oldest trees exceed six hundred years—comparable to those in the Garden of Gethsemane. But enumerators restrict themselves to aboveground wood. As rootstock, or as germplasm, an olive might be much older. And a tree of "only" half a millennium still counts as a survivor of the Little Ice Age, including the Great Winter of 1708–1709. It is a genetic repository of resilience.

Gift-giving olives anchor long-term landscapes and traditional economies, yet their future looks uncertain. An invasive bacterial disease, identified in 2013 in Puglia, the heel of Italy's boot, now spreads across the Continent despite desperate measures to contain it. "We thought they cannot be touched, are immortal," said an Italian whose family has harvested olives for over five hundred years. "Now, we are facing a truth that is a natural truth—that nothing is untouchable."[15] New climate patterns undermine old ways. Although olives are adapted to aridity, traditional groves lose economic viability as summer heat intensifies, and fires rage. Planetary warming—and global competition from lower-cost seed oils—may push the industry further toward intensive, irrigated, flatland operations dominated by agribusiness. It may also push it north of the Alps.

In that event, Mediterraneans may dislodge wizened trees from stony hillsides to create goat pasture. Or, olives may hunker down, revert to thorny oleasters, and outlast the age of humans as shrubs.

GINKGO

Who first noticed a ginkgo and through which sensory organ—nose or eye? It happened long ago in far eastern Asia, somewhere south of the Yangtze. It likely happened in autumn. In that season, a human might have smelled the decomposing seedcoats, with their olfactory resemblance to carrion. Or her eye might have spotted the leaves—bilobed and yellow like no other, with a matchless synchronized drop.

In the past thousand years—roughly 0.001 percent of its life as a species—*Ginkgo biloba* spread from China throughout the world. Or, more accurately, *re*spread. In the geologic past, many species of the ginkgo division prospered throughout the Northern Hemisphere.

These ginkgophytes were, in their evolutionary heyday, the foremost innovators of the plant kingdom. They could shed leaves in winter, go dormant in low-light seasons, switch between stub growth and branch growth depending on conditions, and resprout from lignotubers— energy-storing roots—after disturbances. On a prior planet with relatively few tall plants and no fast-growing angiosperms, ginkgophytes achieved dominance as generalists.

As Darwin said, "rarity precedes extinction," but the duration of rarity varies greatly.[16] *Ginkgo* is a temporal outlier. Ginkgophytes survived multiple mass extinction events and outlived their original seed dispersers, which might have been carrion-eating animals attracted by the sweet-rotten smell of the fleshy seedcoats. After a long period of glory in the Mesozoic era, ginkgophytes declined in the Cenozoic and dwindled to one species by the ice ages. Ginkgoes disappeared from North America, then Europe, and finally Japan, becoming, by the Pleistocene epoch, mountain refugees in China.

In the late Holocene, Chinese people began functioning as *Ginkgo*'s disperser. In a text from the eleventh century CE, a Song dynasty poet described the process by which the "silver apricot" traveled from the highlands to the lowlands: "First it came in silk bags as a tribute"; then the noble prince "brought roots from afar to bear fruit in the capital"; and, by and by, the transplanted tree's first nuts were "presented to the throne in a golden bowl."[17] In short, ginkgo seeds entered Chinese cuisine as imperial appetizers. By the succeeding Yuan dynasty, orchardists grew "white fruit" commercially in the Yellow River drainage. From China, the "duck-foot" tree spread to Korea, and from Korea to Japan.

Ginkgoes cannot serve as crop plants like true nut trees. Their seeds are semi-toxic. Beyond a certain limit, this delicacy becomes a poison. Because of their active ingredients, ginkgo seeds attracted the attention of Chinese doctors, who theorized on their correspondence with earth elements and human organs. By the seventeenth century CE—quite late—the seeds became standard components of Chinese materia medica. Doctors prescribed ginkgo porridge for lung-related ailments that depleted vitality. The leaves of the tree did not undergo comparable medicalization.

During the Edo period of Japan (1603–1868), the seeds, branches, and leaves of the ginkgo acquired new associations. Japanese artists led the way in aestheticizing the unmistakable leaf. The bilobed shape appeared on crests, ceramics, kimonos, and coiffures—including a stylized topknot for sumo wrestlers. Meanwhile, in Japanese cookbooks, ginkgo seeds showed up as common vegetable ingredients. Before or after a big meal, people ate them separately as a digestif. Whereas ginkgo seeds regulated the lungs of Chinese, they regulated the stomachs of Japanese.

Also in Japan, and only in Japan, select branches on ginkgoes corresponded to mammary glands. Old ginkgoes sport hanging growths—aerial roots—called "breasts" (*chichi*) in Japanese. Chichi-bearing ginkgoes became sites of propitiation for women who desired assistance with childbirth or lactation. Not coincidentally, such old ginkgoes typically grew in gardens under sacred guardianship. Following the lead of sanctuaries in China and Korea, the keepers of Buddhist temples and Shinto shrines in Japan landscaped with ginkgoes.

From early modern Japan, knowledge of ginkgo traveled to the West. A German naturalist with the Dutch East India Company observed the species in Nagasaki and bestowed upon it its unpronounceable name. "Ginkgo" was wrong—it should have been Romanized as *ginkio* or *ginkjo* or *ginkyo*—but the error became fixed once Linnaeus adopted it. The Swedish taxonomist didn't have access to the plant's reproductive organs, so he placed *Ginkgo biloba* in the appendix as a "Planta Obscura." Despite the isolationism of Japan, collectors from Europe and the United States obtained seeds and cuttings for wealthy patrons, including the Duke of Weimar. Inspired by the "tree from the Orient," the duke's chief adviser, Goethe, wrote a love poem about the secret of the bilobed leaf—two parts as one.[18]

While gardeners planted ginkgoes, geologists exhumed ginkgo leaf fossils in unlikely places, including northern Europe. The obscurity and secrecy of this tree ran deeper than Linnaeus or Goethe realized. Once Darwin advanced the theory of the Tree of Life, botanists puzzled over the evolution of early land plants: Did ginkgoes branch from ferns? Did conifers then derive from ginkgoes?

The critical evidence arrived in the 1890s, when University of Tokyo botanist Sakugorō Hirase observed the microscopic union of male and female ginkgo. To his surprise, the spermatozoid arrived at the ovum by swimming in fluid. This discovery of motile sperm—an evolutionary holdover from the watery origins of plants—secured ginkgo's status as a primordial species. A decade later, when paleobotanist (and future birth control activist) Marie Stopes visited Tokyo, she witnessed the "grand excitement" at the university during ginkgo's brief fertilization period. Stopes spent three days "hunting *Ginkgo* sperms" under the magnifier. "It is most entertaining to watch them swimming," she wrote in her journal. "Their spiral of cilia wave energetically."[19]

After Hirase's revelation, scientists began speaking of ginkgo as the "missing link" or "connecting link" between ferns and conifers, and a "living link" to the age of dinosaurs. Albert Charles Seward, a noted geologist, applied Darwin's concept "living fossil" to *G. biloba*.[20] It became conventional to call ginkgo the "oldest tree species in the world" or the "oldest living genus." Recent discoveries of fossilized ovulate organs—a better measure of evolutionary change than leaves—suggest that ginkgo has been morphologically stable for some 120 million years.

Do age-old genera contain age-defying chemicals? A German homeopathic company, Schwabe, has profited from this wishful correspondence. In the 1960s, Schwabe developed two botanical extracts. One they advertised with giant sequoia; the other they literally made from ginkgo leaves. Patented as "Egb 761," this extract became the top-prescribed herbal medicine in Germany—where the publicized association with Goethe helped—and later the top-selling supplement in the United States. The efficacy of Egb 761 remains inconclusive. Proponents make unsubstantiated claims that ginkgo trees live longer than a millennium and that ginkgo has been used in Traditional Chinese Medicine for five thousand years.

Chinese horticulturists grow ginkgoes for nuts, not leaves. The nuts are meant primarily for food, not medicine. Bitter when raw, they become tasty with roasting. Fairgoers snack on them during the Mid-Autumn Festival, much like Europeans eat chestnuts at Christmas markets. In China, ginkgo leaf extract is a niche product; discerning consumers prefer the German version.

To meet the global demand for its "mental sharpness" pills, Schwabe operates a plantation in South Carolina, the largest ginkgo forest since the Tertiary period. Every summer, ten million trees are defoliated by machine; every five years, cut to the ground. They receive no sabbatical. Through violent rejuvenation, these piedmont plants have taken the form of subalpine krummholz, with immense root systems supporting leafy stubs. "It's brutal," says Peter Del Tredici, who serves as Schwabe's horticultural adviser. "Morphologically, we've made them into ancient shrubs. That these plants have lasted 35 years has defied everyone's expectations. Ginkgo is astounding."[21]

Ginkgoes even lived through an end of time at the end of a world—Year Zero at Ground Zero. As Hiroshima burned, scores of injured residents who survived the initial impact ran to Shukkeien Garden—1,370 meters from the hypocenter—and perished among skeletonized trees, including an almost toppled ginkgo. Defying death, the tree pushed out new buds and generated a second layer of annual wood, a double ring for 1945. To this day, the leaning ginkgo stands, bearing a bright yellow "A-bombed tree" identification tag. Each autumn, peace activists come to Shukkeien to collect seeds for distribution around the world. A "peace tree" offspring now grows at Oak Ridge, Tennessee, the enrichment site for the uranium that the US military detonated above Hiroshima.

It's no coincidence that ginkgo is longevous on two scales—in evolutionary age as a clade and in biological age as individuals. In the words of Ernest Henry Wilson, preeminent collector of East Asian plants, ginkgo is endowed with "a thousand and one means of maintaining its existence."[22] At the organismal level, it avoids senescence, as recently proven at cellular and molecular levels. A ginkgo's ability to do the stuff of living—growing full-sized leaves, photosynthesizing, generating viable sperm and seeds, producing antimicrobial chemicals—doesn't decline over time. Wood production suffers slightly past two centuries, but not enough to shift a ginkgo from its default mode of immortality. The organism dies from external stress, not internal aging. Alternatively, catastrophic injury can lead to life renewal, thanks to lignotubers and aerial roots (called "stalactites" in Chinese). In Tokyo, landmark ginkgoes regrew after the great fire of 1923 and the firebombing of 1945. Maximum longevity remains undatable, however. Like olives, ginkgoes hollow out, depriving scientists of tree rings and radiocarbon samples going back ten centuries or more. Notwithstanding, it seems reasonable to assume that ginkgoes can be millennials, especially in sanctuaries.

In 2010, the most esteemed ginkgo in Japan, a tree with legendary associations, collapsed in a storm. "So many people came, called and sent e-mails offering their condolences," said the chief priest at Tsurugaoka Hachimangū shrine in Kamakura, Kanagawa Prefecture. "Perhaps the tree fell to draw everyone's attention away from their focus on materialism and money."[23] Nearby, Shinto caretakers planted cuttings from the tree. Out of respect, and hope, they left the giant stump in place. Sure enough, this ginkgo, purported to be 800 years old, generated new growth from its storage roots.

Since the nineteenth century, plant hunters, mainly Westerners, have searched the mountains of China for the oldest, wildest ginkgoes. To the Chinese, "wild" has little cultural resonance, and little practical meaning. All the land below 1,000 meters in elevation was deforested in ancient times. Despite extensive agriculture, a few old ginkgo populations exist—as demonstrated by genetic testing—in highland refugia. One of these remnant groups grows adjacent to an important Buddhist monastery at Tianmushan, Zhejiang Province. Did monks plant these trees, or did monks plant themselves here because of the trees?

At some point in history, ginkgo shifted from endangered to domesticated. Ernest Henry Wilson went so far as to credit Buddhist monks with the survival of the species—a speculation that became a just-so story in popular literature. On safer grounds, Sir Peter Crane, past director of Kew Gardens, has argued that ginkgo is a "good news story: a tree that people saved."[24] Evidence exists on the streets of temperate-zone cities on both sides of the equator. The fad for urban ginkgoes began in Japan, during the Meiji period of modernization (1868–1912). Japanese city planners took a European innovation—the tree-lined boulevard—and made it their own.

Americans later copied this look with source material from Japan. "Stiff and almost grotesque in its early years," wrote Harvard authority Charles Sprague Sargent, ginkgo "does not assume its real character until it is more than a century old." Sargent remarked that it took five hundred to a thousand years for the temple ginkgoes of East Asia to reach their glory. To "plant for posterity," Sargent continued, Americans are "reasonably safe in selecting this tree."[25] That was 1897. Three decades later, when mature female trees began dropping stinky seedcoats, many people revised their opinion. When US street tree wardens came back to ginkgo in the late twentieth century—in appreciation of its tolerance for urban soil and air pollution—they exclusively planted males.

Species in the zone between rarity and extinction cannot rely on humans to keep them going unless they provide something that humans desire. Beyond nuts and Egb 761, people want something special from ginkgo: beauty. No leaf is more attractive, or more distinctive, than ginkgo in its golden autumnal phase. It's tempting to say that ginkgo was preadapted to domestication by an ocular species with an aesthetic sense. However, given that ginkgophytes antedate hominins by some 200 million years, this cannot be an interspecies example of the "evolution of beauty."[26] Rather, a lucky break for *Ginkgo* and a miraculous coincidence for *Homo*. What are the chances that the oldest surviving tree genus on Earth would grow the loveliest leaves in geohistory?

PIPAL

After ginkgo, the most distinctive tree foliage belongs to *Ficus religiosa*— the sacred fig, also called pipal. Its hand-sized leaves have elongated ends, like swallow tails. When they quake in the breeze, they sound like a flock taking flight. Unmistakable long-tipped leaves appear in Bronze Age art from the Indus Valley civilization.

Like those of many *Ficus* species, pipal seedlings may begin their life cycle as epiphytes—nonparasitic hangers-on. They roost in the sunlit canopy of host trees, sending down aerial roots in search of soil. If successful, they use their new vitality to turn on their hosts, enveloping them. Of the "stranglers," the sacred fig is singular because it splits rather than chokes. Like a slow-moving axe, its roots sunder the host trunk vertically.

The original sacrality of this species possibly had something to do with its power to create through destruction. Pipals can split bricks as well as trunks. By building brick architecture, Indus Valley peoples established new niches for pipals. It must have been auspicious to witness a mighty tree beginning its life on the roof or wall of a sacred building. Or perhaps temples were sited near mighty trees. Likely both.

Each species of *Ficus* has a mutualistic relationship with a unique pollinating wasp. One cannot survive without the other. For at least five thousand years, the mammal *Homo sapiens* has joined the insect *Blastophaga quadraticeps* as evolutionary partners of pipal. In return for elongating its life, the tree gives meaning to human life. This material-spiritual relationship seems all the more remarkable given that people neither eat the fruit nor utilize the wood of pipal. Although its leaves and bark feature in contemporary Ayurveda, medicine as such is not the reason for pipal's age-old significance. Unlike any other plant, this species was domesticated by—and for—devotion.

Vedic scriptures associated pipal—*ashvattha* in Sanskrit—with cosmic power, including the power to split and destroy enemies. Later Puranic texts specified devotional practices and divine associations. The Indian epics reinforced this enshrinement. "Among all trees, I am the sacred fig tree," sings Krishna in the Mahābhārata.[27]

When Siddhartha Gautama achieved nirvana, becoming the Supreme Buddha, he did it—according to hagiographies—at the foot of an *ashvattha*. This tree of life and death was also a tree of knowledge. Buddhism did not sacralize the pipal so much as redirect its preexisting sacredness. Not coincidentally, the Exalted One found enlightenment (*bodhi*) at a preexisting sacred place, Gaya, in northern India. Here, Buddha himself began the Buddhist practice of revering the bodhi tree, for he spent the first seven days of his awakened life staring unblinkingly at a ficus.

The early spread of Buddhism, and the bodhi tree, is largely credited to Aśoka, the Indian emperor of the third century BCE who built the first monuments to dharma. Supposedly, Aśoka became a disciple after being an adversary: first he ordered his soldiers to fell, dismember, and burn the tree, only to see it miraculously regrow. In another legend, Aśoka's gift-giving to the ficus was so lavish that the youngest of his wives—assuming that "Bodhi" must be a mistress—hired a sorceress to devitalize the object of the emperor's obsession. Grief-stricken by the withering of his bejeweled tree, Aśoka made additional offerings, and gave proper praise when it recovered. To protect the fig, the emperor built a wall, the first of many.

The site called Bodh Gaya became a destination for pilgrims throughout South Asia and China. According to legends, more than one anti-Buddhist ruler over the centuries felled the ficus in the sacred enclosure.

With or without violence, the main trunk must have perished multiple times—figs are not long lived—but monks always planted offshoots on top of the roots. As part of this process of material and spiritual replication, caretakers added new soil to the tree mound, and, over the first millennium CE, the most sacred fig literally ascended by degrees. An architectural temple—an impossibly steep tower with four straight sides—eventually adjoined the arboreal monument.

As India became culturally Hindu, then dynastically Muslim, Bodh Gaya became isolated, and Buddhists almost vanished from the scene. The bodhi tree remained holy, though not in singular fashion, for Hindus revered pipals categorically and incorporated Buddha as the ninth avatar of Vishnu or a heroic form of Shiva. A Shaivite monastery oversaw Bodh Gaya at the time that British archaeologists "discovered" the "ruins" of the temple of Aśoka.

In their reports, the British noted both the overgrown forest and the decayed condition of the bodhi tree. After a storm knocked it down in 1876, they supervised the planting of seedlings. Raj administrators hardly concealed their philosophical preference for Buddhism over Hinduism—the latter seemed almost Catholic in its ritualism. Moreover, their originalist and prudish approach to archaeology favored older bodhi trees over newer lingam stones.

The British-led restoration of Mahabodhi Temple, completed in the 1880s, had unexpected consequences, for it coincided with new religious revivalism. Anglophone Buddhists from the British Empire, following the example of Protestant mission societies, began to proselytize. They presented dharma as a world religion on a par with Christianity. Their self-styled leader was Anagarika Dharmapala, who founded the Mahabodhi Society after an apprenticeship as a theosophist. He obsessively worked—through agitation, then litigation—to restore Buddhism to Gaya. A belated half-victory came in 1949, when the newly independent Indian state assumed control of the site and delegated management to a committee composed of Hindus and Buddhists.

In British Ceylon, Dharmapala's homeland, a parallel history played out. Historically speaking, there are two "original" bodhi trees—one in

Gaya, India, another in Anuradhapura, Sri Lanka.[28] Or, rather, it is the same tree in two places. The Lankan tree has greater claim to antiquity because Anuradhapura, unlike Bodh Gaya, has been in continuous Buddhist custody for over two millennia. Lanka was the cradle of Theravada, the oldest extant branch of Buddhism, which arrived circa 300 BCE, presaged by the Buddha's legendary visits to the island.

Fittingly, Theravada took root with a ficus, as narrated in the Mahā-vaṃsa, a Pali-language monastic text from the fifth or sixth century CE. Part epic, part chronicle, the story details how a Lankan king asks his ally Aśoka to send his ordained daughter to the island so that an abbey can be established. Along with the gift of ordination, Aśoka's eldest daughter brings a cutting from the bodhi tree. "Cutting" is not literal, for no one would or could take a knife to this being; instead, it agrees to self-sever its southern branch, which floats in the air, emanating celestial rays while growing new roots and branches. The retinue of the reincarnated tree—carried in a golden pot—includes nuns and monks and whole families of specially trained arborists and protectors. After a journey by caravan and boat—including a seven-day detour under the sea to the realm of serpent people—the ficus arrives in Lanka, where the king offers all his gifts, even his sovereignty. The tree plants itself at Anuradhapura, producing rain, temblors, additional miracles, and bonus figs to plant around the island.

For more than one millennium, Anuradhapura served as the royal seat of power. The capital was renowned for its irrigation canals, its domed buildings, and its bodhi trees. Whereas Bodh Gaya featured a temple beside a tree, Anuradhapura—and other locations throughout Lanka—had numerous tree temples: four-sided unroofed structures built around ficuses. Because of the influence of the Mahāvaṃsa, one particular fig next to one prominent monastery won the status of "original" bodhi tree. Long after the kings of Lanka abandoned Anuradhapura—a response to political conflict and climatic change—they episodically returned with processions to signal their dynastic authority and to heap gifts upon the ficus: soil replenishments, ramparts to protect against elephants, and, most spectacularly, special pumps for aerial watering.

Even so, without a large, stable population, the temple complex gradually became derelict and overgrown with sacred figs. When Portuguese, Dutch, and later British occupiers arrived on the island, Anuradhapura functioned as one of those "lost cities" that enthralled Europeans. However, the place was hardly jungled and anything but forgotten. The chronicles of precolonial Sri Lanka are remarkably complete and detailed, and they do not record the death of the venerable ficus. This negative evidence can be interpreted as continuous life, or at least continuous caretaking. It seems likely that a core group of Buddhist devotees always held out at Anuradhapura, serving as spiritual arborists.

In any case, Sir James Emerson Tennent, who served as colonial secretary of Ceylon in the 1840s, declared that the "Bo-tree" was "the oldest historical tree in the world." He dismissed scientific calculations of older trees in Africa, Europe, and the Americas as conjectural and inferential, "whereas the age of the Bo-tree is *matter of record*."[29] It put Gethsemane in perspective.

Fifty years on, anticolonialists appropriated Tennent's words. For Sinhalese Buddhists, the island's ethnoreligious majority, the "oldest historical tree in the world" came to represent their national claim to the island. On a municipal level, the Ceylonese branch of the Mahabodhi Society advocated for spatial segregation at Anuradhapura. The British had turned the ancient capital into a new town, complete with Christian churches, a Hindu temple, and a mosque. After independence in 1948, the Sinhalese-dominated government enacted the Society's vision of two cities—a secular new town divided from an old town restored by and for Buddhists. In 1982, UNESCO affirmed the separation by listing the "Sacred City of Anuradhapura" as a World Heritage Site.

By nationalizing the bodhi tree and its origin story—to the point of adding sacred fig leaves to the flag—the leadership of postcolonial Sri Lanka further marginalized minority Hindu Tamils and Tamil-speaking Muslims. Decades of discriminatory policies begat a generation of civil war. One of the early terrorist attacks by the Tamil Tigers, a militant separatist group of Hindus, took place at Anuradhapura in May 1985, when a busload of shooters opened fire at the transit hub in new town, killing scores of people. They drove on to old town, murdering all the nuns and monks in sight and attacking a sacred fig in the compound. The bodhi tree itself escaped damage.

At Bodh Gaya, World Heritage status, which arrived in 2002, produced a more pluralistic, or at least international, outcome. After an interval of centuries, the temple complex is once again a major Buddhist pilgrimage site—some two million visitors per year—a fulfillment of British imperial statements about the "Mecca" or "Jerusalem" of Buddhism. Tibetan exiles have taken up residence in Gaya, and the Dalai Lama brings international attention to the site. Even terrorist bombings at the compound in 2013 did not deter religious tourism for long. Indian leaders, including Hindu nationalists, promote pilgrimage as good business for a poor region. Certified saplings of India's bodhi tree serve as high-profile diplomatic gifts to other Asian states. In a kind of sectarian competition, donors from various Buddhist countries have erected increasingly lavish temples and accommodations in the sprawling village beyond the sanctuary's walls.

The newest original bodhi tree, reborn the year Queen Victoria became Empress of India, has achieved magnificent size and shape. Few trees anywhere see greater traffic—excluding the lockdown period of Covid-19. Worried for the plant's health, temple managers recently banned electric lights, candles, incense, and offerings of milk from the immediate area; and they planted the next successor nearby. Devotees of the holiest ficus still gild its roots, meditate, pray, chant, circumambulate, and chase any leaves that may fall. Officially, no one may detach a single leaf, yet newspapers have reported allegations that priests sold a whole branch to a Thai millionaire.

Hindus come here, too, but not in mass pilgrimage. Tree worship remains dispersed and localized in contemporary India. Performing puja at a garlanded and colorfully string-wrapped pipal is an ordinary act in villages and cities. Neighbors may hug, massage, and kiss the prana-filled tree—an embodiment of Vishnu—asking it for health, wealth, marriage, children. The most common offering, water, helps explain the ubiquity of figs in a country that has been largely deforested. According to still-living

Vedic practice, pipals cannot be uprooted or killed—even if they sprout on buildings—without inviting misfortune.

Sacred figs occur in far-flung secular locations, too, such as Foster Botanical Garden in Honolulu, an oasis in an ugly concrete city by the beach. This is the "tree museum" referenced in Joni Mitchell's famous song about paving paradise. The garden once belonged to Mary E. Foster, a remarkable woman descended from Hawaiian royalty and haole merchants. From her liminal position, she empathized with the ethnic and religious minorities of Hawai'i, especially its Japanese Buddhists. After the death of her husband, she was wealthy and directionless. She studied theosophy. In 1893, she sought out Anagarika Dharmapala, who stopped in Honolulu on his way back from the World's Parliament of Religions in Chicago. Foster wanted Dharmapala's advice on controlling her anger. That one brief meeting on a steamer led to perpetual financial gifts. Without Mary Foster's fortune, the Mahabodhi Society could not have continued its mission. Twenty years later, Dharmapala returned to Hawai'i to thank his "foster mother."[30]

It's tempting to assume that the *Ficus religiosa* in Foster Botanical Garden came from Dharmapala. Indeed, sources claim that he gifted a cutting from the bodhi tree at Anuradhapura—another epic voyage for a well-traveled tree. This story is too good to be true. Most likely, the tree was already there, grown from seed by the garden's nineteenth-century creator, a German immigrant who wrote the first textbook on Hawaiian flora, and who traveled widely in Asia, including Ceylon, collecting plants.

The exact provenance of Honolulu's "bodhi tree" is less important than what it represents: the globalization of the sacred fig. This species was deliberately spread throughout Southeast Asia by Buddhists, and later around the world by Western gardeners enthused by ancient Eastern religious history. People forgot about the tree's older evolutionary partner, the wasp. In Hawai'i, pollinators arrived by ship or plane in the early years of the current millennium. In short order, Mary Foster's fig began fruiting for the first time. On O'ahu sidewalks, seedlings appeared as splitters and destroyers of concrete.

BAOBAB

The oldest relationship between humans and ancient trees naturally occurs in Africa. The continent's longest-lived tree is the largest, too. Amazingly wide for its height, a mature baobab appears otherworldly. Leafless for most the year—an energy conservation strategy—its branches resemble roots marooned in the sky. According to traditional stories, the original baobab was planted upside down as punishment by gods, heroes, or hyenas.

The "upside-down tree" also goes by "elephant tree." The connection between Africa's greatest megaflora and megafauna goes beyond size. The calloused bark of a baobab is elephantine in color and texture. Botanists speak of pachycauls (thick-stemmed plants) as a cognate of pachyderms (thick-skinned mammals). Moreover, bush elephants consume the bark. In the dry season, tusked males gouge the trunks, peel away strips, and chew their fibrous trophies.

Baobabs heal over wounds that would kill other trees. They are among nature's apex regenerators. Their wood contains a high percentage of living cells, and a high percentage of water—up to 80 percent. Contrary to popular belief, a baobab doesn't perform hydraulic storage like a barrel cactus. Rather, it uses all that watery tissue to prop itself up. The tree is elastic, swelling and shrinking over seasons and stages of life. To keep its spongy mass intact requires a special outer later, like a rind.

From the inside, old baobabs hollow out, producing roomy recesses. Their uses are limited only by imagination. In a haunting Afrikaans-language novel, a woman escapes enslavement by confining herself in a tree: "You, trusty baobab, confidant, home, fort, water source, medicine chest, honey holder, my refuge, my last resort. . . . You protect me. I revere you."[31] The useful emptiness of baobabs had impressed Ibn Battuta, one of the best-traveled persons of the fourteenth century. He saw trees of "great age and girth" on the road to Mali. "I was surprised to find inside one tree, by which I passed, a man, a weaver, who had set up his loom in it and was actually weaving."[32] In Mali and in Sudan, on the western and eastern edges of the Sahara, Africans introduced baobab, and later generations burrowed out giants to form networks of cisterns—infrastructural trees that in time became war targets.

If humans are destroyers, what explains the coexistence of hominins and baobab over millions of years? The longevity of the relationship can be expressed in economic terms as a dynamic of uselessness and usefulness. The tree's absorbent tissue barely qualifies as wood—no good for building, burning, or charcoal-making. Besides, a fat-stemmed succulent is unchoppable. One tree can defeat a bulldozer, as British planners discovered during their abortive East African Groundnut Scheme. Baobab never grew as forests that farmers burned for farmland. Instead, isolated giants of the savannah inspired pastoralists and agriculturalists alike to situate their camps and villages nearby, and to plant future giants in the neighborhood. Humans long ago succeeded monkeys as the main dispersers of the species.

In addition to shade, shelter, and storage, African baobab gives foods, medicines, and textiles. Its velvety seedpods contain roastable seeds surrounded by vitamin-rich pulp that can be eaten raw or processed into meal. (In French, the species goes by "monkey bread tree"; in Afrikaans, "cream-of-tartar tree.") Leaves can be cooked; roots can be nibbled. People peel the bark and convert it into rope for weaving. If debarking is performed properly, and if the tree is allowed to heal, the process can be repeated in future years. All over sub-Saharan Africa, ethnic groups devised customary rules to manage the utilization of this resource that combined the properties of wild organism, crop plant, and sacred tree. Among the Dogon people of Mali, for example, tree guardians wearing terrifying masks patrolled communal baobabs.

The tree's genus name, *Adansonia*, honors French naturalist Michel Adanson. He arrived in Senegal in 1749 and set to work filling his enormous collecting cabinet. He beheld his first baobab on Gorée Island. On another island near Dakar, he observed a huge specimen with overlaid names and dates carved by European voyagers as far back as the fourteenth century—imperial claims of possession. This palimpsest caused Adanson to muse that the lives of these giants "must continue many thousand years, and, perhaps, reach as far back as the deluge." He called baobab both the "largest vegetable creation in nature" and "the most ancient living monument in the earth."[33]

In his magnum opus, *Familles des plants*, Adanson created an idealized growth chart to offer "an idea of the duration of these monstrous trees." Based on diameter and height, Adanson guesstimated that imposing African baobabs might be 5,150 years old—just shy of the age of the Earth according to biblical chronologies. Adanson made no claim of accuracy, but he promoted the idea that a naturalist could, with sufficient growth data, age-date a tree without relying on unreliable "tradition."[34]

Ironically, scientific legend grew up around the inscribed baobab in Senegal. In the enhanced version of the story, Adanson makes an incision into the cambium layers adjacent to the dated tree graffiti. By measuring the widths of the excised tree rings, he calculates the annual growth rate for mature baobabs. Separately, he observes the growth rate of young baobabs. These data combined with the tree's height and girth—exactly measured, of course—allows the naturalist to reliably estimate the age: 5,150 years. Thus, the ingenious Frenchman offers proof of extreme longevity in trees, a topic of speculation since the Greeks and Romans.

This apocryphal story felt veracious to European scientists enraptured with metric precision. French dictionaries and encyclopedias of the post-revolutionary period listed baobab as the "Thousand-Year Tree" (*L'arbre de Mille Ans*). On the strength of Adanson, Alexander von Humboldt—the most esteemed naturalist of the nineteenth century—called baobab "one of the oldest inhabitants of our globe." Adanson's "discovery" lasted in textbooks in spite of debunking by US botanist Asa Gray: "The vitality of an erroneous statement is truly wonderful," he wryly observed. Charles Lyell sustained this error in his groundbreaking *Principles of Geology*.[35]

Darwin read Lyell on the *Beagle*, which made first landfall at the Cape Verdean island São Tiago in January 1832. While traversing a valley near Praia, the port city, Darwin beheld his first giant tree, a baobab—a species he knew from Humboldt. "This one bears on its bark the signs of its notoriety—it is as completely covered with initials & dates as any one in Kensington Gardens," wrote Darwin. He returned to the site with measuring instruments. He doubted any plant could endure 6,000 years, though this one "strikes the beholder that it has lived during a large fraction of the time that this world has existed."[36] However, the erosion of the valley that sheltered the tree must have taken much longer, he noted. As of 1832, baobab grew at the border of biblical and geological time.

Today, it's unclear if *Adansonia* naturally occurred in Cape Verde before the fifteenth century CE. The islands had no human population before the Portuguese turned them into a human trafficking hub. The presence of baobab, introduced or not, marked the islands as African spaces. At São Tiago, like Gorée, enslaved persons awaiting their trans-oceanic uprooting would have seen these trees as vestiges of home. Adanson observed that Wolofs in Senegal wore tobacco pouches on their necks with second pockets for baobab seeds and other treasures. On the Middle Passage, slaves may have carried such pouches, while slavers may have stocked pulpy seedpods as anti-scurvy edibles. On the other side of the Atlantic, in the Caribbean islands, displaced Africans planted slow-growing baobab, specimens of which have survived to this day as landmarks of cultural tenacity.

The baobab diaspora across—and alongshore—the Indian Ocean started earlier and lasted longer. Historic *Adansonia* specimens, up to eight hundred years old, exist from Zanzibar to Oman to Iran to India to Indonesia. Baobab followed the movements of traders and later of captives taken by Arab, Portuguese, Dutch, French, and British enslavers. In contemporary India, many baobabs co-occur with the shrines of Muslim saints. Some of these shrine trees are maintained by the Siddi, an ethnic group descended from East African Bantu speakers. Some baobabs have been Hinduized. One of India's oldest grows near the Ganges outside Allahabad, an ancient pilgrimage site. Seemingly out of place, this *Adansonia* gets confused with *Ficus*.

Arab traders also brought African baobab to Madagascar. A remarkable specimen grows in the port city Mahajanga, in the center of a roundabout, meters from the beach. Its planting there was a kind of reunion, for six distinct baobab species grow endemically on the island. Seedpods from proto-baobabs likely crossed the Indian Ocean eons before Malagasy canoes and Arab dhows. Humans are not the only long-distance dispersers. Migratory animals and ocean currents work wonders, too, as suggested by western Australia's single outlier species of baobab, which genetically might be the oldest.

The validation of extreme longevity in African baobab only recently moved beyond the musings of Adanson. Because of their great girth and unusual morphology, baobabs have never been well suited for tree-ring dating. Radiocarbon works better, provided that inner tissue can be obtained. Firm evidence that baobabs can live one thousand years came in the 1960s, when two things coincided: the calibration of the radiocarbon dating method, and the construction of Kariba Dam on the Zambezi River, a megaproject that required mass demolition of megaflora.

More recently, a Romanian chemistry professor secured funding to research the maximum age span of angiosperms. The project got off the ground because one of the most famous baobabs—the Grootboom of Namibia—had just collapsed, allowing access to ancient inner wood. The investigation appeared to be prophetic, or cursed: wherever the professor went looking in southern Africa for the oldest baobabs, he found fallen giants. (Strangely enough, he moonlights as an authority on paranormal phenomena.) In 2018, in a prestigious academic journal, the professor reported that of fourteen known baobab millennials, ten had buckled or perished in the twenty-first century, including iconic specimens in South Africa and Botswana. The outlier had, before its sudden death, reached approximately 2,500 years, quadruple the age of ancient olives.

This limited, emblematic sample made climate crisis headlines around the world. "It's sad that in our short lives, we are able to live through such an experience," said the primary investigator. A coauthor called the mortality cluster "a canary in the mine."[37] Botanists questioned aspects of this study, but no one doubted the takeaway: southern Africa will continue to get hotter and drier, shrinking the habitat of baobab.

Long before the global media attention, Zimbabweans spoke anecdotally of baobab decline. According to some, the mysterious "black soot" on trunks was the doing of spirits angered by misbehavior in the present and the past. A Black nationalist, pro-government media source suggested that the "curse" began in 1855, when David Livingstone reputedly carved his name on one or more big baobabs near Victoria Falls (which he brazenly named). These profaned trees held ancestral spirits of the Tonga people. According to microbiologists, the unsightly fungus is endemic, and not by itself fatal. It may signal stressors like drought and overuse. In Zimbabwe, fly-by-night bark collectors have damaged baobabs in violation of customary rules.

The African baobab seems destined for further domestication and, in cities such as Dakar, street tree status. Programs are underway—combining lab science and Indigenous knowledge—to create superior cultivars for propagation as cosmetic and pharmaceutical material, and as crop trees. This is an example of "conservation through utilization." Baobabs take decades to reach sexual maturity, so the fruit of this research will arrive in the future. In the meantime, the food industry has discovered baobab pulp, another "superfood" additive for discerning consumers of smoothies in the Global North. The harvesting of seedpods, outsourced to village women, brings cash to poor locales, though the market may not be sustainable.

For as long as ecologists have studied baobab—just one century—they've noted a paucity of young trees as well as centurial gaps between mature cohorts. This is characteristic of various slow-growing, long-lived trees. What explains it? The multiyear co-occurrence of atmospheric and soil conditions needed for seedling recruitment simply happens rarely. Now that people and livestock have greater footprints and hoofprints—and now that industrial countries have irrevocably altered the climate of the planet—who knows when that optimal sequence will recur in Africa?

It required human assistance for baobab to expand beyond its original habitat; and now it requires human care for baobab to remain. Twentieth-century Africa inspired two remarkable tree-planting NGOs: The Men of Trees led by Richard St. Barbe Baker, and later the Green Belt Movement led by Wangari Maathai. Of the two, Maathai's Christian feminist

Indigenous environmentalism seems capable of longer life. Maathai understood the intersectional possibilities of stewarding land, empowering women, and reforming government. As the Nobel laureate once said: "You have to nurture it, you have to water it, you have to keep at it until it becomes rooted so it can take care of itself."[38]

She was talking about a tree, and she was talking about so much more.

II.

MEMENTO MORI

TREE OF DEATH — CHURCHYARD BOTANY — ARBORETUM
NATION — CHURCHYARD HISTORY — TREE OF TIME

The first plant subjected to sustained scientific investigation in regard to longevity was European yew, a species that had already been used to death in much of Europe. In the British Isles, however, specimens by the thousands survived into great age because of their natural ability to regenerate, and their contingent location in churchyards. In the eighteenth and nineteenth centuries, these consecrated trees, which held religious meanings at the local level, acquired literary and scientific meanings at the national level. The deathly, deathless yew became the most poetic plant in the early United Kingdom. Gradually, the churchyard yew lost cultural resonance with the rise of garden cemeteries and urban crematoriums, and the decline of Anglicanism. In the late twentieth century, dendrophiles renewed the efforts of Victorians to catalog and date Britain's elderflora. Today, in the context of urbanism and secularism, the ancient yews of "God's acre" are both famous and unremembered, protected and insecure.

TREE OF DEATH

Taxus baccata (European, English, or common yew) may be the oldest extant tree species native to the landmass called Europe. It occurs from the Caucasus to the British Isles and in a few island environments in mountainous North Africa. As dated by the molecular clock, the *Taxus* family evolved around the end-Cretaceous extinction event. In other words, yew species have lasted through the entire age of mammals. *Taxus* does well in oceanic climates, and it thrived in the Tertiary period, when the planetary north was mild and humid.

Then came the Quaternary, when extreme climatic oscillations dried, iced, thawed, and re-iced the European subcontinent. Once a hotspot of conifer diversity, Europe was repeatedly, progressively deconiferized. During deep freezes, Europe's yews retreated to Mediterranean refugia. With each interglacial, the species faced stiffer competition from faster-growing angiosperms, particularly beeches. *T. baccata* requires decades to reach sexual maturity; and then, to reproduce, requires male and female members as well as avian seed dispersers. In the Holocene, the current (and possibly terminal) interglacial, the lowland Mediterranean grew too hot and dry for yews. The species' advantages—tolerance for shade, endurance over time—count for more in clement, stable climates.

Once established, a shrubby yew outlives—or outdies—almost anything. Under duress, it can slow or stop its growth, or grow sectionally. After catastrophic injury, it can restart life from the roots or from epicormic buds in the trunk—even from the stump. If a yew bough falls to the ground, the organism can initiate a new vertical lead. Most impressively, in a process called "layering," downward-growing branches root themselves, then grow new leads upward. A single old organism can thus comprise a tiny grove. A hollow specimen can even layer from the inside, filling its void with a new branch-cum-trunk that fuses with the old shell. Through layering, fusing, and hollowing, one tree can assume varied forms over centuries. Besides thickness and hollowness, there is no classic morphology to a long-standing yew. Each appears different, and differently eldritch. A yew can even switch its sex—male to female,

female to male. No other perdurable has so many ways to rejuvenate. No timeworn being can be so young.

The earliest evidence of the *Homo-Taxus* relationship is archaic, utilitarian, and deadly. The Paleolithic "Clacton Spear"—the oldest known woodworked object, uncovered at Clacton-on-Sea, Essex, in 1911—dates back some 400,000 years. That time was a prior interglacial stage when *Homo heidelbergensis* and *Homo neanderthalensis* walked the shores of future Albion. In the Holocene, sapiens displaced other hominins. Cave paintings, the first masterpieces of European art, imply yew in the form of weapons held by hunters. From bogs in northern Germany and Denmark, archaeologists have dug up hundreds of yew shafts and bows from the Neolithic. The ancient man dubbed Ötzi—mummified in Tyrolean ice for five thousand years—carried a stave of yew.

From archaic to medieval times, European bow makers preferred *Taxus* because its fine-grained wood combined resilience and tensile strength. The best bows were carved from staves containing two kinds of growth: the cream-colored sapwood, thin and bendable, and the honey-colored heartwood, thick and durable. With such a contrivance, a long-bowman could, without it cracking or breaking, create great potential energy in the yew, transferred kinetically to the arrow.

As domesticated animals gained economic value, Europeans had another reason to take heed of yew. *Taxus* is toxic. Every piece of the plant, save one, can poison ruminants, horses, humans—and human cancers, as now evidenced by Taxol, the brand-name bark-based drug. The exception is the aril, the fleshy seed-cup that turns bright red in fall. (Yews lack cones, despite being conifers.) The somber foliage—the most chemical part of the tree—occasionally shows up in Greco-Roman sources, and in pathology reports, as a means to suicide. In *Macbeth*, the Third Witch adds these ingredients to the cauldron: "Liver of blaspheming Jew / Gall of goat, and slips of yew."[1]

Folkways of yew grew throughout premodern Europe and most fully beyond the English Channel. In Britain, the plant's cultural richness reflected the coniferous poverty of the postglacial landscape. Only two forest conifers and barely more evergreens (including box, holly, and juniper) are native to the isles. Unlike Scots pine—restricted to the far north—yew occurred widely, excepting the marshy lowlands of the east. Before early modern conifer introductions, yew was the sole British species that offered Britons year-round shelter from rain and wind. All the colonizers of Hibernia and Albion—Celts, Romans, Scandinavians, Saxons, Normans—fashioned implements and meanings from yew.

In contemporary Britain, unplanted yews occur rarely. In soils too chalky for farming, *T. baccata* can take the form of eerie monospecies stands—canopies of understories. And on isolated cliffs too steep for human land use, individual yews can hang on indefinitely. More commonly, yews appear in parks and gardens, often in the form of hedges. With its dense, form-fluid growth, the species can be trained into any shape. Most strikingly, in churchyards, amid broken headstones, yews planted centuries ago survive as elements of parochial landscapes. This Irish and British tradition also occurs on the peripheries of France and Spain (Brittany and Normandy; Galicia and Asturias), a distribution that suggests, if not proves, a Celtic determinant.

Churchyard yews, like holy wells, represent historical accommodation between paganism and Christianity. The Church's crusade against sacred trees—symbolized by hagiographic stories of totem-fellers such as Saint Barbatus of Benevento, Saint Boniface, and Charlemagne—was never complete. From the fifth through the seventh centuries, at multiple synods, Church leaders reiterated that making prayers, vows, and offerings to trees and stones remained illicit, idolatrous, fanatical, foolish. A half millennium later, missionaries in Anglo-Saxon England expressed the same enduring disapproval. Eradication took so long that eradicators ended up appropriating: they sanctioned pilgrimage to landforms rededicated to saints.

Celtic Christians attended to yews in Ireland and Wales well before the conversion of England. An early witness was Giraldus Cambrensis (Gerald of Wales). In 1187, in Ireland, he observed more old yews than he'd ever seen—and the archdeacon had traveled widely, as far as Paris and Rome. According to Gerald, "holy men" from "ancient times" had planted yews at burial grounds and other sacred places, an Irish custom continued by abbots.[2] In medieval Wales, law protected such trees. The penalty for felling the "yew of a saint" was one pound—compared to fifteen pence for an undedicated yew.[3] Some offenders paid a higher price. Gerald reported that during the recent Norman invasion of Ireland, archers with Henry II laid "violent hands in the most irreverent and atrocious manner" upon some churchyard yews near Dublin. For their offense, God smote the king's bowmen with pestilence.[4]

In England, Norman rule ushered in an outstanding period of church building, including city cathedrals and far more numerous village churchyards. A consecrated enclosure typically contained a rectory, a chapel, a cross, a yew—and, crucially, the remains of Christian ancestors in the ground. Only in the High Middle Ages did it become common for the English to congregate their dead in churchyards, which they began to call "God's acre" (adapted from German, *Gottesacker*). The favored spot for interment was the sunny south side of the chapel, by the yard's main entrance, beneath the primary yew.

The new churchyard trees of England were semisacred, protected by custom and law, but not inviolable. A royal statute in 1307 alluded to controversies between vicars and villagers over yews, and noted that trees had often been planted to defend chapels from gales. Edward I clarified that such yews belonged to the Church, meaning their management fell under canon law. Nonetheless, the king prohibited clergymen from removing trees except if chapel repairs so demanded.[5]

Even as the Church guarded thousands of yews in parish churchyards, the Crown imported millions of staves from Continental forests. The Normans favored longbows—at the Battle of Hastings, from a distance, they shot the last Anglo-Saxon king in the eye—and the subsequent House of Plantagenet extended this lethal enthusiasm. With expertise from conquered Welsh bowmen, the English marshalled a new and devastating force: whole armies of high-volume longbow killers, men whose bodies deformed from the repeated stress of pulling the arm while twisting the spine. From the thirteenth through the sixteenth century, to the end of the House of Tudor, the Crown compelled its male subjects to own bows and practice archery. "The eugh, obedient to the bender's will," eulogized Spenser in his poetic catalog of trees.[6]

After the interstate carnage of the Hundred Years' War, the English returned to internecine violence in the Wars of the Roses. For each new conflict, they needed further supplies of what Shakespeare called the "double-fatal yew."[7] Munitions came from mid-elevation forests from the Pyrenees of Spain to the Carpathians of Poland, where *T. baccata* grew as an understory plant. In 1472, having perceived that a "great Scarcity of Bowstaves is now in this Realm," Edward IV imposed a special duty on ships from Venice and other ports that had previously supplied yew to England: for each ton of merchandise, four staves due.[8]

English demand for the tree of war inspired early instances of forest capitalism and forest conservation. Centuries before "rare species" became the lingo of conservation biologists, Europeans applied to yew the concept of rarity, or species-level scarcity—a limited supply of a strategic reserve. In Poland, in 1423, the king and his parliament agreed to a set of land laws, including one that recategorized yew as the exclusive resource of landowners, and redefined warrantless yew-cleavers as poachers.[9] In the neighboring Holy Roman Empire, the Crown granted limited fee-based privileges to private companies to harvest yews. Boats laden with two-meter staves floated down the Vistula, the Oder, the Elbe, and the Rhine, on their way to London. Controlled markets did not prevent depletion or waste. In the early 1500s, Bavarians, Tyroleans, and Austrians debated whether to ban or further regulate the monopolies, given the collateral damage to the overstory. By the time the first English-language archery book, *Toxophilus*, appeared in 1545, England had effectively introduced yew shortages to central Europe. Here, the species never recovered from rarity.

Toxophilus was dedicated to Henry VIII, a toxophilite who pushed old-fashioned technology on his subjects. For his own soldiers, though, the king provided firearms from Birmingham factories. In short order, the nation of longbowmen developed the original gun culture. The infamous King Henry also inaugurated the English Reformation. The dissolution of abbeys—many adorned with yews—was one topographical consequence of the break from Rome. To encourage national self-sufficiency, the House of Tudor encouraged naval buildup and the introduction of nonnative conifers from Europe and North America. At the dawn of the seventeenth century, the depleted state of Britain's forest cover was proverbial: "Let them that live longest, fetch their wood farthest."[10] Forward-thinking estate owners began tree planting.

Dissenters, the most protesting of England's Protestants, further reformed the landscape. In particular, Puritans engaged in widespread religious vandalism. They toppled crosses and attacked the iconic Glastonbury Thorn, but they didn't touch churchyard yews, even though parishioners used *Taxus* spray as fronds for Palm Sunday—the kind of ritual that Nonconformists despised as popery. Perhaps ancestral bones enwrapped by roots gave would-be desecrators pause. More likely, British yews no longer retained enough heathen or saintly associations to be worthy of iconoclasm. In Brittany, by contrast, the Catholic bishop of Rennes caused regional controversy by ordering the uprooting of all diocesan yews, the subject of ongoing popular traditions he considered paganish and contrary to the Counter-Reformation.[11]

England's off-and-on six-hundred-year history of internal warfare culminated in the English Civil War. In its aftermath, John Evelyn, diarist and virtuoso of horticulture, studied wood supply as a security issue. National power was naval power was tree power. In *Sylva*, his groundbreaking treatise from 1664, Evelyn urged the planting of useful "exotics" for the afforestation of the kingdom. His book—the first to bear the imprimatur of the Royal Society—followed his polemic against urban air pollution from coal burning. Evelyn's preference for wood was geopolitical, medical, sentimental—and ironic, given that his inherited wealth derived from gunpowder. The dendrophile had a soft spot for native conifers. "Since the use of *Bows* is laid aside amongst us, the propagation of the Eugh-tree . . . is likewise quite forborn; but the neglect of it is to be deplor'd," he wrote.[12] He praised its usefulness in woodworking—the source of "everlasting" axles—and for its shade, perennial verdure, and applications as hedges. Because of its plasticity, yew was ideal for topiary. Adapting French style, English royal gardeners bent the gloomy tree into fancy peacocks and scenes from the life of Jesus.

In a different section of *Sylva*, on extraordinary growth, Evelyn provided a short list of "monster" yews in British churchyards.[13] In the coming two centuries, gentleman scholars would become preoccupied with these slow-growing curiosities. Even as Britain completed its revolutionary transition from an organic to a mineral economy—leading the world into a fossil fuel future—Britons developed new relationships with elderflora. Through all the yew-filled wars following the Norman Conquest, and through the dismal entirety of the Little Ice Age, Norman-era churchyard yews and their Celtic Christian precursors had kept on living, dying, rejuvenating, thickening. By the time the coal-powered United Kingdom entered its domestically peaceful imperial phase, it owned the world's greatest collection of old-growth *T. baccata*—not forest trees, but consecrated plantings.

As Britons began to perceive themselves as moderns, they discovered new things in familiar yards: arboreal antiquities.

CHURCHYARD BOTANY

Being a barrister and a bachelor, Daines Barrington had both money and time to devote to his hobby, natural history. A fellow of the Royal Society, Barrington in 1767 introduced the "Naturalist's Journal"—a preformatted logbook, with columns and rows, one week per page, to facilitate daily recordkeeping of weather and other phenomena. Quotidian observation was the foundation of modern science. Through patient accumulation of data, the laws of nature might be revealed.

When it came to interpreting data, Barrington had more enthusiasm than erudition. On the problem of Britain's native trees, he took the position that yew came from "foreign" sources. "Every church-yard, indeed, proves that this tree hath been for many centuries introduced into England; it seems, however, very extraordinary that we should have no account when, or for what purpose, this so very general a practice hath so long prevailed with us."[14]

In the eighteenth century, European naturalists speculated on the origins and ages of British yews. Initially, the inquiry combined two scientific approaches—a Baconian search for rare, extraordinary, and monstrous life-forms; and a Linnaean study of species characteristics within a system of life. With yews, both perusals required precise measurements, specifically of the girths of trunks.

The first and greatest botanist to write about yews was Augustin Pyramus de Candolle. Based in Geneva, he founded the city's botanic garden, and held the university's first chair in natural history. Unfortunately, he never finished his life's work, a natural system of *taxonomy*—a word he invented—based on relationships among plant species. Candolle labored in the shadow of Linnaeus, who had successfully completed his artificial system based on sexual morphology. (Michel Adanson, the proponent of multimillennial baobabs, was another Francophone botanist who hoped to supplant Linnaean classification.) Candolle infrequently left Geneva but traveled the world in his mind through his vast network of correspondents.

In 1831, Candolle published a treatise on arboreal longevity, a work that remains remarkably fresh. To paraphrase the botanist: A tree is a plant that is two things at once, an individual and a collective. Trees don't die of old age; rather, they eventually succumb to accidents—so it stands to reason that some have reached extraordinary ages. Scientists should endeavor to find and date such trees. Just as people preserve documents and coins from antiquity, they should preserve ancient trees, for evidence as well as sentiment. By determining the ages of the oldest living things, scientists might be able to fix dates on the "last revolutions of the globe."[15]

Anticipating the science of dendrochronology, Candolle insisted that annual growth rings must be measured, not just counted. Using slips of paper, he marked with pencil the boundaries between tree rings. "The collection of these slips, not unlike those in the shops of tailors," allowed Candolle to tabulate average girth increase, decade by decade, in various species. For some trees, he had more data than for others. Yews were too rare. Based on three Continental samples, Candolle suspected that *T. baccata* had extreme age potential, beyond that of olive and Lebanese cedar, and second only to African baobab.

Through his readings, Candolle had learned of four enormous British yews, including a hollow specimen in Crowhurst, Surrey—roomy enough for a congregation, equipped with a door—that had succeeded a similarly monstrous yew in Hardham, Sussex, as a curiosity. Since Candolle couldn't travel to Britain, he solicited colleagues across the Channel to verify girth measurements and to extract sections of trunk for tree-ring analysis. No single person could track down every monumental tree: "The life of man is too short."

In Britain, Candolle found a champion in John Henslow. This Cambridge don—Darwin's favorite teacher, the one who recommended him to the captain of the *Beagle*—transformed the university's Renaissance physic garden into a modern botanical garden. In his other role as a reverend, the professor had access to parish records, where he found exact planting dates for certain churchyard yews, which allowed him to refine Candolle's girth-to-age formula. Henslow argued that age estimates for old yews ought to be reduced by one-third.[16]

Candolle's system was further improved by John Eddowes Bowman, a minor Mancunian banker who retired early to geologize and botanize. Bowman (whose surname suggested an ancestral yew connection) approached churchyard trees as data sources, and argued for a uniform system of girth measurement that excluded burls and other excrescences. Somehow, he got permission—or didn't—to remove multiple pieces from the boughs of two enormous specimens, just as Candolle suggested. For his tools, Bowman used a frame saw and a custom-built bone saw, or trephine (similar to the increment borer later used by dendrochronologists). Through his surgical work, Bowman documented exceeding irregularity in yew growth. Rings from the same year varied in width from tree to tree, and side to side. Based on his "experiments," Bowman concluded that young yews grew faster than assumed, old ones slower. In other words, he thought Candolle overestimated the young while underestimating the old. For many decades, Bowman's 1837 article was the last word on yew longevity.[17]

Georgians and later Victorians carried out a second inquiry into yews, this one about local history—or, in the terms of the era, antiquarianism. The timing of this scholarship can be determined with various editions of *Popular Antiquities*, a reference book compiled before the study of "vulgar customs" became professionalized as folklore studies. The original 1725 edition by Henry Bourne said nothing about yews, nor did the annotated edition prepared by John Brand in 1777. As expanded by Henry Ellis of the British Museum, the 1813 edition included a lengthy section on yew lore, derived primarily from literary sources.

This arbo-antiquarian turn was anticipated—indeed, precipitated—by graveyard poetry. The first half of the eighteenth century witnessed a vogue for sorrowful lyrics, especially elegies, set in churchyards with yews. At burials, rectors began reading self-composed elegies. Sentimentality entered British culture as the emotive power of evangelical revivalism halfway softened the spirit of Calvin. Leafy feelings *about* death superseded arboreal associations with *means* of death—bows, arrows, poisons. The plant that Chaucer called the "shooter ewe" became, in Dryden's phrase, the "mourner yew."[18]

Two foundational poems set the tone: Robert Blair's "The Grave" (1743) and Thomas Gray's "Elegy Written in a Country Churchyard" (1751). A century later, Tennyson's "In Memoriam" marked the apotheosis of evergreen melancholia. By that point, the shorthand "Old Yew" summoned a set of night thoughts and gloomy images, several of them immortalized by Gray. As a cultural phenomenon, his elegy has never been bested. It became meta-canonical as the most quoted, most memorized poem in English history. In it, the "rude forefathers of the hamlet" sleep in "many a mould'ring heap" beneath the "yew-tree's shade."

The boundaries between poetry and history, botany and geology—also amateurism and professionalism—appeared indistinct in Britain until the late nineteenth century. The Victorian bourgeoisie, at least the male half, took on scholarly pursuits reserved for elites in other nations. Although excluded from the Linnaean Society of London, citizen botanists could read papers before any number of regional societies and local clubs devoted to natural history. Britain's unprecedented modernization—a fossil fuel economy, a large and literate middle class, an urban majority—permitted an unparalleled study of rural nature. Descriptive research on the countryside filled libraries and parlors.

For example: *Magazine of Natural History* in 1833 ran a piece on churchyard botany. The author, a clergyman from Warwickshire, applied a Byronian phrase about the Colosseum—"a noble wreck, in ruinous perfection"—to a churchyard yew in Kent. This "vegetable ruin" was an "interesting relic": "Oh! Couldst thou speak!" Following scientific protocol, the rector-naturalist wrapped a measuring tape around the trunk, in contrast to the boys of the day who mutilated old trees for "sheer sport and mischief" in the "spirit of modern Vandalism." In a lengthy, digressive footnote, the naturalist ended his piece with a hoary question: "Was the yew brought to the church, or the church to the yew?"[19]

Victorians had answers—a surplus of notions they presented as "theories." They held up pagans, Christians, Celts, and Romano-Britons as possible instigators of the British yew tradition. Maybe churchyard yews marked the locations of ancient springs or barrows, or the miracles of saints, or the cells of anchorites. Maybe Christians planted them for spiritual protection against witches or sanitary protection from corpses. Perhaps churchyards served as munition storehouses for archers. Or possibly it all began with the absence of fronds for Palm Sunday processions.

The two lines of scrutiny—on the longevity of *T. baccata*, and on the origin of churchyard yews—crossed at Fortingall, a Scottish hamlet in Glen Lyon, Perthshire. Here, a far northern yew had been put on the academic map by Daines Barrington, who measured it twice over twenty years, as it decayed from a monstrous 52-foot circumference to a bifurcated tree tunnel. By the nineteenth century, the hollow was wide enough for funeral processions. The split trunk became a shell, then split further—one tree transformed into many, like a fairy ring—all the while producing new spray. Well-read travelers cited Candolle to validate the "oldest tree" or "oldest vegetable remain" or "oldest living thing" in Europe, the greatest antiquity of Scotland. Inevitably, locals began pilfering branches and re-selling them to tourists. To discourage souvenir hunters and pyromaniacs, the parish erected an iron railing, and later a stone enclosure.

Fortingall gained competitors. Influenced by poets, antiquarians, and naturalists, rectors started claiming that *their* churchyard yew was the oldest in Britain, or the only tree listed in the Domesday Book, or another relic from the time whereof the memory of man runneth not to the contrary. Fortingall's parson played this game. Taking advantage of Victoria's midcentury holidays to the Highlands, he penned a poem, the *Queen's Visit*, which named Fortingall the birthplace of Pontius Pilate. However weakly, this local legend about Roman Scotland connected the sempiternal yew to Jesus Christ, defeater of death. No yew could top that.[20]

The former stature of Fortingall can be measured in the extracurricular activities of Sir Robert Christison, professor of medical jurisprudence. In 1879, this towering figure in Scottish academia—forensic expert on arsenical poisonings and staunch opponent of female education at the University of Edinburgh—delivered a three-part lecture on the "exact measurement of trees." The octogenarian toxicologist, who had recently climbed a 3,000-footer in the Highlands with a boost from coca leaves, presented a literature review on plant longevity, followed by an analysis of new data, personally collected, on the growth of churchyard yews. Christison concluded his heroically pedantic discourse with a new calculation of the age of the senior Caledonian: 3,000+ years, on a par with recent reports about monster trees in California.[21]

Along with arboreal antiquities, Britons reevaluated architectural monuments, especially Gothic ones. Centuries after dissolution and plunder, Cistercian abbeys underwent cultural and physical renovation. In 1767, an affluent connoisseur of gardens and ruins purchased Fountains Abbey, North Yorkshire, and remade it into a must-see sight on the national tour—the domestic answer to the grand tour of Europe. Trained by graveyard poets, tourists saw monastic trees with new eyes. The row of large yews at Fountains Abbey—supposedly the shelter of the original monks—became known as the "Seven Sisters."

These garden-like sisters had a wild-growing complement: the "fraternal Four" that stood side by side on a mountainside above Borrowdale, Cumbria. This alliterative name came from Wordsworth, a great promoter of Lake District tourism.[22] Practically in his lifetime, the image of England's mountainous northwest transformed from repellent to attractive. To Wordsworth's romantic eyes, these unconsecrated yews—a natural temple—were even worthier of note than a nearby churchyard yew, the pride of locals. As the fame of the poet grew, the yews he christened became Wordsworthian attractions in themselves.

It was a sign of modernization that Europeans gave so many new meanings to European yew in such a brief span of time. Ironically, in the same decades that *T. baccata* became important in British science and literature, Britons created novel landscapes and modern traditions that would render yews anachronistic.

ARBORETUM NATION

As the metropolis became denser and smoggier, the countryside grew more piney. An unprecedented afforestation occurred in Britain, the least wooded Western nation, though not in the way John Evelyn had imagined. Modern Britain imported colonial and Scandinavian timber for naval ships, while extracting domestic coal—an underground forest of geological wood—for industrial factories. As peasants moved to the city, becoming proletarians, merchants moved to the country, becoming manor owners. The enclosure movement, which converted communal lands into hedged-off estates, obviated woodland forestry as traditionally practiced by rural commoners. Unlike modern France and Germany, Britain did not train state foresters as replacements. For arboreal assistance, landowners turned to private landscape gardeners.

In the process, educated Britons embraced a new type of private-public pedagogical space, the arboretum. The archetype existed in Paris, where French revolutionaries converted the Jardin du Roi into the Jardin des Plantes. As grand display, the arboretum borrowed from the royal garden; as encyclopedic collection, it borrowed from the physic garden (*hortus medicus*). Serious tree collectors grouped their specimens either geographically or taxonomically, following the Linnaean system.

The preeminent British arboretum was Royal Botanic Gardens, Kew, where public education mixed with imperial science. Drawing on the far-flung network of "botanical discoverers," and the patronage of Sir Joseph Banks—veteran of Captain Cook's first voyage on HMS *Endeavour*, adviser to George III, president of the Royal Society—Kew assembled the world's greatest herbarium and living collection of plants under the successive father-son directorships of William and Joseph Hooker. With the papers of Linnaeus already in London (purchased in 1783), the British capital became the nineteenth-century center of economic botany as well as botanical taxonomy.

Meanwhile, on country estates owned by the gentry and the nouveaux riches, landscape gardeners created arboretums with conifers from Lebanon, California, Chile, and New Zealand. Because Britain had climatic "hospitality" for "exotics," it hosted an "experiment in arboriculture without parallel."[23] Tree connoisseurs—using riches extracted from slaves and coolies, cotton and tea, and bituminous coal—competed for the most species, the best specimens, and annual tree-planting medals from the Society of Arts. For their conifer collections, they used a pretentious coinage: "pinetums." These were sites of scientific curiosity and aesthetic consumption, not economic production.

Starting with Capability Brown, British landscape gardeners of middling class status became heroic figures for ornamenting the nation with greenery. Of these figures, William Barron loved yews best. He advocated for "British winter gardens"—coniferous landscapes that never lost their foliage. He called *T. baccata* the most beautiful evergreen in the Western world, especially now that nurseries stocked golden and variegated yews, and an "Irish yew" cultivar that grew in thick, upright form, ideal for hedging. Thanks to Barron, vegetable sculpture, which had been ridiculed by Alexander Pope, returned to style in the mid-nineteenth century. At Elvaston, Derbyshire, on former priory ground, Barron embellished a Gothic Revival castle with hundreds of yews clipped into cones, and a topiary tunnel that snaked around monkey puzzle trees from Patagonia. Barron specialized in transplanting old yews—using custom-built wheeled machines—to add Gothic romance to the estates of private clients.

Public pinetums existed, too, with John Loudon as their great champion. His oversized, illustrated, eight-volume *Arboretum et fruticetum Britannicum* (1838)—a disaster for his personal finances—was a garden in itself. It contained technical descriptions, practical tips, and innovative drawings of every tree and shrub thought to be tolerant of Britain's climate. This reference work could never have been finished without his wife-amanuensis, Jane Loudon, who had previously authored her own multivolume book, *The Mummy! A Tale of the Twenty-Second Century*, a speculative fiction about a revivified living antiquity.

For John Loudon, arboretums were key markers and preservers of modern civilization—Britain's answer to Egyptian pyramids (and the mummies therein). An outdoor tree museum was a place of both education and reflection. According to Loudon, the new industrial working class needed access to leafy landscapes to improve their knowledge, taste, manners, and morals. Never idle, Loudon spread his ideas through his own periodical, *Gardener's Magazine*, and various how-to books.

His philosophy benefited from timing. Right when horticulturists were advocating new green spaces, public health reformers were advocating the replacement of urban churchyards with garden cemeteries. This landscape change represented nothing less than a social revolution.

For roughly one millennium, the final destination for nearly every Briton was the churchyard. Here, the yew signified the inevitability of death, the shared future of all. In Welsh and in English, *sleeping under the yew* had popular meaning. At burials, bits of the consecrated tree accompanied the body's descent to root level. One of Shakespeare's characters thus imagines his end: "My shrowd of white, stuck all with Ew."[24] According to common law, everyone deserved funeral rites in their parish yard, regardless of church attendance or class status. Within the sacred space, a local community existed across long temporalities. As Thomas Hardy expressed in rhyme: "Portion of this yew / Is a man my grandsire knew."[25]

Given the close association between living yews and dead bodies, Britons naturally wondered about mutual effects. In the 1660s, one physician speculated that the original purpose of yew planting was botanical sanitation: the toxic tree, being attracted to "poysonus vapours," would, via its roots, imbibe "gross oleaginous gasses" emitted by putrefying bodies.[26] Thus, the hurtful became beneficial. In Blair's famous elegy, *Taxus* is the "Cheerless, unsocial plant! That loves to dwell / 'Midst sculls and coffins, epitaphs and worms."[27]

In early nineteenth-century London, churchyard decomposition became societal phobia. A representative proponent of exurban cemeteries portrayed urban burial grounds as "unroofed charnel-houses of putridity and corruption."[28] The miasma was moral as well as medical. According to critics, London's overladen churchyards attracted depraved persons— body snatchers, grave robbers, corpse composters, necrophiliacs. Charles Dickens wrote *Bleak House* (1852–1853) amid the sanitation crisis. In a mortuary scene, the narrator piles on ghastly adjectives to describe the profaned space: *pestiferous, obscene, malignant, beastly, savage, reeking, poisonous, shameful.* A character asks: "Is this place of abomination, consecrated ground?"[29]

Through the gospel of sanitation, Britons addressed their anxieties about immigrants and strangers living and dying cheek by jowl. The foremost reformer, Edwin Chadwick, developed an antiseptic dictum: "All smell is disease."[30] He viewed mortal remains not as a religious matter but as septic material. The solution would come from experts, not the clergy. An outbreak of cholera in 1848–1849 accelerated the rise of urban planners. A series of laws passed by Parliament in the 1850s decreed that London would henceforth exclude the dead. Deconsecrated burial grounds would become parks and playgrounds, as well as commercial properties.

London could close its churchyards because the city had gained a cemetery greenbelt between 1832 and 1841—the "Magnificent Seven." These garden cities of the dead promised health and pleasure for the living. The cemetery was everything the churchyard was not: a civic, cosmopolitan space; a commercial and nondenominational institution; a tourist attraction. The new memorial landscape encouraged visitors to think of death individually, not communally. Paris, a city that violently overthrew the old regime, invented the modern cemetery with Père Lachaise, a memorial park filled with celebrities and full of celebrity seekers. Great Britain, which had many more industrial cities, soon surpassed France as cemetery capital. Glasgow, Newcastle, Leeds, Manchester, Liverpool, and Birmingham each acquired exurban burial grounds.

Some cemeteries, notably Abney Park, were explicitly designed as arboretums, with taxonomic information labeled on trees for the edification of the public. Taking John Loudon's advice, cemetery designers lined curving paths with evenly spaced, neatly trimmed yews. To complement *Taxus*, they planted *Cupressus*, the funereal genus of the Mediterranean. The combination of cypress and yew—the "cypress of the north"—reinforced the historical claim that the modern cemetery shared a genealogy with the ancient necropolis. Loudon preferred prim, sober, salubrious evergreens to deciduous trees and flowers. However, he approved the adornment of individual graves with weeping willow—a showy introduced ornamental that symbolized grief.

In their effort to reform British burial practices, hygienists and horticulturists found allies among Nonconformists who hoped the growth of garden cemeteries would hasten the disestablishment of the Church of England. At the same time, certain dissenting Protestants agitated for the civic right to be interred in Anglican churchyards, for they liked the idea of waking up on Judgment Day beside their ancestors. Following contentious, drawn-out debate, a parliamentary amendment in 1880 separated religion from final disposition. Before, God's acre was the reserve of parishioners; afterward, just another cemetery.

Even within the "old mother church," death customs had been changing with the times. Memorials became more permanent and sentimental, with biographical information, not simply *Hic jacet* (here lies), *Memento mori* (remember death), or *Ut hora, sic vita* (life is as an hour). When the eldest daughter of Charles and Emma Darwin died of scarlet fever at age ten, her parents commissioned a headstone with a heartful message: A DEAR AND GOOD CHILD.

The Darwins buried two more of their children under the impressive hollow yew at St. Mary's Church, down the road from the family home in Downe, Kent. Charles had considered becoming a rector, for married rural life—supplemented by a voyage around the globe—suited his work habits. In 1881, less than a year before he died, the empire's greatest naturalist sent a letter to his friend Joseph Hooker at Kew. "I am rather despondent about myself," he wrote, "for idleness is downright misery to me, as I find here, as I cannot forget my discomfort for an hour. I have not the heart or strength at my age to begin any investigation lasting years, which is the only thing which I enjoy; and I have no little jobs which I can do. So I must look forward to Down graveyard as the sweetest place on earth."[31]

When death at last relieved him of pain, the state displaced his remains. Charles Darwin no longer belonged to St. Mary's in Downe but to Westminster Abbey in London. There, his body still rests in soilless splendor, alongside monarchs and heroes, safe from roots and worms. His wife, who did so much to create and maintain her husband's work environment, did not receive the same regards. Emma's gravesite is parochial, familial, and arboreal—another "yew-arched bed."[32]

Over Darwin's forty years in Downe, Anglicans became a minority in the United Kingdom, and urbanites became a majority. Considering the number of Britons living (and dying) in India and Africa—not to mention all the Anglo citizens of self-governing colonies—the poetic image of the rural English yew-side funeral was outdated by the Edwardian era, even before cremation, legalized in 1885, became common. The horrific Great War further disturbed burial practices. In the conflict's immediate aftermath, the Imperial War Graves Commission struggled to inter and remember hundreds of thousands of servicemen who died far from their places of home and worship.

The hollow yews of churchyards remained in place, yet grew out of time—the social time of townspeople, living and dead, and also geological time. By creating garden cemeteries and arboretums, Britons had both accelerated the postglacial return of European conifers and redirected evolution by introducing non-European conifers. Although the British Isles had more yews than at any point since the last interglacial, the elders, except a small fraternity of the famed, lost conspicuousness.

CHURCHYARD HISTORY

Compared to Victorians, twentieth-century Britons couldn't agree on the meaning of ancient yews, or whether churchyard yews were ancient after all. Over the decades, select English yews acquired the historical significance of English oaks, witness trees of the royal past. Synchronously, through a sequence of geographical dictionaries, emblematic yews got younger, younger still, and finally old again.

In 1897, the first such gazetteer appeared. The author was John Lowe, avid gardener, fellow of the Linnaean Society, soon-to-be Physician Extraordinary to His Royal Majesty, and reader on the fishes of Norfolk—the kind of gentleman scholar the British Empire produced like hothouse flowers. But Lowe's work could not have been composed without help. The dedication, "to my wife," concealed more than revealed, given that research correspondence addressed to "Mrs. Lowe"—for a planned second edition—kept arriving years after John Lowe's death.[33] The research couple relied upon a pastoral network. In response to requests, local priests made girth measurements and supplied descriptive and historical information from churchyards and parsonages.

Through their correspondence, the Lowes compiled an annotated list of yew sites. After centuries of commentary, this was the first serious attempt to catalog these common rarities. (The Ordnance Survey, a key project of the modern state, had mapped every churchyard but rarely identified individual trees.) John Lowe referred to his indexed yews as "notable," not ancient. In his opinion, age-dating formulas derived from Candolle ranged from "very fallacious" to "entirely fallacious." The historical importance of *Taxus* required no embellishment. Equal to the oak, the yew had served "to place England in her present exalted position."[34]

Quercus and *Taxus*—"the manly oak, the pensive yew" in Walter Scott's pairing—functioned differently in British memory.[35] In former royal forests, modern Britons individuated ancient oaks and turned them into heritage sites. They imagined that specific trees had witnessed events leading to the Glorious Revolution, the origin of constitutional monarchy. Historic oaks stood for royalism and also constraints on royal power. As part of their historicization, trees received personalized names: the Parliament Oak, Wallace's Oak, the Oak of Reformation, Queen Elizabeth's Oak, the Royal Oak. Such legendary trees (alongside Shakespeare's mulberry and Newton's apple) functioned for English Protestants not unlike how saints' yews once functioned for Irish Catholics. They imbued the landscape with venerable meaning.

For some modern Britons, olden oaks had additional associations with mysterious priests from antiquity. Druidomania affected England in the eighteenth and early nineteenth centuries, when antiquarians mused that Stonehenge had been a temple. A series of speculations became just-so stories: Druids venerated oaks, and this veneration dated back to the Oak of Mamre mentioned in the Torah. Druidism was the Ur-monotheism that anticipated Judaism and Christianity, the primordial religion of the patriarchs, the long-lived figures who existed between Adam and Abraham. Keepers of the pure patriarchal religion migrated to Gaul, then Britain, bringing oaken traditions with them, building Stonehenge within a sacred grove, before their religion degenerated into idolatry, and their oaks disappeared.

In 1827, at the tail end of Druidomania—before biblical scholarship and geological theory stifled it—one author added *Taxus* to the list of Druidical holies. In a wild book of pseudo-scholarship titled *Celtic Druids*, a Yorkshire magistrate elaborated on the notion that the original Druids had been ancient Indians, for whom the Levant marked a midpoint in their diasporic history. It was no coincidence, he said, that the Irish word *iubhar* (yew) sounded something like the Hebrew word *YHWH*. The "longest lived tree in the world" was in fact the tree of God.[36]

In the mid-nineteenth century, a *Taxus* wildwood in Surrey—one of the few such microhabitats in Britain—acquired a Druidical varnish. Located in Norbury Park, a country estate with public footpaths, this tangle of yews attracted ramblers, who mused on time and history while visualizing priestly tree circles on the shrouded slope. Inspired by his walks in "Druid's Grove," Surrey-based novelist George Meredith wrote a poem that begins and ends: "Enter these enchanted woods, / You who dare."[37] The "grove" eventually became so popular that Surrey County purchased it as parkland.

In contrast, consecrated yews didn't attract legends of Druids and monarchs—with one royal exception. A thick, fluted specimen by the Thames at Ankerwycke, near a former Benedictine priory, was reimagined by William Thomas Fitzgerald, aka "the small-beer poet . . . who has written more verses against Buonaparte than any man living."[38] In

Fitzgerald's patriotic verse, this yew became the possible location where rebel barons convoked before signing the Magna Carta, and where Henry VIII later "woo'd the illstarr'd maid" Anne Boleyn.[39] Thus historicized, the poetic plant earned a place in guidebooks, though without a name like "King's Yew" or "Boleyn's Yew."

From the nineteenth century onward, Britons identified yews by place-name, typically the name of the associated village: Ashbrittle Yew, Buckland Yew, Crowhurst Yew. As a species designation, the word *yew* could be treated as English heritage: a metonym for the longbow. As archers sing out in one of Arthur Conan Doyle's novels: "So men who are free / Love the old yew-tree / And the land where the yew-tree grows."[40] But as individual plants under which national history happened, ordinary yews had far less witnessing power than rarified oaks.

Rather, they witnessed the parochial and the quotidian. To contemplate a yew in historical time meant to imagine the daily, weekly, monthly, and yearly cycles that comprised the common lives of parishioners, from christenings to burials, across generations. This was the tree where locals gathered for mystery plays and end-year celebrations; where pastors pruned "palms" for Lent and red-on-green ornaments for Advent; and where children snacked on sweet-and-slimy "snottle-berries," being careful to spit out the seeds.

In 1912, London schoolteacher Walter Johnson produced the first study of yew customs since *Popular Antiquities*. Influenced by German academics, Johnson put antiquarianism in new terms: "prehistory" for the ancient past, "folk-memory" for its resonance today. In *Byways in British Archaeology*, Johnson summarized Victorian scholarship on yew longevity—"plethoric," he called it. He couldn't resist adding to that girth, though his main interests were beliefs and traditions, such as casting off devils and warding off witches. Johnson argued that emblems and landmarks provided access to "latent," "unconscious," or "subliminal" memories of racial groups. In a nod to James George Frazer's *The Golden Bough*, Johnson concluded with his "instinctive feeling" that the modern English reverence for churchyard yews derived from "customs of ages exceedingly remote."[41]

The strongest indications of pre-Christian sacred yews in the British Isles came from Ireland, not England, as evidenced in place-names and poetry. Early Irish manuscripts contained references to *bile*, a type of sacrosanct tree. As the site for royal inaugurations, a bile demanded veneration from allies, while attracting violation from rivals. A bile could be an ash, or oak, or yew. There was scant evidence for yew's preeminence in Celtic tradition, and no evidence for Druidical rites beneath the poisonous branches. Whereas John Lowe dismissed the latter notion outright, Walter Johnson reserved judgment.

The ambiguous, heterodox possibilities of yews impressed Marie Stopes, paleobotanist-turned-sexologist. In 1922, the scandal-seeking activist delivered to Anglican bishops *A New Gospel to All Peoples*—the good news of mutual orgasm and non-procreative coitus, written in God's voice. In the prefatory note, Stopes explained that she received the revelation word for word on a day spent alone "in the cool shades of the old yew woods on the hills."[42]

Outside the woods, and inside the churchyard, yews could just as easily represent conservative values in the twilight years of the Empire. During World War II, the independent scholar Vaughan Cornish, son and brother of rectors, assembled a new gazetteer. Over his life, Cornish had transitioned from geomorphology to landscape aesthetics, becoming a leading advocate for British national parks. Prior to that, he achieved prominence as an expert on imperial geography and "race theory." In 1925, he argued that maintaining the global proportion of white people was "a eugenic movement tending to the well-being of the world."[43] This ethical imperative would require each married white woman to give birth to least four children. Cornish himself had no children, and no regular employment, supported as he was by his wife, Ellen, an engineer.

In his slender volume *The Churchyard Yew and Immortality* (1946), Cornish searched for existential beauty in the countryside of England, Wales, and Normandy. For him, girth and age didn't matter as much as aesthetics. He was interested in timeless values, not measurable ages. He saw the yew as an ideal Christian symbol, and approvingly noted that Anglo colonists had transplanted the tradition to Tasmania. In the home country, churchyard trees stood as topographical relics of medieval land-scapes. By sending questionnaires to Anglican bishops, Cornish enumer-ated more historic sites than Lowe—roughly five hundred.

In the period around the two world wars, botanists created new meanings for unchurched yews that likewise went beyond age-dating. Post-Candolle tree-ring science had found a home in the semiarid North American West, where long-lived conifers rarely hollowed out from rot. British scientists such as Arthur Tansley transitioned from yew dendrology to yew ecology, building on the work of German and Polish researchers. Tansley and his journal, *New Phytologist*, represented the post-Victorian professionalization of botany. Rather than individual trees and flowers, Tansley studied plant biogeography and vegetative communities. He cherished the *Taxus* wild-wood at Kingley Vale, West Sussex, which he brought to the attention of Charles Rothschild, founder of the Society for the Promotion of Nature Reserves. Through Tansley's ecological advocacy, the state preserved this microhabitat in 1952 as a Site of Special Scientific Interest, using the new authority of the National Parks and Access to the Countryside Act.

In this postwar time of budding environmentalism, a new reference book on yews appeared. The author, an 88-year-old mycologist and for-mer natural history curator, was even more dismissive than John Lowe of girth-to-age calculations. The longest-lived yews may be 400 years young, he said. Very few exceeded 1,000 years. He observed that vicars used hollow yews as sheds for coal, coke, diesel, or tools, and as recepta-cles for clippings, weeds, and trash.[44]

A generation later, when yews reappeared in British consciousness, they became ancient again, and sacred again—only differently so. For neopagans and modern Druids, who espoused a kind of magical anti-modernism, yews joined oaks as coequal emblems. The pivotal yew reviv-alist was Allen Meredith, a fey and shadowy figure. An infantry veteran of the Royal Green Jackets, Meredith pedaled around the countryside in the 1970s and 1980s, guided by dreams, channeling the voices of yews. On a practical level, he wanted to produce an authoritative update of Vaughan Cornish's gazetteer. He supplemented his sleuthing in church-yards and archives with intuitive, psychic, and visionary "research."

To Meredith, *T. baccata* wasn't just a Eurasian conifer but a cardinal species on Earth. It was the Tree of Life in Judeo-Christian cosmology. It was Yggradsil, the World Tree in Norse cosmology. The oldest specimen—5,000 years, he claimed for Fortingall—served as a genius loci and an axis mundi. Meredith's mix-and-match mysticism borrowed from William Stukeley, James George Frazer, Mircea Eliade, Carl Jung, Arthurian legends, Green Man images, runestones, and the Bible. He posited an exalted role for the British Isles in world religion. He believed that Druids transported yews across the Channel to protect the species. Britain was meant to be the sanctuary for the world's most sacred tree. Planted by ancient guardians, these plants now guarded over modern Britons.

Despite being disheveled and reclusive, Meredith had a gift for missionizing. He appeared in dailies and magazines and on public television as a "tree historian." The self-made expert warned that Britain's millennial yews were critically endangered. He made the fact-free claim that of one thousand ancient yews that existed as of World War II, half had been lost. When developers made plans for property near the Anker-wycke Yew, Meredith persuaded journalists and activists that the Magna Carta had been signed under that very tree. In due time, Druids and Wiccans began coming to Ankerwycke, offering ribbons, crystals, and stones during Samhain and Beltane.

New Agey assessments of yew longevity spread through establishment figures. Meredith's allies included Alan Mitchell, the leading dendrologist in Britain. Most consequentially in the age of TV, Meredith received endorsements from Robert Hardy and David Bellamy. Hardy was a respected historian of the longbow as well as a beloved actor (who would go on to play Cornelius Fudge, Minister for Magic, in the *Harry Potter* film franchise). Bellamy, a tele-naturalist, regularly appeared on BBC. Partnering with *Country Living* magazine, Bellamy started a yew campaign in 1988. Following the spirit if not the letter of Candolle, he asked readers to mail in girth measurements; in return, he would mail out certificates—cosigned by Hardy and the Archbishop of Canterbury—giving calculated ages. Bellamy's estimations were as outsized as his personality. He baited readers with the exuberant statement that Crowhurst might be 4,000 years old, and Fortingall, 9,000 years old.

Bellamy didn't stop. Through his charity, the Conservation Foundation, he convened a "Green Magna Carta" at the Ankerwycke Yew in 1992, the year of the Earth Summit. And in 1996, he launched "Trees for the Millennium," a yew-planting drive cosponsored by the Church of England. As the clock counted down to Y2K, the Conservation Foundation took thousands of cuttings from "ancient yews," propagated them, and, after ecclesiastical blessings, distributed them to parishes with certificates of age. Purportedly, the scions provided living links to the time of Jesus.

The only question was an age-old question: Was it all too good to be true?

TREE OF TIME

The latest, best gazetteer goes by the name Ancient Yew Group, an interactive website built on Google Maps. Notwithstanding the collective name, maintenance of the website falls to one self-effacing volunteer named Tim Hills.

A math teacher by education, a musician by inclination, Tim surprised himself by becoming Britain's expert on yew geography. Trees meant nothing special to him in his first five decades. But when his children approached adulthood, Tim and his wife, Gaye, went looking for hobbies. They joined the Bristol Naturalists Society and met a fellow member who wanted to compile a master list of yews in Somerset and Gloucestershire. Intrigued, Tim read *The Sacred Yew*, a compilation of Allen Meredith's research.

Alarmed by the thought that Britain had lost half its ancient yews in his lifetime, Tim straightaway drove to a Somerset churchyard to begin measuring and photographing. There he bumped into a man in his late eighties who wanted to know why Tim was interested in the parochial landmark. After Tim held forth on millennial yews, the elderly parishioner posed the simplest, hardest question: "How do you know?" And he challenged the idea that 1,000 years should be the threshold for arboreal antiquity. "The conversation had a profound effect on me," Tim recalled, "and I determined that I would become as questioning and skeptical as he was."

At age fifty-five, Tim quit his teaching job to research yews full-time. He had become disillusioned with the classroom following the National Curriculum for England. His wife's income as a doctor allowed him to turn his arboreal hobby into a full-time occupation. Over the next fifteen years, Tim visited over two thousand churchyards marked on Ordnance Survey maps. He worked alone, except for the companionship of BBC Radio 3, the classical music station.

Tim's conversion to skepticism was right for the times. Yew dating, stuck in the 1830s for a century and a half, improved in the 1990s, when new radiocarbon and tree-ring data permitted refinement of the girth-to-age formula. A recent scientific appraisal puts the majority of churchyard yews—of approximately two thousand total—under one millennium, with an upper limit of two millennia, which is still "ridiculously old," to quote one of the investigators.[45] The Ancient Yew Group uses three categories based on probable longevity: *notable* (300+ years), *veteran* (500+ years), and *ancient* (800+ years).

The churchyard functions as Tim's second office, a strange if appropriate outcome for someone who rejected Christianity. His father, a major in the Salvation Army, sold "life assurance" without compensation, and Tim as a boy helped collect penny dues, door to door. He wore a Salvationist uniform: a roll-neck red sweater with BLOOD & FIRE emblazoned on the chest. Today, Tim respects believers, including neopagans, despite considering religion nonsensical.

Nonsense is a favorite word—something Tim loves to hate and tries to correct. He initially included a forum for comments on the website, but quickly shut it down when it attracted visionaries, pseudo-shamans, and mad interpreters of King Arthur and the Knights Templar. Although Tim knows that the website will outlive him, he worries that future webmasters may not have the time, resources, or inclination to maintain his high standards. The ephemerality of digital information haunts him.

When I visited Tim's home in Bristol in 2017, he spoke of two personal milestones—his seventieth birthday and his twentieth anniversary with *Taxus*. He referred to the seventies as a decade fewer people leave than enter. He complained about his guitar hand seizing up. He wanted to record himself performing Heitor Villa-Lobos while he still could. "Soon I'll need a prop," he joked, "just like an old churchyard tree." His funeral music had been preselected: *The Unanswered Question* by Ives.

With Radio 3 playing, Tim drove me to Wales, past used-up coal mines, to a boarded-up church with a spooky, unpruned yew overgrown with ivy. Gazing at the empty churchyard, Tim offered me his own unanswered questions: "Does an ancient tree have any value on its own? If no one visits it, does it really exist?" The villagers and their memories had departed this place. Without the living, the deathless yew and the human dead both fall out of communion.

Wales is the *loco centri* for ancient specimens of *T. baccata*. Of yews in this country, approximately 95 percent grow in churchyards or former sites of worship, compared to 55 percent in neighboring England. In the coming decades and centuries, these two yew populations may experience divergent histories due to legal differences. In England, Anglican churchyards belong to individual parishes, not the Church of England; in Wales, the Property Department of the Church in Wales manages the old churchyards, which began losing parishioners well before deindustrialization. Indeed, Nonconformists (particularly Methodists) have been a Welsh majority since the nineteenth century, now followed closely by "nones," people who claim no religion. The dwindling Church in Wales has recently shut down hundreds of redundant properties, selling those it could. Hundreds more closures are imminent.

What will private owners of repurposed churchyards do with "their" yews? Could they create a new kind of country living that resacralizes the deconsecrated? The full answer will come in the early twenty-second century, when the hundred-year legal waiting period for closed cemeteries expires. "That's the crunch time," Tim told me, after which old bones can no longer protect old trees. Unlike historic sites and buildings, old yews are not "listed" under British law, though local planning authorities can issue individual "tree protection orders."

Tim is philosophical about the potential loss of churchyard yews. Although he would love to see Parliament pass a law that categorically protects the country's elderflora, he accepts that landscapes change as values change. He simply wants Britons to act informedly, deliberately, choosing to continue or discontinue this landscape tradition. "Do we really want to be the generation that ends this relationship with yews?" Tim asks. "If so, so be it."

Throughout rural Britain, three institutions anchored modern village life: the post office, the pub, the church. In the postwar period, many villages lost one, then two, then everything. Today, over 80 percent of Britons live in cities, and the percentage will likely rise. Currently, Anglican churchyards, like orchestral halls, are spaces for older white citizens. Amid the interminable negotiations around Brexit, Prime Minister Theresa May regularly attended Sunday services, and press photographers captured her striding past a yew as she entered St. Andrew's Church in Sonning, Berkshire, a wealthy village on the Thames, upstream from the capital. It was a perfect picture of anachronistic conservatism. Tories could no more revive Anglican village life than resurrect the British Empire.

Online, at least, parochial trees have the veneer of world heritage. Nearly all of Britain's churchyard yews are accessible as geocoded data in the cloud. On the ground, however, most of these trees lack informational signage—and visitors. Within a half-day's drive of London, scores of world-class trees exist, yet cosmopolitan Londoners, the kind of people who voted "Remain," are more likely to know Muir Woods outside San Francisco. To see a yew in Worcestershire or Wiltshire, one must motor down unmarked roads—narrow, twisting, hedged, and canopied—to destinations unrecognized by satnav.

A handful of yews have become "celebritrees"—chiefly those at Fortingall, Crowhurst, and Ankerwycke. For the Golden Jubilee of Elizabeth II in 2002, a charity called the Tree Council honored this trio among "50 Great British Trees." Dame Judi Dench, who has portrayed multiple queens onscreen, visited Crowhurst in 2017 for her BBC documentary

My Passion for Trees. Looking into the camera, this legend of the screen cheerfully admitted her tree-hugging habits, and spoke of pagans and Druids.

For secular, urban Britons, venerable yews may be more alive in fantasy than geography. According to J. K. Rowling's official *Harry Potter* website, *Taxus* wands are as rare as they are powerful. When buried with its owner, a wand of yew sprouts into a tree that guards the grave. Possessing the power of life and death, such wands choose to be handled by heroes and villains. Yews chose Harry's future wife, Ginny, as well as his mortal enemy. Voldemort carried a toxic wand, and *Taxus* grows in the graveyard where the Dark Lord returned from the dead.

There can be something off-putting about British uses of *Taxus*, from bowmen to topiarists to witches and wizards. How can one remarkable plant attract so many flawed supporters? Allen Meredith may be the extreme case. The year after *The Sacred Yew* appeared, he went to prison for sexually abusing minors. During the years he found an audience as a "tree historian," he worked as a housemaster at a boarding school for the children of Scottish servicemen. Meredith took favorite boys, ages 9 to 12, on walks in the woods and excursions to churchyards.

Long out of public view, Meredith reappeared in 2012 with a co-written book and a fantastical claim: a churchyard yew in Defynnog, Wales, was the true oldest tree in Europe at 5,000 years.[46] The "God Tree" sported an odd branch with yellowish spray—a genetic mutation interpreted as a mystical return of the golden bough described in the *Aeneid*. When media outlets publicized Meredith's age assessment, the Ancient Yew Group disputed the number, hoping to prevent the outcome at Fortingall. There, in view of signage about "Europe's—and possibly the world's—oldest living thing," visitors regularly snatch twigs as souvenirs. As with all heritage sites, there's a fine line between inadequate attention and too much.

Despite the depredations of tree lovers, the Fortingall Yew has avoided certain death for nearly three centuries, and may continue not dying indefinitely. The wall erected to protect its shell now constrains its growth. In 2015, astute visitors noticed arils on the crown of the tree. One branch of the ancient male had transitioned to female. Even familiar yews can surprise humans with their multitudinousness.

The shape-shifting ability of *T. baccata* is a boon to conservation biologists. The Royal Botanic Garden Edinburgh recently replaced its row of holly with a genetically diverse yew hedge composed of hundreds of tagged samples—each a slightly different shade of dark green—collected throughout its range. Part of the International Conifer Conservation Programme, this super-hedge in the Scottish capital is a living archive in miniature, a vast genotype in a compact area. Fortingall grows here, entwined with rarities from the Atlas Mountains. Perhaps no plant species can be genetically "backed up" so efficiently. The caretakers of the conservation hedge honor British garden history—topiaries and pinetums—as they mind the future.

Meanwhile, in West Sussex, thousands of carefully cleaned and canistered yew seeds lie in dry cold storage at the Millennium Seed Bank. This cryo-conservation bunker—the first such vault to prioritize noncrop plants—ranks as the world's greatest biodiversity hotspot. The building was designed to last five hundred years, halfway to the goal. The seed bank sits within an estate, Wakehurst, owned and managed by Kew Gardens. The Wakehurst mansion is Elizabethan, but its garden is both younger and older. Sequoias introduced in the nineteenth century tower above a yew planted in the fourteenth. Beyond the garden, on the manor's Rock Walk, yews of indeterminate age swallow sandstone ledges with tortuous roots that look positively earthen.

A limited but expressive range of English adjectives has for centuries been affixed to yew: *dark, dismal, gloomy, poisonous, fatal, sepulchral, solemn, melancholy*. The *OED* lists "symbols of sadness" as one of its leading definitions. But as trees of death, trees of life, yews symbolize something more: timefulness.

With apologies to T. S. Eliot, the moment of the yew and the moment of the rose are not of equal duration.[47] Yew time is slower, longer, weirder. Even at low-end age estimates, churchyard yews span historical divides—before and after the Black Death, the Reformation, the slave trade, empire, coal. Why and however this landscape tradition started, it benefits the species, for Britain's widespread churchyard yews cross-pollinate with isolated yew wildwoods, benefiting gene flow. Assisted longevity means greater chronodiversity means enhanced fitness at the species level. With or without God, the peoples of the Isles have reasons to care for God's acre "till the wind shake a thousand whispers from the yew."[48]

English-language prose on *Taxus* cannot match the poetry: "A living thing / Produced too slowly ever to decay; Of form and aspect too magnificent / To be destroyed."[49] Nonetheless, moments of fair wisdom exist in the amateur pages of Victorian natural history publications. My favorite comes from Leopold Hartley Grindon, a Mancunian cashier who taught botany on the side. In one of his books on plants, he likened yews to whales and bats—intermediary organisms that could serve as conceptual bridges. *Taxus* was a link to longer temporalities. "We often hear of 'railway time' and of 'sidereal time,'" Grindon wrote, referring to carbon-burning engines and hydrogen-burning stars. "The yew tree helps to enforce upon us the grandeur of the idea of 'tree time.'"[50]

The modern study of tree time started with *T. baccata*, and soon moved beyond it. Although the species has proven unfit for the data demands of tree-ring science, it remains ideal for long-term thinking. People never needed dendrologists to know that yews can be extraordinarily old. Throughout premodern Europe, proverbs listed the age potential of creatures in relation to each other; and in Ireland, these lists of old-and-older animals ended, remarkably, with a plant: "Three lifetimes of the salmon for the yew / Three lifetimes of the yew for the world from its beginning to its end."[51]

Proverbial tree time could be adapted for the present unperfect, when the planet that has been is being undone.

For the Holocene, beginning till end, six lives of hallowed yews.

All the Cenozoic, one lifetime of *Taxus*.

III.

MONUMENTS OF NATURE

HUMBOLDTIAN FLORA — WATER LORDS — HOMELAND PROTECTION — PROTECTED AREAS — THE NATIONAL TREE

After Spanish conquests in the Americas—foreshadowed by the takeover of the Canary Islands—scientific discussions about plant longevity included more kinds of trees. The most spectacular was ahuehuete (Montezuma cypress), a species of pan-Indigenous significance. One gigantic specimen, El Árbol del Tule, was eventually crowned "oldest living thing," a title it inherited from a dragon tree in the Canaries. For Alexander von Humboldt, the Prussian naturalist who made his fame in Latin America, landmark trees of the New World were "monuments of nature." In German lands, foresters and conservationists adapted and institutionalized Humboldt's idea as small-scale nature care. Some Germans cared to extremes. Before and after the ruination of World War II, industrial nations and intergovernmental organizations codified systems of protected areas, including natural monuments. Like all heritage, arboreal heritage is subject to politicization. In dictatorships as well as democracies, the interaction of state conservation and local care has produced varied outcomes for peoples and trees—from long life to total death. Mexican conservationists know the score.

HUMBOLDTIAN FLORA

In *Cosmos*, the all-encompassing multivolume magnum opus that crowned a career of universal acclaim, Baron Friedrich Wilhelm Heinrich Alexander von Humboldt reflected on the childhood inspirations that compelled him to become a globe-trotting naturalist-philosopher. Chief among them was a colossal dragon tree in an old tower of the Botanic Garden, Berlin. It and the greenhouse palms "implanted in my breast the first germ of an irrepressible longing for distant travel," he wrote.[1]

For Europeans, "dragon tree" (*Drachenbaum* in German) referred to a spectacular succulent—an arborescent asparagus—endemic to the Canary Islands, Cape Verde, Madeira, and isolated sites in the Anti-Atlas Mountains of Morocco. The rediscovery of the Canaries, the Azores, and Cape Verde in the 1300s induced "insulamania" in Europeans, who recalled ancient stories of the Isles of the Blessed, Hesperides, Atlantis, and other lost worlds.

The Crown of Castile claimed the Canaries by right of conquest—defeating and enslaving Guanches, who had migrated from North Africa centuries earlier. Castilian soldiers crushed the last resisters the same decade that Columbus plundered his way through the West Indies. All four voyages by the Admiral of the Ocean Sea included stopovers in the Canaries, the initial "New World." The words Columbus chose to describe Taínos—*naked, godless*—mirrored descriptions of Guanches, and vice versa. On both sides of the Atlantic, Spanish colonizers and their accessories—guns, horses, sugar, disease—turned aboriginal homelands into unfortunate isles.

For Europeans beyond Iberia, the Canaries remained for centuries semi-mythological, and *el drago* joined the symbology of late medieval art. In paintings, engravings, and tapestries, this exotic plant represented Eden, or Egypt, or the Orient. Its unmistakable upturned shape—a multiplying candelabra atop a single spongy trunk—appeared in works by Bosch and Dürer. As an icon, a drago was the vegetal equivalent of a unicorn or a griffin, only real. Seeds and cuttings found their way into cabinets of curiosity. Royal gardeners in Lisbon and Madrid tried to acclimate dragos—an effort foreclosed by the Little Ice Age.

The name of the plant, and a portion of its fame, derived from the "milk" or "liquor" tapped from its tissue, one of various red resins sold as "dragon's blood" in apothecaries in Britain and Europe at a premium price. Rare and valuable as living dispensers of medicine, dragos also had life-protecting value as war material. The fibrous trunks could be made into blade-swallowing shields, a resource controlled by the Castilian Crown.

A century after they invaded the Canaries with swords and cannons, Spaniards began writing histories of their exploits, and sharing descriptions of a lost tree on the island El Hierro—an aboriginal sacred site from which fountains of water miraculously sprayed. On more solid footing, chroniclers claimed that a gigantic drago on Tenerife, the main island, had hosted Canarian convocations. According to subsequent folklore, this tree in the village Orotava had been used as an altar by pagans, then Catholics, and served as a boundary point for dividing land dispossessed from Guanches.

Regardless of the truth of those particulars, this landmark predated the conquest, survived the upheaval, and ultimately became far-famed as El Drago de la Orotava. In a globalizing world, it became a secular axis mundi. In 1776, surveyors used it as a triangulation point for establishing the precise location of the Canaries on navigational charts. Its eighteenth-century owners, a noted family, did not extract dragon's blood; this plant had greater value as a garden curiosity. As Tenerife turned into a holiday destination, the tree El Drago joined the volcano El Teide as the island's must-see sights. In Orotava, tourists climbed a ladder to a viewing platform built within the succulent's stout branches.

Because of location and history, the Canary Islands could be imagined as Mediterranean or tropical, African, or even American. When Alexander von Humboldt visited in 1799, carrying an all-access royal passport stamped by Charles IV of Spain, he wrote home to his diplomat brother about his "state of ecstasy at finding myself at length on African soil," surrounded by palms and the "growth of a thousand years." "I could almost weep at the prospect of leaving this place," he emoted, "yet I am scarcely out of Europe."[2]

Barometer in hand, Humboldt raced to the thin-aired summit of the volcano, though not before he took his tape to Orotava, where he measured the prodigious plant's circumference as 45 feet (*Parisian* feet, not English nor Prussian). Monumental trees have always evoked feelings in people, he noted, but only in recent times has care been taken to numerically determine their sizes and ages. Without abundant data, the great naturalist made the categorical suggestion—later adopted by textbook authors as fact—that longevity corresponded with enormity.[3]

Humboldt was a genius of bold thinking and rushed inference. In his *Personal Narrative*, the best-selling travelogue of his five-year journey to and through the Americas, he confidently placed the dragos of Tenerife with the baobabs of Senegal as the "oldest inhabitants of our globe."[4] More generally, he expressed conviction that the Earth sustained trees as old as Rome, Greece, and Egypt. There might even be trees, he conjectured, that had witnessed the most recent physical revolution, by which he meant a catastrophic transition from one geological epoch to another.

In the pictorial atlas that complemented his instant-classic narrative, Humboldt included, as the final plate, an overimaginative rendering of Orotava's landmark.[5] Although presented as a scientific illustration, the visual language of the engraving derived from older religious iconography. Being the first mass-produced view of the monster succulent, it caused a sensation. Humboldt cemented the tree's status as the most illustrious single plant of early modern Europe. A distant competitor was the much larger "Hundred-Horse Chestnut" (*Il Castagno dei Cento Cavalli*) near Mount Etna, a waypoint on nineteenth-century grand tours, and the subject of legends and exaggerated age estimations. In 1745, in one of the earliest bureaucratic acts of tree protection, the royal viceroy of Sicily had signed a conservation order for this *monumento dell'insigne naturale portento* (monument of distinguished natural wonder).[6]

Even after El Drago lost half its crown in a hurricane in 1819, it retained wondrous stature in the imagination. It enchanted Charles

Darwin and filled him with "burning zeal" to do field science. He copied Humboldt's descriptions of Tenerife and read them aloud to friends and family in 1831. From Cambridge, he described his "tropical glow" to his sister. "In the morning I go and gaze at Palm trees in the hot-house and come home and read Humboldt: my enthusiasm is so great that I cannot hardly sit still on my chair," he gushed. "I will never be easy till I see the peak of Teneriffe and the great Dragon tree." To others, Darwin spoke of his "Canary ardor" and his "Canary scheme." "Nothing will prevent us seeing the Great Dragon tree," he told his mentor at Cambridge.[7]

In fact, Darwin never set foot on the island—though he came tantalizingly close the next year. As the *Beagle* prepared to drop anchor, Spanish functionaries approached by boat: they ordered a twelve-day quarantine as precaution against cholera. Captain FitzRoy promptly sailed on. Gazing from deck at his long-wished-for object, Darwin took measure of his gloom. "Oh misery, misery," he scratched in his diary.[8] Consolation came just days later, when the *Beagle* docked at Cape Verde. The young English naturalist marveled at volcanoes and at baobabs—another kind of subtropical megaflora he recognized from Humboldt.

Like baobabs, dragos inspired misinformation as much as research. By the time of Humboldt's death in 1859, travel writers stated, falsely, that the polymath had calculated the age of the Orotava specimen at 4,000, 6,000, even 10,000 years. When this "living antiquity" collapsed in a storm in 1868, foreign commentators castigated the Spanish for not having propped it up, though the earliest photographs clearly show props. An Italian horticulturist portrayed the demise of this "historical monument" as proof of the "general carelessness that Spanish people and the Spanish Government have for whatever concerns botany and natural beauties."[9]

Today, the word *monument* mainly connotes impressive, place-defining structures that are intentionally commemorative. In European languages, the word once had wider and subtler meanings. A monument could be anything from the past that offered, like parchment or coinage, reminders to the present.

The former spaciousness of the word can be seen in Humboldt's *Personal Narrative*. The naturalist spoke of languages as the original monuments of nations, and posited that philology—in the absence of built monuments—could allow scholars to infer ancient migrations in the past. In another passage, he referred to the edifices of Olmecs, Toltecs, Zapotecs, and Aztecs as monuments, though he called them "barbarous" or "semi-barbarous" monuments, meaning that they were historically valuable as evidence of the advancement of the species, but not aesthetically valuable. By contrast, he considered Greek ruins to be monuments of art *and* history.

Compared to his regard for Mexico, whose pre-Hispanic temples and gardens he frequented, Humboldt expressed specific disappointment in Venezuela. Here, he said, the grandness of nature was unembellished by monuments of antiquity. When Humboldt barometrically measured the summit of the coastal range above Caracas, and determined it lower than the highest points in Iberia and Tenerife, his local hosts expressed their own letdown. Humboldt responded with condescension: "How can we blame that national feeling, which attaches itself to monuments of nature, in a spot where the monuments of art are nothing?"[10]

Humboldt was a complicated figure—an anti-imperialist who used diplomatic connections to travel throughout New Spain, where he surveyed landscapes with imperial eyes, extracting local knowledge in service of global science.[11] At the same time, he gave significance to local features and ecologies in a way that inspired national feelings in Latin America and Europe. For example, Humboldt's most enduring statement on monuments—one that would go on to influence the development of international conservation law—concerned a Venezuelan tree.

In early 1800, the baron went to visit some local dignitaries at their haciendas near Maracay, roughly sixty miles west of Caracas. Traveling the main road from the colonial capital, past sugar plantations, in the valley of Río Güere, he encountered—"discovered," he pompously wrote—a tree that looked like a "tumulus" or "umbrella" or "vegetable roof." Humboldt liked to measure everything, and he calculated the diameter and circumference of this amazingly wide and symmetrical saman, or rain tree, an arborescent member of the pea family. It had persisted unchanged since the conquistadors, he said. El Samán de Güere was as old as El Drago de la Orotava, he said. Humboldt reported that villagers, especially Indigenous ones, held this rain tree in veneration, and that a farmer had been tried for removing a branch. To accompany that local information, Humboldt made a universal assertion: in countries destitute of monuments of art, people give protection to monuments of nature, and severely punish their violators.[12]

Although attributed to Humboldt, the phrase "monuments of nature" predated him by a generation or two. It entered the vocabulary of eighteenth-century French savants who investigated what became known as geology. Early geologists explicitly compared fossils to coins and documents—historical evidence that could be read and interpreted to assemble a chronological story—a *geo*history. The renowned naturalist Comte de Buffon even proposed a schema for *les monuments de la nature*: subfossil bones in the soil, fossils in surface rocks, in subsurface strata, and so on.[13] In characteristic fashion, Baron von Humboldt borrowed this geological idea, applied it to botany, and made it his own.

Adding substance to Humboldt's usage, the nineteenth-century botanist Augustin Pyramus de Candolle proposed a policy of municipal preservation of arboreal monuments. "Old trees are coins of a different kind," he wrote. Living antiquities should be identified by naturalists, then placed in public trust. The scientific question of tree longevity needed to be solved "while there is time," urged Candolle. Population growth and industrialization threatened to destroy old trees to the far ends of the Earth. Compounding the threat, "changes in religious opinions, and the decay of some notions worthy of respect, though superstitious, tend to diminish the veneration which certain trees formerly inspired in the people." Naturalists could fill the spiritual void by ascertaining dimensions and ages. This was a perfectly Humboldtian approach: nature veneration through instrumental measurement.[14]

In Humboldt's senior years, as he labored on *Cosmos*, his orbit shrank by degrees to Berlin and Potsdam, among the forests, gardens, palm houses, and tree-lined boulevards of the Hohenzollern dynasty. Among the fairest fruits of European civilization were the abilities to import and grow exotic plants, and to connect emotionally to faraway nature by studying natural history and landscape art, said the baron from his privileged vantage.

Before he died, a parade of far-flung visitors called on the Prussian eminence, whose international fame rivaled Goethe's and Napoleon's. In 1858, Pál Rosti, a Hungarian nobleman-cum-cameraman, came to Brandenburg to deliver an inscribed copy of his album of American views—the first-ever photographs of Venezuela and Mexico—which he made in Humboldt's footsteps. After showing the dedication, Rosti turned the leaves to reveal El Samán de Güere. According to Rosti, an astonished Humboldt said something to this effect: Look at me, so close to the grave, and this tree appears unchanged from sixty years ago![15]

Just as well that white-haired Humboldt didn't see an albumen print of the *drago milenario*. At Orotava, the crypto-religious habit of relic-taking worked against science-based protection. In an 1858 book on telescopic investigations from Tenerife, an astronomer included stereographs of a hollowed-out, bricked-up wreck of a garden plant, its disintegration hastened by English health tourists who hacked off pieces. Toeing the line of disrespect, the scientist inquired skeptically: Where is the proof this "asparagus stalk" is the oldest tree in the world—or a tree in the first place? Being a monocot like a palm, a drago produced no "woody substance" and no tree rings to count.[16]

Here, as with so many ancient trees in modern times, science awkwardly resembled belief. In the astronomer's shrewd words: "Such is the weight deservedly attached to every utterance of the great Humboldt in his immortal 'Personal Narrative' that the simple words have, nearly everywhere, been received as fact."[17]

WATER LORDS

For seven months in 1803, Baron von Humboldt had lived in the Valley of Mexico, absorbing information from its scholars, libraries, pre-Hispanic artifacts, and volcanic landscapes. He frequented Chapultepec, the forested hill to the west of the capital, where he saw two large, beautiful cypress trees that dated to the Aztec Empire.

These were specimens of ahuehuete (Montezuma cypress), one of the great conifers of the Americas. At well-watered sites, the species grows to stupendous size. Ahuehuetes favor the margins of rivers and marshes, plus seemingly arid upland sites with subterranean water. In Indigenous Mexico, the species symbolizes the life-giving power of springs. Water once gushed from the porous volcanic aquifer at Chapultepec. Toltec elites used this sacred hill, with its powerful combination of caves, springs, and towering trees, for ceremonial purposes, as did their Aztec successors. In classical Nahuatl, a Uto-Aztecan language, "cypress" was a metonym for "royalty." Today, *ahuehuetl* is typically translated as "old man of the water" (*viejo del agua*), though "water lord" is closer to the original meaning.

For centuries, the advantage *and* the disadvantage of the Valley of Mexico was water—too much or not enough. After choosing to erect their capital, Tenochtitlán, on an island in a lake, the Mexica people went to extraordinary lengths to deliver freshwater from Chapultepec by aqueduct. Their lakeside rivals-turned-allies, the Acolhua people, had their own impressive capital, Texcoco, and their own sacred hill. On it, the Acolhua poet-king Nezahualcóyotl created a botanical garden flanked by ahuehuetes. To reinforce dynastic symmetry and topographic duality, he designed a complementary resort for his Mexica cousins at Chapultepec, on the opposite side of Lake Texcoco. Here, formal lines of ahuehuetes grew alongside dams, tanks, and canals. Moctezuma II made additional improvements. Commoners couldn't visit the canopied pleasure park, but they knew ahuehuetes from the "floating gardens" (*chinampas*) that sustained Tenochtitlán. Mexica farmers created island-like plots by piling mud into the shallow lake and anchoring the corners with cypresses.

Hernán Cortés came to the Valley of Mexico one century after the Aztec engineering boom and beheld temples and trees in imperial splendor. In 1520, the year before taking down Tenochtitlán—in a siege that began with demolition of the aqueduct—the conquistador and his Native allies came close to defeat. Barely escaping the forces of Moctezuma II, Cortés stopped in the night to count his losses. Chroniclers later named this moment *La Noche Triste* (The Night of Sorrows). Supposedly, Cortés cried beneath an ahuehuete—what Spaniards called a *sabino*, a word for

juniper. In the nineteenth century, after Mexican independence, El Árbol de la Noche Triste became a landmark of arbonationalism.

In the immediate post-Conquest era, the ahuehuetes of Chapultepec,
guardians of the lifeblood of Tenochtitlán, retained meaning as Aztec
metonyms. A Nahua annalist of the early 1600s, a convert, recalled a
broad ahuehuete in the Valley of Mexico that had been planted sixty-one
years after the birth of Jesus and that had lived one millennium—1,008
years, by ancestral count—before collapsing. Christian leaders felled
more. With crude zeal, Franciscans converted a regal sabino from Chapultepec into a towering cross for the first Native chapel in Mexico City.
According to a Spanish chronicler, local Indians gave special reverence to
this cross, believing it derived from a "deified thing."[18] The colonists of
New Spain showed no such respect for water lords. Gold seekers tore up
the sacred hill of the Aztecs in 1615, knocking down cypresses, making
barrenness from a garden.[19]

After Tenochtitlán fell, Chapultepec officially remained royal—a
hunting and recreational domain for Spanish viceroyalty. In the eighteenth century, Spaniards erected a hilltop castle. After the United Mexican States won its freedom in 1810–1821, the castle became a military
academy. This college proved no match against the militarism of Mexico's
sister republic to the north. In 1847, US Marines stormed Chapultepec in
preparation for their march on the "Halls of Montezuma." A group of
Mexican cadets refused orders to retreat from the hill, and five of these
young men died pointless, heroic deaths. National indignities would
continue at Chapultepec: After a subsequent French intervention, newly
installed emperor Maximilian I set up residence in the castle. During
his brief reign, Maximilian and his wife, Carlota of Belgium, hired
Austrian gardeners to ornament the former Aztec garden with exotic
plants.

With the republic restored in 1867, Mexican nationalists reimagined
ahuehuetes, as well as pre-Columbian ruins, as republican patrimony.
Ironically, El Árbol de la Noche Triste caught on fire in the midst of the
ten-year anniversary of Cinco de Mayo, in 1872. Throughout the valley,
on a suddenly sorrowful night, people saw the landmark burning like a

chandelier. Thanks to firefighters, the cypress survived—barely—minus large parts of its crown. Newspapers blamed the fire on vandals and described the damage as barbarism and profanation.

In the 1870s and 1880s, arbonationalists lavished attention on the capital's other outstanding ahuehuete, the pride of El Bosque de Chapultepec. Called *El Árbol de Moctezuma* or *El Centinela* (The Sentinel), and later *El Sargento* (The Sergeant), this tree was a multitrunked marvel. Fanny Calderón, the Scottish-American socialite married to Spain's minister to Mexico, felt duly amazed by its "venerable and druidical look."[20] The plant became the subject of romantic paintings and the site for annual commemorations of the cadets who died fighting Yankee aggressors. Next to this monumental cypress, the state erected treelike obelisks to *Los Niños Héroes*—"the child heroes" and martyrs of the nation—thus giving Mexican cadets the symbolic standing of Aztec lords.

Chapultepec hosted the presidential residence during the protracted reign of Porfirio Díaz, who brought "order and progress" to Mexico along with authoritarianism, cronyism, and repression. Even as officials in Mexico City co-opted Aztec heritage and celebrated a modern mestizo identity, they enacted policies that disempowered Indigenous communities. Urbanites fared better than campesinos in the Porfiriato (1876–1911). Mexico City became a European-style capital, with trolleys on tree-lined boulevards leading to spa resorts. Porfirian engineers dug thousands of artesian wells around the capital while draining the remainder of Lake Texcoco. For a luxurious moment, the Valley of Mexico seemed emancipated from its historic water problems. In this second golden age of Chapultepec, well-to-do urban dwellers swam and relaxed in the shade of sabinos. The tree of the premodern past was the tree of the modern future.

Hydraulic plenty begat hydraulic profligacy, and, by 1900, the once-regal forest looked haggard. As the water table fell, springs dried up, and the famous "Moctezuma's Bath" went dry. With lower humidity, Spanish moss fell from cypresses, then cypresses dropped branches. Insects and pathogens took advantage of stressed-out trees. In the midst of the dieback, one of Díaz's *científicos*—as he called his European-trained advisers—redesigned and enlarged El Bosque de Chapultepec.[21] To succeed native sabinos, foresters turned to mass plantings of drought-resistant eucalypts from Tasmania. In time for the national centennial in 1910, a regreened hill boasted a modern zoo and superior attractions to Central Park. However, the nation's park was no longer prime habitat for the national tree, an honorific given to ahuehuete in 1921.

To see a cypress in picture-perfect health, Mexicans went to Santa María del Tule, a Zapotec village near Oaxaca de Juárez, the capital of Oaxaca. The Spanish word *tule* derived from the Nahuatl word for the native sedge, another water-loving plant. At El Tule, in the midst of sedgy springs, grew a tree of preternatural, almost supernatural circumference. By any system of tree measurement—and there have been many—its girth was, and still is, the largest ever documented.

The earliest Spanish description, from 1590, indicated that Indians gathered at the colossus for "ceremonies, dances, and superstitions."[22] In this era, the rulers of New Spain coerced people out of the Valleys of Oaxaca—from Mixtec, Mixe, and other Indigenous communities—to labor in the mines. Despite enslavements and epidemics, and despite Christianization, Native dances at the tree persisted into the early nineteenth century, and later restarted as an annual celebration for locals and visitors. The front doors of the colonial-era church faced the ahuehuete and swung open toward it.

Like a time-lapse video, travel accounts portrayed a dynamic organism whose exceptional habitat allowed for prodigious growth and regrowth. A seventeenth-century hollow that accommodated groups on horseback became an eighteenth-century tree cave, accessible by ropes, that sheltered nocturnal wildlife. In the middle decades of the nineteenth century, spring water issued from the saturated trunk. When the first US minister to Mexico reported the tree's circumference to the American Philosophical Society, skeptical savants in Philadelphia requested (and received) a remeasurement.[23] In 1844, the distinguished US botanist Asa Gray implored the "next intelligent traveler who visits this most ancient living monument" to cut a lateral section to reveal the growth of recent centuries.[24] Instead, visitors carved names or inserted boards with poetic messages into the spongy bark. *El Gigánte de los Árboles* healed so quickly, it swallowed inscriptions within years. On one illegible plaque, late-nineteenth-century visitors believed they could make out letters spelling Alexander von Humboldt's name. To this day, fakelore that the baron measured El Tule in 1803 remains unshakable.

This much is true: A parade of Humboldt-like figures passed through Santa María del Tule between stops (or looting expeditions) at Monte Albán and Mitla. One was the French archaeologist Désiré Charnay, who took the first photograph of the cypress in 1859. According to Charnay's amanuensis, Eugène Emmanuel Viollet-le-Duc—the great authority on French monuments—Santa María's villagers took protective care of their organic landmark, regulating visitors, prohibiting branch removal, and sweeping its base every day.[25] On the basis of the combined technique of rephotographing trees at Mayan ruins and counting tree rings therefrom, Charnay argued later that woody plants can enjoy more than one growth period per year, for "in a tropical country nature never rests."[26] This insight might have caused intelligent travelers to reconsider the age of El Tule, but the plant's overwhelming enormity in comparison to the dragon tree of Tenerife encouraged extreme estimates.

Low-cost lithography and new-fashioned photography added to the stature of trees in the Americas. In 1863, a German forester published a reference book on megaflora that ended with a comparative chart—a Humboldtian infographic—showing famous trees from around the globe, including El Drago, El Tule, and the "Mammoth Tree" of California, arranged against prominent domes, steeples, towers, and obelisks of Europe.[27] To the satisfaction of Mexican arbonationalists, encyclopedias in the era of Díaz listed El Tule as the world's oldest living thing, with the Golden State's giant sequoias the only serious competition.

In 1903, on the centenary of Humboldt's apocryphal visit, Hermann von Schrenk, a noted plant pathologist from St. Louis, made his way to Oaxaca. Assuming the role of the scientific hero summoned by Asa Gray, he planned to "make old Tule give up the secret of his age." Introduced to Santa María's mayor as a representative of the US Bureau of Plant Industry, Schrenk initially received a courteous welcome. But when the

Yankee announced that he needed a core sample and pulled out an auger, the mayor raised both hands. El Tule is sacred, he protested; locals will defend it with their lives. In Schrenk's recollection, "simple Mexicans" stood "between science and sanctitude," defending the latter.[28] Undeterred by his failure to obtain tree-ring data, the pathologist guesstimated the tree's age between 4,000 and 6,000 years.

Almost two decades later, after the Mexican Revolution, an expert with better social credentials—the director of the botanical garden in Oaxaca de Juárez—estimated the age downward, at 1,500 to 2,000 years. Worryingly, he observed new artesian wells all around the church. Talking to locals, the director heard "popular legends" about El Gigánte: Quetzalcoatl, or an ancient Zapotec prophet, had planted this cypress.[29]

A Oaxacan poet collected such "fables" from Santa María's Indigenous elders and rewrote them in the 1920s. His mythopoetic anthology combined traditionalism and modernism, Humboldt and hamadryads, and references to Nobel Prize winners Maurice Maeterlinck and Rabindranath Tagore. By way of preface, he described the village's annual fiesta: pilgrims danced by the tree and received gifts of twigs—though not from El Gigánte. Anyone who damaged the sacrosanct plant was subject to arrest and community service.[30]

The same year the poet published his work, a Ukrainian immigrant named Zelig Schñadower arrived in the port city Veracruz, Mexico. He spoke Yiddish, not Spanish, and carried all his savings in his pocket. With his last pesos, he bought a roundtrip rail ticket to Oaxaca, the only place in Mexico he knew something about. Back in third grade, his teacher, a Russian, had taught a unit on Humboldtian geography, including natural wonders of the world. When the teacher described El Árbol del Tule, it produced an indelible image in Schñadower's mind. After an improbable journey from the edge of Europe, he beheld the tree. Decades later, as an elderly man in Mexico City, he recalled: "Suddenly, it stood before me, as if it were waiting for me, as if the centuries had fed its enormous veins in anticipation of this magical moment when I would arrive to realize my childhood dream." Impulsively, the Jewish émigré called out to all the villagers and tourists in sight. How many would it take to encircle the giant? Using pidgin Spanish, he organized a circle of wonderment.

"We were able to join fifty-seven pairs of hands," said Schñadower, "and when we were all standing there, interlaced, I was overtaken by a sense of completion, and emotion, over what I had accomplished."[31]

HOMELAND PROTECTION

If a dragon tree in Berlin had compelled Humboldt—and followers—to travel to Latin America to encounter monuments of nature, the Humboldtian current recirculated back home, where Europeans created new protections for monumentalized trees. Landmark by leafy landmark, people at the fin de siècle attempted to reconcile—where reconciliation seemed possible—industrial capitalism with nature preservation, nationalism with localism. The designation and care of organic monuments were expressions of the modernity of the state and projections of the imagined community of the nation.

In particular, modern Germans cared about the future of the "German forest"—a place-idea that emerged after the Protestant Reformation and that merged with nationalism in the decades before and after the unification of the German Empire in 1871. Just as the forest long ago aided Germanic tribes in repelling Romans, so the analogy went, Germany's forest served as a shield during the Napoleonic invasions. The notion that Germanic pagans formed an exceptionally profound relationship with the primeval forest—passed down to the *Volk* as forest-mindedness (*Waldgesinnung* or *Waldbewußtsein*)—was likewise an invention of nineteenth-century scholars, poets, painters, composers.

Before they romanticized the primeval forest, Germans, and also the French, pioneered the silvicultural forest—a habitat of science and economics. Silviculture arose in response to early modern European anxieties about wood supply and sustainability (*Nachhaltigkeit*, a word that existed in German long before English). Whether or not shortages were imminent or dire, "scientific foresters" successfully—if contentiously—argued for their jurisdiction at the expense of other users, even after economies shifted from charcoal to coal. The new forest was the domain of state experts, not royal merchants, and especially not peasant graziers and wood harvesters. In Germany, state-trained foresters often worked for landed nobility (*Junkers*), who conserved private forests as aristocratic hunting reserves. As a result, German lands contained more forests in 1900 than in 1800.

On straight paths through straight trees, forest engineers surveyed their stock with military discipline. They classified trees by age cohorts so that they could harvest mature specimens—typically fast-growing, nonnative pines and firs—at optimal times on sustained-yield rotations. Theirs was the "high forest" of plantation timber, not the "low forest" of coppiced firewood. Instead of trees with individuality, silviculturists worked with fungible "normal trees," whose future productivity could be geometrically calculated.

A favorite tool of forest engineers, later picked up by tree-ring scientists, was the growth drill (*Zuwachsbohrer*). This elegant, T-shaped instrument allowed removal of pencil-thin core samples, ring-by-ring records of annual growth, without significant injury to the organism. Because a Swedish manufacturer captured the American market, US foresters came to know the device as the "Swedish increment borer," though a Dresdenite had perfected the design.

The age-class coniferous forest coexisted uneasily with the older mixed deciduous forest. Even as wardens and technicians policed against proletarian "wood thieves," they fought a cultural "forest war" against fellow moderns—middle-class back-to-nature types who believed that pine plantations were monotonous, soulless, commercial landscapes, an affront to nature rather than a sustainable partnership.

In France, this conflict played out at Fontainebleau, the former royal forest south of the capital, newly accessible by train. Charles Baudelaire may have mocked the cult of "sanctified vegetables," but many of his Parisian friends fell in love with the historic forest and objected to the designs of foresters, who planned to replace oaks with conifers.[32] Landscape painter-activist Théodore Rousseau petitioned Napoleon III regarding ancient oaks: these "monuments of nature," these "venerable souvenirs of past ages," deserved no death except from old age.[33] A noted art critic, in literary conversation with Rousseau, submitted that the postrevolutionary state, being more immortal than oaks, had a duty to nurse these long-lasting beings.[34] In 1861, the French government—in the person of the emperor—set aside the core of the mixed forest as an "artistic reserve."

Across the Rhine, in relatively decentralized Germany, the contradiction between forestry and forest-mindedness was partially resolved through the protection of "remarkable trees," the inverse of "normal trees." While forest engineers segregated timber into age cohorts, they marked extraordinary specimens on their cadastral maps and placed stones around them—relicts to be preserved in the midst of rationalization. The vegetal category of the remarkable had originated in the early modern period, when it suggested freak forms of growth suited for cabinets of curiosity.[35] By the time of German unification, the remarkable included the biggest, the oldest, the most historic—trees associated with kings, or Luther, or Goethe, and ideally situated near spas and hiking paths.

In their acts of secular reverence, modern Germans, like their French rivals, prioritized oaks, the totem of Germanic and Gallic pagans. A new category of antiquity, the "Thousand-Year Oak," came into being. The titular instrument of Mozart's *The Magic Flute* derived from the heartwood of such a tree. Looking to the long-term future, many townspeople celebrated the birth of the German Empire by planting oaklings meant to thrive coevally with the state. The practice of inventorying remarkable oaks migrated from forestry departments to Heimat clubs, which published regional tree guides, *Baumbücher*.

In late-nineteenth-century Germany, *Heimat* was an evocative yet nebulous word for one's small homeland, or subnational place of spiritual belonging. Those who loved Heimat feared its loss; they worried about materialism, urbanism, and alienation from nature in the midst of industrialization. Modern life threatened to uproot everyone, even those who never moved. Across the political spectrum, Germans searched for alternative modernities, including "life reforms" like homeopathy. Sometimes, longing caused by the loss of localism led to exoticism—for example, costumed "Indian play" in the woods.

On the right, meanwhile, Volk-minded defenders of Heimat favored prohibitions on landscape disfigurement. The forceful coinage *Heimat-schutz* (homeland protection) originated with Ernst Rudorff, composer and musicologist from Dresden. Starting in the 1880s, he and various historical and antiquarian clubs called for legislation to register historic and prehistoric structures, following the leads of France and England. In Germanic fashion, Rudorff wanted to supplement built monuments with objects of nature, including venerable trees.

Legal precedents for registries came from geoconservation. In various German kingdoms, duchies, and principalities, early modern rulers had protected marvelous caverns and peculiar rocks by fiat. Then, in Switzerland, in the 1860s, the first systematic program of nature protection took place. Swiss societies of natural history led an effort to inventory glacial erratics—monuments of the recently named "Ice Age" (*Eiszeit*)—and safeguard them from crushed rock quarrying. On municipal and cantonal land, local governments banned exploitation of erratics. Protecting such blocks on private land meant purchasing them, so societies raised funds.

By the opening decade of the 1900s, with industrial capitalism reaching a crisis point, small-scale nature protection in central Europe became a large-scale enterprise. A key figure in the nationalization of the subnational was Hugo Conwentz, a paleobotanist turned bureaucrat. Born in Danzig in 1855, he stood out as the stable—and incessantly serious—member of a Mennonite family marked by mercuriality and mental illness, including a mother who compulsively knitted socks. At university, Conwentz developed a lifelong interest in yews, the region's longest-lived trees, as vestiges of the primeval "German forest." He went on to become director of the West Prussian Provincial Museum in Danzig, one of hundreds of Heimat museums that opened in German lands between unification and World War I. A tireless proponent of outreach, Conwentz made thousands of school visits over the decades.

Conwentz's concern for European yew as a "dying species" pushed him from educational to governmental work. For him, the category "remarkable" was scientific more than aesthetic or cultural. Instead of kingly oaks, he gravitated to rare plant associations that provided clues about former climates. He surmised—correctly—that microhabitats such as heaths and bogs could have outsized ecological importance. Endangered biota deserved protection irrespective of beauty or recreational value, he argued.

After writing the first volume of a comprehensive report on the forest trees of West Prussia—the foremost *Baumbuch*—Conwentz received an assignment from the central Ministry of Culture to devise an empire-wide register of geologic and biotic peculiarities. His memorandum, published in 1904, became a technical classic.[36] Conwentz revived the Humboldtian phrase *la monument de la nature* (in German: *Naturdenk-mal*), gave it legalistic meaning, and promulgated it to civil servants as best practice. Two years later, Conwentz became director of the world's first state agency devoted to protected areas: the Prussian Office for the Care of Natural Monuments.[37] Although a minor player in Berlin, Conwentz became internationally famous among conservationists in the years before World War I, when "international conservation" became a field of policy. The intensive European natural monument was admired as an alternative to the extensive US national park, exemplified by Yellowstone. There, US Army Rangers had instituted state control, pushing out Indigenous users and the future possibility of local managers.

Nature love or nature protection—which followed which? This was a serious question, and Conwentz felt certain of the answer. Enduring care came from local participants, not corporations or states. He didn't fault capitalists for ecological destruction in Germany; instead, he criticized German educators for insufficiently nurturing place attachments. By design, his office did not literally protect anything. Having no mandate or money for land acquisition, much less land management, it focused on fact gathering through citizen science. After mailing out thousands of questionnaires to provincial experts, Conwentz and his staff collated the returns and redistributed the information to museums and schools. A centralized database of objects registered for care would, they believed, encourage a decentralized practice of caregiving.

Conwentz wanted German settlers to find Heimat in West Prussia—that is, partitioned Poland—and wishfully imagined that ethnic Poles could be fraternized on the basis of topophilia. Stopping just short of ethnonationalism, his attitude resembled the "civilizing mission" that defined so many settler-colonial projects. Conwentz felt no compunctions attaching Prussian heritage tags to trees within Polish communities and designating former "beehive pines"—a Polish land-use tradition outlawed by German foresters—as monuments. His blind spot about belongingness came across in a sneaky suggestion. To safeguard an ancient yew or other remarkable tree in a Catholic region, he said, a forester could simply hammer a cross or icon to the trunk.[38]

In a deeper sense, the remarkable tree and the sacred tree intertwined at the root. A century before Conwentz, Baron von Humboldt had interpreted Indigenous and mestizo veneration of monumental trees as evidence of civilizational retardation. Humboldt's characterization of Venezuelans aligned with the nineteenth-century German scholarship that more or less invented "religion" as a discrete category of research. Professors placed tree worship (*Baumkult*) at the primitive, animistic end of religious evolution. As codified by Conwentz and the Ministry of Culture, however, care for an arboreal monument (*Baumdenkmal*)—a sacralized tree of the state—indicated civilizational progress.

This gospel of care was inherently conservative. Natural monuments represented a bourgeois compensation to, not a radical confrontation with, industrial development. Advocates of maximal home protection lampooned Conwentz's conventional approach using the wordplay *conwentzionellen*. A well-known Heimat writer dismissed the care of natural monuments as *Pritzelkram*.[39] This idiomatic insult is difficult to translate—akin to "cutesy botanizing" or "small potatoes," with connotations of fiddling while Rome burns or rearranging deck chairs while the *Titanic* sinks.

Professionally, Conwentz's reputation rose highest outside Germany. A frequent guest speaker in neighboring lands, the botanist found allies on all sides. In England, he worked with Cambridge University Press on a revised, translated edition of his report, and found an appreciative audience in Arthur Tansley, fellow yew enthusiast and founding editor of the *Journal of Ecology*. Plant geographers and ecologists embraced monument designation as a culturally resonant legal instrument for protecting rare species, relict populations, and pristine pockets. They wanted to preserve microhabitats with buffer zones, just as historic preservationists pushed for perimeters around registered buildings. Tansley gave Conwentz a platform to share Prussian talking points: nature protection was home protection; the care of natural monuments, like the support of science, was the duty of every nation.[40]

Before World War I blew everything apart, great-power rivalry aided Conwentz's cause. In France and Germany, national-level organizations had arisen: La Société pour la protection des paysages et de l'esthétique de la France (1901) and the Bund Heimatschutz (1904). Strategically, Conwentz framed nature protection as heritage preservation, a legal realm wherein the German Empire had fallen behind France.

From the Revolution onward, the French state had emphasized the protection of built monuments. Revolutionaries and later the French Commune defaced or destroyed many edifices; in response, the nation asserted its powers of classification and preservation over royal and ecclesiastical monuments, reimagined as patrimony of the people. Analogous to German foresters registering ancient trees while modernizing the forest, French planners created architectural protection zones while Haussmannizing the city. This top-down effort resulted in the seminal 1887 French law on the "conservation of monuments and objects of art of historical and artistic interest," a blueprint for listing sites and establishing perimeters.

The most sophisticated statement on monuments—though initially not the most influential—came from Europe's leading multicultural empire. The Austrian state appointed Alois Riegl, a lawyer-philosopher-curator in Vienna, to draft legislation on historic preservation. In his theoretical preface from 1903, "The Modern Cult of Monuments," Riegl suggested that the state, through monument care, could inspire the kind of social cohesion formerly supplied by the Church.[41] The "religion" of monuments (*Denkmalkult*) went beyond Heimat, beyond ethnonationalism, to someplace transcendent.

Riegl assigned pluralistic meanings to monuments: the intentional "commemorative value" from the past, a subjective "historical value" legible to the educated, and an "age value" accessible to all. Of these, Riegl considered age value the most modern, visual, direct, and emotional. Because anything could get old, any landscape feature—including the vernacular, the unintentional, the organic—could become an index of time, thus gaining memorial value.

In modernity, a condition of newness, the worth of age was the aura of becoming and passing. From the contemplation of temporariness came wisdom and reverence—an ethical education. Riegl wrote in near-religious terms about the redeeming power of decay. The temporal *was* the spiritual. He looked forward to a future age of altruism when societies protected all timeful things, not just the rarest and oldest. He thought the age threshold for legal registration should be sixty years rather than a century or a millennium. Caring for a building corresponded to respecting a tree: letting it age slowly through noninterference. Riegl imagined a secular sacred landscape permanently graced by long-term cycles of impermanence.

In short, monumentality meant many things—not all of them compatible—as the twentieth century dawned. Basel-based Paul Sarasin, intellectual forefather of the World Wildlife Fund (WWF) and the International Union for Conservation of Nature (IUCN), spoke of Africa's megafauna *and* its aboriginal peoples as "natural monuments" that required protective intervention. In France, in 1906, the state codified a distinctly French notion of natural monuments of *artistic* character. In the United States, the Antiquities Act of 1906 combined the archaeological and the geological in its definition of "national monuments." Internationally, the proto-ecological Prussian system of small-scale nature reserves was the most influential, serving as the model for the Netherlands, Sweden, Italy, Russia—and, more surprisingly, Japan.

Manabu Miyoshi, a botanist trained in Leipzig, helped to establish the (Japanese) Society for the Preservation of Historic Sites, Places of Scenic Beauty, and Natural Monuments in 1911. In general, the society's trustees were ex-samurai or descendants of samurai. Conservationists gained the upper hand because the Meiji state had fixated on joining the imperial league of industrial nations. The society advocated for a national preservation law, passed in 1919, adapted from European, especially Prussian, precedents. Echoing Conwentz, Miyoshi wrote: "That which is not endangered is not considered a natural monument."[42] By World War II, Japan listed nearly eight hundred monuments—more than any state outside Europe—with ancient trees well represented.

Another surprising conservation nation was Poland, given that the Polish Republic, conquered three ways, officially didn't exist. Compared to the German and Russian partitions, Poles in the Austrian partition had greater freedom to be Polish. At universities in Cracow and Lviv, botanists and paleobotanists worked to inventory and register natural monuments—a term they borrowed from Conwentz, whom they admired as a colleague. One of Galicia's leading intellectuals considered *Heimat* such an efficacious concept that he cleverly translated this untranslatable into Polish.[43] Not unlike Germans and the "German forest," Poles celebrated their deep connection to the primeval "Lithuanian forest," another modern symbol. In 1834, in *Pan Tadeusz*, the national epic, patriot-in-exile Adam Mickiewicz had eulogized "our monuments" of the woods devoured by the Muscovite axe.[44]

During World War I, the German Army rolled into Białowieża, the former Polish-Lithuanian royal forest. Forest-minded Germans exploited the old growth, indiscriminately logging for war materials. The destruction might have been worse had not Conwentz gone to the eastern front in 1916 to persuade military officials to spare the heart of the forest. After the armistice, with the republic restored, Polish nature protectors decolonized Prussian natural monuments and converted "German" yews into "Polish" patrimony. To his credit, Conwentz coordinated with counterparts in Warsaw and Gdańsk (Danzig). He died in 1922 with his magnum opus on yews unfinished.

The Weimar Constitution (1919)—a symbol of all that might have been, and all that was—marked the high point of Conwentzian preservation. In the section on education and schools, the charter for a democratic Germany affirmed that monuments of nature enjoyed the protection and care of the state. It did not, however, authorize Berlin to enforce this provision; the safeguarding remained the duty of an educated citizenry.

The essence of Heimat—a nation of provincials protecting local landscapes—would seem more conducive to bioregionalism than ultra-nationalism.[45] In German lands, the Prusso-centric turn to eco-fascism was not inevitable. But the nationalization of nature always created politically combustible material. This could be seen in Venezuela, too, at the very site where Humboldtian natural monuments symbolically originated.

The rain tree that Europeans sometimes called "El Árbol de Humboldt" became emblematic in postcolonial Venezuela by association with another great—arguably greater—figure from the period, Simón Bolívar. An 1823 epic by the nationalist poet Andrés Bello (who, like Mickiewicz, wrote from Paris), compared El Liberator to El Samán de Güere.[46] Given that the approach to Caracas passed by the famous tree, Venezuelans could easily imagine that Bolívar, on his "Admirable Campaign," had laid down his cot, drawn plans of attack, and proclaimed to his troops beneath the generous canopy. In 1851, the local government preserved this prototypical monument of nature in memory of Bolívar.

After independence, Venezuela struggled to become republican. The Liberator himself vacillated between democracy and dictatorship. One century after Bolívar's death, Venezuela was a petrostate ruled by Juan Vicente Gómez and his cronies. The strongman called himself the father of Venezuela, likened its citizens to children, and compared his fatherliness to a saman—a crown providing shade to all.

Gómez went further, presenting himself as the protector of the specific plant that stood for Bolívar and the fatherland. In his youth, he claimed, he had dreamed of paying respects at the rain tree where the Liberator had suffered for the nation. When he first visited, he felt dismayed to see the machete marks of relic hunters.[47] In 1926, as president-for-life, Gómez staged a ceremony to show his care. A fence made of bayonets and flag-colored cannons encircled the cement-patched tree; an archway granted access to the enclosure, where a bust of Bolívar and a marble engraved with an arboreal sonnet greeted patriotic pilgrims. To demonstrate that Venezuela was a modern nation, the demagogue designated El Samán a national monument.

Had any Germans been paying attention, this authoritarian tree protection might have warned of the dark potential of nationalist topophilia.

PROTECTED AREAS

In the twilight of the Weimar Republic, many German conservationists threw in their lots with the National Socialist Party despite their allegiance to an older *sub*national tradition of Heimat. Walther Schoenichen, who inherited Hugo Conwentz's directorship in Berlin, exemplified this devil's bargain. He joined the party in 1932 and facilitated the warping of Conwentzian internationalism into blood-and-soil anti-Semitism.

Under Nazi rule, the "German forest," already associated with the ancient past, became a symbol for the nation's eternal future. In 1935, when the party enacted a new nature protection law, including the right of the state to seize natural monuments, conservationists around the world applauded. The next year, the Führer welcomed the world to Berlin for the Olympiad, where gold medalists received potted oak saplings. The Olympiastadion—a coliseum of stone intended to last one thousand years—had risen from a deforested section of the Grunewald. Nazi planners preserved a single oak, and erected the gateway towers adjacent to it as complementary landmarks.

Even before the invasion of Poland, National Socialists destroyed more nature than they safeguarded. They drained marshlands in pursuit of autarky; they razed woodlands for war materials. The most disconcerting instance of fascist "forest protection" occurred at Ettersberg, the mountain above Weimar. This had been a favorite haunt of Goethe, who, as a young man, lazed in a cabin with Schiller, drinking and contriving forest comedies. In 1827, as an old man, with *Faust, Part Two* nearly complete, Goethe went to the highland one last time. Breakfasting on partridge and wine—drunk from his golden chalice—the genius surveyed the glories of the world, and felt free, he said, as nature intended.[48]

Thuringian surveyors had mapped the forest in 1797 and inventoried remarkable trees, including an olden oak.[49] Preserved by ducal foresters, this oak still existed fourteen decades later, when the SS took over the property for an early concentration camp. Unlike the death factories later constructed in occupied Poland, camps within the Reich served as slave labor facilities, where resisters and nonconformists died from exertion, starvation, illness, neglect, and abuse. Local elites in Weimar, protective of their city's classical reputation, requested a name other than Ettersberg. The head of the Totenkopf division suggested *Hochwald* (High Forest); Himmler chose *Buchenwald* (Beech Forest). At gunpoint, the earliest prisoners, a who's who of German intelligentsia, cleared all the trees in the camp perimeter. All except the oak. According to instant legend—whether it began with guards or prisoners remains unclear—Goethe composed his second "Wanderer's Night Song" here at the tree. Anyone who damaged the "mighty oak" would get twenty-five lashes.[50]

On their design plans, SS architects simply labeled it *Dicke Eiche* (Thick Oak). In spring 1939, as horror stories began escaping from Buchenwald, Jewish Austrian expatriate Joseph Roth drafted—just days before he died in despair in Paris—a satirical piece on a nature protection law that saved the historic oak.[51] In fact, its preservation required no special legal mechanism from the Reich. The late-nineteenth-century impulse to care for arboreal monuments had been internalized by Germans. In 1942, when state cartographers updated the public map of Thuringia, they identified the big oak, while rendering the camp around it invisible.[52]

As the one green thing in a place of utter darkness, the oldest living being in a place of constant death, the tree by the laundry became precious to many prisoners; they called it the "Goethe Oak." Fraternally, they referred to it as the original inmate of Buchenwald. It reminded novelist Ernst Wiechert of the lost goodness of German culture. From its base, at just the right angle, he could glimpse freedom in the valley below.[53] Goethean legends multiplied here. At this very bough, the poet once sat with Charlotte von Stein. In the shade of its canopy, he composed "Walpurgisnacht" from *Faust*.

German prisoners weren't the only ones who knew Goethe by heart. Warsaw native Edmund Polak translated the "Wanderer's Night Song" in an attempt to keep his humanity on the "malicious mountain." After the war, he recalled that a Polish professor had convened a literature club in the barracks and lectured on Goethe's life. Inmates from the east related their own legends and prophecies. The Goethe Oak had begun to wither after Nazis hung prisoners on it, they said. If and when this cursed monument fell, so would the Reich. Each spring, Polish inmates felt the tiniest rush of hope when they counted the foliage. By August 1944, a single leaf remained, wrote Polak.[54]

SS guards probably never used the oak as a gallows, though they did practice arboreal sadism. With routine capriciousness, they took prisoners beyond the barbed wire and gun turrets, and left them strung up on beeches, feet scraping the ground for hours. Wails along with birdsong carried into the camp from the forest of death. Near the crematorium, a tree trunk cemented in place facilitated public torture—"hanging on the tree." Within the interrogation building, the SS practiced additional "tree hanging" on walls. The master sergeant, Martin Sommer, earned the sobriquet "Hangman of Buchenwald."

On August 24, 1944, when US planes targeted the neighboring armaments factory, an errant incendiary bomb landed in the camp, igniting the defoliated oak. One anonymous Pole—almost certainly the microbiologist Ludwik Fleck—remembered prisoners reacting with secret joy. He painted a scene of resistance: although he and others had buckets, they denied water to the burning tree. Fleck had recently been transferred to Buchenwald from Auschwitz to assist with production of typhoid vaccine. He survived his postdoctoral enslavement and returned to Lviv to search for his wife and son. For Fleck, the beastly oak—preserved by Satan—had no redemptive quality. Goethe was dead; Himmler had extinguished him.[55]

Fleck's allegorical viewpoint was hardly universal. After the SS pulled down the charred oak, prisoners scrambled to collect wooden relics. Nico Pols traded cigarettes for a chip, which he carried in his hand on the death march down the mountain, and back home to the Netherlands.[56] The German artist Bruno Apitz secured the largest fragment of the Goethe Oak and concealed it in the pathology office. Working in secret, in mortal danger, Apitz carved a tender wooden death mask, a collective monument to the victims of the anti-forest.[57]

The Soviets who liberated Ettersberg used the mountaintop to torture and kill National Socialists and other enemies. In the mid-1950s, when it came time to officially memorialize KZ Buchenwald, the East German government ignored the war crimes of the Red Army and focused on the depravity of the anti-communist SS. The Goethe Oak had been a monument of humanity for inmates, read the guidebook from the German Democratic Republic, while the piety of Nazis for this oak was like a mass murderer caressing a dog.[58]

In West Germany and Austria, the idea of the small homeland survived, even flourished, amid the ruins of the Reich and the expulsions of peacetime. Displaced people went to the cinema for the escapism of *Heimat-filme*. For his part, Walther Schoenichen, the former Nazi officer of nature care, retreated to a professorship and authored a tree book.[59] But Heimat was unredeemable in international context: no longer associated with a spectrum of subnational movements, the word now connoted genocidal *Lebensraum*. Among the Atlantic alliance, no one talked about Germanic contributions to the first international conservation movement. White US Americans, always eager to claim exceptionalism, came to believe they had invented modern nature protection with their system of "wilderness" parks, starting with Yosemite Valley and the nearby "Big Tree Grove," both reserved in 1864. They told this exculpatory story so insistently—leaving out the violent displacement of Indigenous peoples—that it felt true.

When conservation restarted in the late 1940s against the specter of bombed-out cities, North Atlantic elites set a "one world" agenda that was both cosmopolitan and neocolonial. The World Bank, the IMF, and the FAO formed one (strong) axis of the postwar order; the WWF, the IUCN, and UNESCO formed another (weaker) axis. The IUCN originated in 1948, at Fontainebleau, with support from Julian Huxley, UNESCO's director general.[60] Today, the IUCN—a hybrid of NGOs and governmental agencies—is best known for its "red lists" of threatened species (including the Canary Islands dragon tree). It also collects data and sets guidelines about habitat protection. Since 1978, the IUCN has promulgated standard categories of protected areas, including the "natural monument," or Category III.

Looking past the neighborhood-scale Prussian Naturdenkmal, the IUCN modeled Category III after the US national monument, which, like everything in America, was extra large. Over time, European states standardized and Americanized their schemas of nature protection. In reunified Germany, in 2010, the updated Federal Conservation Act added an awkward neologism—*Nationales Naturmonument*, a Category III designation that excluded most of the old monuments of nature, now defined clinically as protected areas less than five hectares in size. This bureaucratic move followed a cultural shift. As postwar Germans reckoned with their shameful past, they separated *Gedenkstätte* (memorials, especially Holocaust memorials) from *Denkmale* (historic structures), leaving ecological monuments conceptually homeless. The evocative compound *Naturdenkmalpflege* (natural monument care) gave way to the global English "protected area management," literal and utilitarian.[61]

Meanwhile, cultural energies that once went into natural monuments have been redirected into UNESCO's Convention Concerning the Protection of the World Cultural and Natural Heritage, better known as the World Heritage program. In force since 1975, this convention has become one of the most successful treaties of all time in terms of number of signatories. The shared idea is that select sites—evaluated by apolitical experts following academic protocols—deserve recognition for "outstanding universal value." Initially, UNESCO imposed a Eurocentric conception of heritage—castles, cathedrals, old towns. As non-Western and postcolonial states fought for inclusion on the list, they insisted that their own evaluations of heritage should be respected by the international community. By the politics of reciprocity, every nation became entitled to listed sites. Ironically, a program that began with "world citizenship" ended up serving nationalistic and parochial interests—including cynical efforts to boost tourism to the detriment of sites themselves.

Whereas Hugo Conwentz and his bureaucratic colleagues had believed that well-schooled locals would protect national heritage, Julian Huxley and his technocratic successors believed that science-based NGOs would safeguard global heritage. At the scale of the planet, postwar internationalists tried to reconcile development and preservation. In 1992, a half millennium after the Columbian encounter, UNESCO and IUCN appeared to achieve success at the Earth Summit in Rio de Janeiro. The Convention on Biological Diversity, a UN treaty eventually signed by 168 nations (though unratified by the US Senate), led to strategic plans, including targets for 2020. Signatories promised that at least 17 percent of the Earth's terrestrial area (excluding Antarctica) would be conserved through systems of protected areas.

The world has come close to that goal, at least on paper. Global governance does not in fact ensure local care. In 2018, *Science* magazine published an analysis of satellite imagery: Of "protected" land, fully one-third (six million square kilometers) fell under intense human pressure.[62] In addition, many reserves had been "PADDDed" through "protected area downgrading, downsizing and degazettement."

Conservation biologists, notably E. O. Wilson, warned that 17 percent was insufficient to prevent catastrophic biodiversity loss, considering the compounding effects of climate change. Wilson pressed for "half-Earth" conservation. Getting halfway to half—25 percent—seems feasible yet chimeric because not all protected land is equal. Signatories to the convention started with the easy: areas of low population density and low economic value. Instead of areas of special importance to biodiversity, they prioritized total surface area—a more easily quantified target.

Island-like refugia can have ecological value out of proportion to their surface area. For example, in Britain, centuries-old hedgerows serve as shelters and corridors for wildlife, including invertebrates. From the 1950s through the Thatcher years, farmers in pursuit of intensification and efficiency removed hundreds of thousands of linear miles of high-maintenance hedges. Unwittingly, they produced depauperate landscapes ungraced by birdsong. Nineteenth-century Heimat advocates such as Ernst Rudorff had championed hedgerows as cultural monuments; their urge to maintain tradition was, from an ecological standpoint, right for the wrong reason.

Like hedgerows, small protected areas of trees rely on volunteer caretakers, not park rangers or state foresters. Some of the planet's rarest arboreal habitats exist on sanctified nonstate land. In Britain, "God's acre" in Anglican churchyards—cumulatively a vast area of seminatural yew-shaded habitat—gives refuge to creatures great and small. Likewise, the church forests of northern Ethiopia (protected by monks and priests) and the sacred groves of the Western Ghats (protected by tribal villagers) are critical habitats outside formal governance. Belatedly, in the early 2000s, the IUCN recognized the importance of established religion and Indigenous spirituality by issuing guidelines on "sacred natural sites."

Recognized or not, sacred land can be a feature of secularism, too. An example is the "Hambach Forest" (*Hambacher Forst*), located immediately next to Germany's largest brown coal strip mine, in the Rhineland near Cologne. Starting in 2012, climate activists from across the European Union staged occupations of this mixed deciduous microhabitat, the property of an electric utility. "Hambi stays," they chanted, just as coal must stay in the ground. This woodland, though expressly endangered, was not primeval in the literal sense. When tree-sitters spoke to journalists from their perches about a community of trees untouched for 12,000 years, apologists for big coal scoffed at the false belief.

In fairness, the notion that Hambi could be as old as the Holocene seemed *just as serious* as the prior notion of the immortal Alexander von Humboldt that the dragon tree of Tenerife could have dated to the last revolution of the Earth. For nineteenth-century Humboldtians, the exactness of any one measurement was less important than the total practice and comparative applications of measuring and discovering. Some of the emotive spirit of Humboldt abides at Hambi—an anti-fascist "German Forest," a place of ecological patriotism without blood-and-soil nationalism. Only two generations after Nazis appropriated the tradition of homeland protection, Greens reappropriated it. Come 2020, the German parliament enacted a national phase-out of coal. At Hambach Forest, citizen-naturalists, in defiance of state police, had discovered something rare, remarkable, even cosmic in the local. Here, activists for the planet took the measure of a nation in relation to ordinary old trees and vastly older monuments—layers of lignite, as yet unviolated—resting beneath their roots.

THE NATIONAL TREE

Traced through trees, the history of nature care in Mexico differs from that of Germany, Japan, the United States, and other countries that industrialized in the nineteenth century. The Porfirian law on monuments, enacted in 1897, covered only archaeological sites. Following the Mexican Revolution, the remade state—showing the enduring influence of France—added protections to "artistic monuments" and "areas of natural beauty." But Mexico skipped Prussian-style natural monuments and went straight to national parks during the Depression-era presidency of Lázaro Cárdenas, who combined land reform with nature protection. Compared to the US model (natural wonderlands dispossessed from Natives and managed for tourists) or the Swiss model (strict nature reserves managed by scientists), the prototypical Mexican national park was more democratic. In upland forests outside the capital, the state hoped to assist campesinos to live and work sustainably, caring for the nation's trees.

After World War II, the promise of sustainability through social justice faded as Mexico became a one-party oligarchic petrostate. In Mexico City, urban ecologists of the 1950s recognized the daunting challenges of an industrializing city pulling in hundreds of thousands of impoverished migrants from rural hinterlands. An unplanned megalopolis in a lakebed, the capital experienced floods along with dust storms along with streams of raw sewage—and ever-worsening smog. Gridded plots of Aztec-era ahuehuetes that dotted the former lakeshore dried up, died, and became ordinary firewood.

Hard times came to Chapultepec, too, beneath the presidential palace that Cárdenas had turned into a national museum. The urban poor cut down desiccated cypresses for charcoal production, and homeless people lived in the hollows of large survivors. A midcentury drought added to the toll. In the 1960s, the number of mature sabinos fell to the low hundreds, and the most iconic, El Sargento, passed away. The nadir was yet to come. In the 1980s, acid rain, erosion, and inundations of trash and rats devastated "the lungs of Mexico City."

In the northwestern neighborhood of Tacuba, municipal officials did their best to keep El Árbol de la Noche Triste on life support in the 1960s. The tree received guy wires, fertilizer treatments, and irrigation works. Its life still ended prematurely, in 1980, when vandals lit the monument on fire. After the attack, this landmark of the Spanish Conquest—and of Indigenous resistance—looked like a blackened artillery cannon upended by a bomb. Adding insult to death, arsonists returned with gasoline the following year. From Los Angeles, the leading Spanish-language newspaper in the other United States mourned the news: "The old tree, about which so many writers have dedicated thousands of pages, is now the one who cries at the savagery of humans."[63]

In faraway Santa María del Tule, El Gigánte experienced its own periods of crisis. Road crews routed the Pan-American Highway through the village, barely beyond the churchyard, covering and compacting the sabino's outer soil. Making things worse, aquifers fell as economic modernization came to the Central Valleys of Oaxaca. New factories used more water than old farms. By 1950, El Gigánte faced death by thirst. An area resident who was both a history devotee and a vice president of an equipment manufacturer saved the day by organizing a special irrigation system for the tree.[64]

By the 1980s, Santa María contained a small industrial corridor, including factories within blocks of El Árbol del Tule. When the giant's health again declined, investigators discovered that locals had clandestinely rerouted the irrigation—an episode that generated distrust between educated activists in Oaxaca de Juárez and the small-town wardens of the national icon. A city-based organization called Mi Amigo el Árbol advocated for World Heritage status for the tree, and for protected area management by NGOs or federal agents.

This proposal came in context of UNESCO's listing of Oaxaca's colonial center in 1987. The tourism that followed World Heritage boosted the local economy, while exacerbating water shortages and social inequalities. Foreigners enjoyed hotel showers during the dry season, when ordinary townspeople relied on bottled water or rainwater stored in backyard tanks. The canopy of Oaxaca's old town—unmentioned in the UN citation—suffered from insufficient watering. In this context, tree-loving Oaxaqueños took it upon themselves to do something reminiscent of central Europe one century before: they conducted inventories, created lists, installed plaques, and published tree books.[65] In the 2000s, the governor's office recognized these efforts with a registry of officially declared "Árboles Notables." This monumental tree protection program has served as a model for other states in the republic. Oaxaca's provincial inventory showed that El Tule was unique in size only. Almost every village in the valleys and mountains of central Oaxaca has a *viejo del agua* by the church, in the zocalo, or in the cemetery.

Today, Santa María del Tule—practically a borough of urbanizing Oaxaca—retains day-to-day management of its monument, despite periodic accusations of fiscal and arboricultural malpractice. The Pan-American Highway was rerouted in the 1990s, allowing caretakers to expand the compound with flower beds and fountains. Up to half a million visitors come here each year. The fortunate ones arrive on the second Monday of October, when the inner fence comes down, allowing them to hug one of the world's largest organisms. Children sing "Las Mañanitas" (happy birthday), well-wishers leave flowers and incense, the local priest presents the icon of the Holy Mother, and Indigenous dancers perform. Nearby vendors sell corn on the cob, fried grasshoppers, and paletas.

Contemporary visitors at El Árbol del Tule are more likely to be Mexican than foreign; but, in Oaxaca, "Mexican" doesn't necessarily mean *mestizo*. El Tule remains sacred to Zapotec, Mixtec, and Mixe communities. The world-famous tree has become a pan-tribal symbol, even a Pan-American Indigenous symbol. In 2016, a Peruvian activist held a wedding ritual at (and with) El Tule with the aid of a shaman. Previously, the activist had "married" trees in other Latin American nations to call attention to unlawful logging and dispossession of Native lands.[66]

Not every Montezuma cypress can be lucky like El Tule. Most grow along rivers, not in church compounds. Riparian forests (*bosques de galería*) are increasingly threatened by water diversions, legal and illegal. According to the Mexican constitution, the state controls all riverways, yet environmental enforcement is weak. As a result of global warming,

Mexico—a semiarid nation already in water crisis—will contend with more intense droughts. A warning from the past came in the 2000s, when tree-ring scientists compiled the first year-by-year record of Mexican climate history, using increment borings from millennial ahuehuetes growing in a gorge outside the Valley of Mexico. The data showed several megadroughts in the pre-Hispanic past, including one during the decline of the Toltec state, plus a sustained dry period in the initial decades of the Spanish Conquest. A future megadrought, in combination with human consumption of surface and subsurface water, could doom old sabinos, if not the Mexican state. To address the endangerment of the national tree, scientists and environmentalists held the first nationwide conference on ahuehuetes in 2017, in Veracruz, with subsequent meetings in Oaxaca and Mexico City.

One of the coordinators was Rodolfo Alfredo Hernández Rea, a forest engineer at Oaxaca's technical college. The care of ahuehuetes has become his life calling. He speaks about it deliberately, with quiet passion. I met with him in March 2018 at the city's decommissioned rail depot, recently converted into a children's museum. The adjoining neighborhood, originally a separate pueblo, falls outside the World Heritage district. Tourists don't flock here. Hernández led me to the end of the platform, where one of his students held a hose to the bole of a fenced-in cypress.

This railyard tree has aura. It has age value. Otherwise, it appears to be El Tule's opposite: unfamous, unassociated with Humboldt, barely huge, desiccated, injured. Its gaping hollow is halfway filled with cement, allowing passage of bats. In the last century, this tree survived three fires—one caused by lightning, another by a holiday firecracker, and one by railroad sparks. Hernández pointed out thick iron nails high in the tree—relics of a railyard pulley system. Then he directed my eyes beyond the wall, across the road, to the pueblo church of Santa María del Marquesado. When Spanish colonists erected this building, three monumental sabinos grew here by the Río Atoyac. According to Hernández, Indigenous people celebrated *la danza del arbol* under their commingled branches.

The single survivor of this trio is quite possibly older than El Tule. A clue comes from Hermann von Schrenk, who returned to Oaxaca in

1933, three decades after his original Humboldtian trip. This time, the Yankee received a consolation prize: permission from state officials to drill into the railyard tree, from which he cranked out a thousand-year core. Still thinking like Humboldt, Schrenk extrapolated that El Tule, many times thicker, was many times older.[67] In 1937, Oaxaca-based botanist Cassiano Conzatti got consent in Santa María del Tule to remove a branch for tree-ring analysis. A few years later, he showed the sample to US dendrochronologist Edmund Schulman—the expert who would go on to prove that extreme age in trees does not strongly correlate to largeness. Schulman agreed with Conzatti: El Tule's age had been overestimated.[68]

Schrenk's tree-ring boring failed to make an impression in Oaxaca de Juárez. The millennial tree was just a tree, hidden in plain sight, with people bustling past it. Not many paid mind to its withering, especially after the privatized railroad stopped serving Oaxaca in 2004. Things began to change when a local artist-activist included the cypress on a tourist map (later a guidebook) of the city's emblematic trees.[69] Street tree enthusiasts began speculating that this ahuehuete might have been planted by vanguard Aztecs or a Zapotec prophet.

Once the city took over the railyard in 2013, Hernández took it upon himself to become the tree's caretaker. He named it "Ahuelito"— grandpappy of the water. He's the poor parent, he told me; El Tule is the rich one. He hopes Oaxaqueños learn to love them equally. As of 2018, the cypress still looked battered, though the foliage had regained its viridity, the bark its sponginess. For optimal health, the tree needs five hundred liters of water per day—more than many Mexican families use in a month. Individual tree protection may seem like "First World" or "full-stomach" environmentalism, but Hernández disagrees. "If there are no trees, there will be no water, neither for people nor for trees," he said.

His relationship with the city's "oldest living being" goes back decades. With a graduate degree and no prospects, he moved to Oaxaca de Juárez in 1995—a year of deep recession following the Mexican peso crisis. Unemployed, he walked the city, searching for remarkable trees and green spaces. When he saw the half-wrecked sabino in the railyard, a door to his childhood opened in his mind.

He's nine or ten. His mother is watching the news on television. Onscreen, El Árbol de la Noche Triste smolders. "They burned it," his mother says despondently. "Those ignorant people burned the national tree." The boy doesn't know the meaning of ahuehuetes, doesn't know the word, but he reads his mother's ashen face. Somehow, the boy would forget this moment, even after matriculating to forestry school. Under the tree in Oaxaca, the memory flooded back.

Hernández teared up when he related this story. Standing next to him, I felt viscerally what I knew intellectually: In the modern world, ancient trees exist in multiple media on multiple scales. Especially since Humboldt's journey to the dragon tree, megaflora and elderflora have become fixtures in the global imaginary. Through small-scale protection of such natural monuments, moderns have tried, with partial success, to compensate for large-scale destruction of nature. As a worldwide phenomenon, the modern cult of arboreal monuments could conceivably serve as a cross-cultural foundation for geotemporal thinking. However, any given monumental tree is a hyperlocal object subject to state law. It manifests tensions between scales of belonging—to one's subnational Heimat, to one's country or homeland, and to everyone's home, Spaceship Earth. Tellingly, the saman in Venezuela that inspired cosmopolitan Humboldt later inspired populist Hugo Chávez.

Mexican heritage looks different from German heritage, yet some of the same questions apply: Who defines national nature, and who takes ultimate responsibility for its care? Better than anyone, Hernández knows that ahuehuete decline is a nationwide problem. Because of economic instability, political corruption, and cartel violence, he cannot count on Mexico City for solutions. Reminiscent of Hugo Conwentz, he prioritizes the provincial: gathering data, marking sites, educating children, and presenting awards to caregivers and caremakers. The first went to the founder of Mi Amigo el Árbol.[70]

Hernández ticked off a list of other notable and emblematic ahuehuetes, including a recently rehabilitated one in the center of the Zapotec village San Pablo Güilá, where Mixe pilgrims from the Sierra Norte attend an annual Lenten mass dedicated to the tree, and where parishioners once painted the Virgin of Guadalupe on roots that emerge from a pool of spring water. Hernández mentioned Humboldt, of course. And he told me about Chalma, one of Mexico's most popular Catholic pilgrimages, which includes a stop at a sabino by a spring-fed pool. First-time pilgrims to Chalma bathe and dance before the tree; would-be mothers make offerings to it, and women who have successfully borne children leave dried-up umbilical cords in gratitude.

Finally, Hernández told me something that went beyond fact or falsehood: In October 1968, Chapultepec's *viejos del agua*, guardians of Mexican culture, died all at once. That's the belief, he said. It was God's way of trying to change humanity.

Much later, I learned that this popular story originated with a controversial "historical novel" about Mexico City, 1968—a time of student protests and state killings prior to the Olympics. The novel portrays the deaths of hundreds of demonstrators not as an atrocity but as a sacrifice by a human-arboreal collective. One hundred *viejos* on Chapultepec follow El Sargento into death even as four hundred latter-day *niños heroes* give up their lives at Tlatelolco. The ritual passing of the last Aztec tree lord signals the coming of a new era, a new consciousness—the rebirth of the nation.[71]

The language of *la nueva mexicanidad* is far removed from IUCN white papers. It's more emotional and more combustible. Arbonationalism—like homeland protection—may not be scalable upward to global environmental governance. But Rodolfo Alfredo Hernández Rea made an undeniable point when he said: "If you can't build a Mexican environmental consciousness around ahuehuetes, then around what?" Placing his hand on the unmonumental monument of nature by the disused tracks, his eyes glistened and his voice cracked: "I talk to my students about this tree; I've had some five hundred of them, and they talk to their family and friends. It multiplies. This tree's not in a hurry. We're not either."

IV.

PACIFIC FIRES

AOTEAROA —— CHILE / CALIFORNIA —— JAPAN / TAIWAN / JAPAN —— CALIFORNIA / CHILE —— AOTEAROA

In extensive populations of plants, bigness and oldness rarely co-occur, except on the Pacific Rim. At least that used to be true. In Chile, California, New Zealand, and the Empire of Japan, modern peoples encountered sky-holding canopies of ancient conifers—and burned them, or cut them, in spite of marveling at them. The burning and the cutting increased exponentially in the long nineteenth century with colonialism, capitalism, and motorized tools. Settlers cleared forests as a matter of course; timber corporations followed. Appreciation of arboreal age sunk in too slowly to arrest large-scale demolishment. In the twentieth century, state foresters made new claims on remnant stands of big old conifers, as did environmental activists. Indigenous peoples made ancestral claims. The territorial enshrinement of relict old-growth as relics— private, public, and tribal parks—is a pan-Pacific legacy of colonialism.

AOTEAROA

Some 385 million years ago, in the Paleozoic era, when animals had barely crawled from the seas, the original forests of megaflora—strange vegetal forms we wouldn't recognize—dominated landscapes. In the Mesozoic, by contrast, big reptiles coexisted widely with big conifers that would appear familiar to us. After the K/T extinction—a diminution as well as a defaunation event—small mammals and birds inherited habitats where coniferous trees still achieved peak size. Come the late Cenozoic, when our species began making art and telling stories, a mythopoetic convergence occurred: the zenith of large mammals, the dusk of large gymnosperms. Overall, megaflora fared better than megafauna in the early age of humans. Twilight lasts longer in tree time. In a few refugia—remote islands and island-like habitats with climates moderated by the Pacific Ocean—the sovereignty of conifers prevailed.

From a floral point of view, the most perfect Cenozoic forests occurred in Zealandia, the geological name for a mostly submerged microcontinent that includes New Caledonia and New Zealand. Except for the highest mountains, the entire landmass of the latter was forested in unique assemblages when humans arrived mere centuries ago. On the North Island, some 85 percent of native plant species were endemic. They included relicts from the breakup of the supercontinent Gondwana as well as later dispersals from across the Tasman Sea. By water and by air, seeds and birds—and seeds in droppings—had made the voyage to outer Zealandia. Mammals had not. Plants and pollinators of the North and South Islands thrived in isolation from the mammalian class, excepting a couple of species of ground-hugging bats. Giant birds strutted in the shade of giant conifers.

The longest-lived and largest-growing of New Zealand's endemic conifers is kauri.[1] Its genus used to be present in the Northern Hemisphere, and Patagonia, too. In the Pleistocene epoch, kauri retreated to the North Island's north, where glaciers never reached. The species does well on infertile soils, including steep slopes and low-lying bogs. Its prodigious shedding makes the soil more acidic and less fertile, conferring additional advantage. Kauris take root—and later die—in cohorts, their life cycle defined by infrequent, large-scale disturbances like cyclones and eruptions. They are poorly adapted to fire. In recent millennia, Northland's climate featured abundant rain and little lightning. Before *Homo sapiens*, a species characterized by pyromania, kauris in the Holocene reached maximum life potential: 1,500 to 2,000 years.

In morphology, millennial kauris resemble mega-dragon trees, only more miraculous. Charles Darwin in 1835 measured one girth of over 30 feet, and heard reports of greater. He described a forest of parallelism: rank upon rank of "noble trees" with smooth cylindrical boles, untapered to heights of 60, even 90 feet, topped with a burst of stout, upward-facing branches.[2] The trunks, textured like hammered metal, oozed resin. The thin bark and the olive-like leaves seemed all out of proportion to the organism.

An etiology of the tree, translated from Māori into English in the 1850s, has become well-known throughout Aotearoa—the modern indigenous name for New Zealand. The story concerns Whale and Kauri, the largest creatures of sea and land. Come away with me, says Whale. Don't be fooled by the smallness of humans; they can cut you down for canoes. Kauri declines to move, yet fatefully exchanges its original bark for oleaginous whale skin.[3]

The major migration of Polynesians who became Māori occurred around 1300 CE. Their claim to the North Island originated with their mythic ancestor Māui, who pulled up the landmass with his fishing hook. To prepare the forested island for sweet potato farming, Māori used another one of Māui's gifts: fire. An agricultural technique developed in Polynesia—slashing, burning, planting—had an exaggerated effect in Aotearoa, where nutrient cycles took decades instead of years. Layers of charcoal—and corresponding layers of black carbon preserved within Antarctic ice sheets—bear witness to the fire-bringers. In a geological instant, Māori burned half of the North Island and converted low-lying east coast forests into tussock, bracken, and scrub. In highlands and on the wetter western coast, mixed kauri forests remained unburned, but not unchanged, for Māori swiftly hunted moas to extinction and introduced rats and dogs. Another round of Polynesian swiddening followed the introduction of American potatoes via Europeans.

Had imperialists, starting with Dutch explorer Abel Tasman, avoided "Nova Zeelandia," kauri might have stood strong into the present. Traditionally, Māori valued the species for its resin (useful for tattoos, torches, and medicine) more than its timber. They compared big kauris to great chiefs and vice versa: they had *mana*, or spiritual prestige. To breach the *tapu* (spiritual restriction) on a mana-rich tree demanded a ceremony and an appropriate purpose—like the carving of a war canoe.

Kauris grow thin, tall, and straight in their youthful first century, before shifting, in their extended maturity, to crown maintenance and girth growth. A French crew in 1772 recognized that "olive-leaved cedars" could serve as naval-class spars for man-of-war frigates. They set about felling a pair of kauris, ignoring a tapu. Local Māori retaliated, killing many crew members and dealing a blow to would-be "France Australe." Following the American Revolution, the British Empire looked to the South Pacific for naval supplies. A new kind of triangular trade developed: British ships offloaded convicts in Australia, crossed the Tasman Sea to load kauri "rickers," then returned home via Cape Colony.

For decades, the Crown relied on Māori laborers to procure this resource. In one recollection, a Briton described the effort of moving an 80-foot mast from forest to water: Scores of men tugged on ropes, in spurts, as directed by their leader, who stood with his staff weapon on the neck of the tree—decapitated, but honorably decorated with flowers and feathers. The leader called out, and the men responded as one, while children ran ahead to smear mud on the skid path.[4]

Māori controlled the spars and wanted things in return: iron tools, nails, bullets, muskets, powder. With their sophisticated war culture, Māori immediately grasped the benefits of British weapons. The initial arms advantage went to *iwis* (tribes) whose homelands included kauri forests and sheltered bays where ships could dock. Equipped with muzzle-loaded long guns from Birmingham, Māori warriors settled old scores.

Amid the intra-Māori "Musket Wars," British missionaries arrived with the intention of staying. Unlike seasonal whalers from Nantucket, they used the language of pioneering and improvement, as illustrated by Reverend Samuel Marsden in 1819. Under the shade of a kauri, he mused about being "surrounded with cannibals," "literally at the ends of the earth." He felt fearless and purposeful, for he "had been sent to labour in preparing the way of the Lord in this dreary wilderness," this "dark and heathen land."[5]

Formal colonization of New Zealand came relatively late, when Māori still held numerical and military advantage over *Pākehā* (foreigners, white people). In 1840, at the Bay of Islands, a proxy for Queen Victoria signed the Treaty of Waitangi. The document sailed around the country, gaining signatories and rejectors in equal measure. Crucially, most signatories marked the Māori-language version of the treaty, which guaranteed their chieftainship over villages, lands, and all treasured things (*taonga katoa*). In return, chiefs gave the Crown the right to purchase land. To mark the birth of its newest colony, British officials erected a kauri flagstaff on Maiki Hill, overlooking the bay, and raised the Union Jack.

Conflicts over the meaning of the dual-language treaty began almost immediately after land alienation began. A few symbolic moments suggest the larger struggle. In 1845–1846, a Māori resister felled the flagstaff on Maiki Hill and defiantly toppled the next three spars the British re-erected.[6] In the late 1850s, iwis from central North Island tried to coordinate against Pākehā rule. Some favored the "King Movement" (*Kīngitanga*) and a sovereign authority equivalent to Queen Victoria. A crowning ritual for the first such king included a Bible, a kauri flagstaff, and a pan-Māori flag.[7]

British settlers approached land purchases as legal transactions devoid of spiritual and ancestral dimensions. They made payments to extinguish Māori titles to "properties"—the English misunderstanding of "all treasured things." Twenty years after the Treaty of Waitangi, Māori had become a double minority: numerically as inhabitants and proportionally as landowners. Ancestral forests had become surveyed properties and then "waste lands"—a legal category introduced from Britain that allowed the Crown to make fee-simple grants in forty-acre blocks to soldiers and emigrants. To create the "Britain of the South"—in contrast to Australia, a former penal colony—the Crown awarded free passage to emigrants it considered morally fit.

Charles and Mary Hames fit the bill. In 1864, after arriving in Auckland, the married couple, a tailor and a dressmaker, devout Methodists, received the deed for Lot 41—part of a paper city called Albertland—on Kaipara Harbour, the heart of kauri habitat. Mary, traveling in hoop skirts, carrying an infant, looked forward to making a home on the land grant. After a one-month walk—a distance now made by car in 1.5 hours—the family arrived at Lot 41, downhearted. "With God's help, we will one day see green fields here like those we left in England," Charles related to his diary. "First how to get these big trees down, I never having felled a tree in my life, nor used an axe."[8]

At a pace that astonished themselves, settlers and their Māori contract laborers transformed "the bush" into a simulacrum of the conifer-poor English countryside. As Darwin had anticipated, the ecosystems of New Zealand reeled under invasion by herbaceous plants from Eurasia, plus all kinds of mammals, feral and domesticated. Anthropogenic fire compounded disturbances. Settlers soon learned that resinous kauri burned hot: "I was always reminded of the smell of frankincense and myrrh," wrote a witness of 1860s infernos.[9] Fifty years later, after the agitation of flames and hooves, John Muir, the great naturalist of California, visited "frowzy farm & dairy country" around Kaipara Harbour. "Not a tree survives," he penciled in his diary. Sawyers told Muir that "smoke is sufficient thing" to kill a kauri, exaggerating only slightly.[10] The whale-like tree could have used the thick bark of giant sequoia and coast redwood, fire-adapted species that Muir knew well.

A political debate in New Zealand about forest destruction and its effect on climate and soil came and went in the 1870s. To the consternation of certain provincial officials, unlicensed loggers had set up illegal sawmills in the Northland, with or without Māori permission, beyond the reach of the state. Cosmopolitan members of parliament called for state forests, following models of central Europe and the British Raj. They cited *Man and Nature* by US authority George Perkins Marsh, warned of erosion and desertification, excoriated barbarous improvidence, and recognized that "in the matter of forests, the Anglo-Saxon is the last man in the world that ought to be left alone."[11] The adversarial majority, standard-bearers of "progress" and "improvement," prevailed. They accepted the settler dogma that kauri would go extinct, and posited a crude kind of Darwinism: by the "same mysterious law," Indigenous peoples and native plants lacked the vigor to resist displacement.[12]

In the end, Parliament passed the Forests Act of 1874, which outlined the potential for lightly regulated logging on Crown lands reserved for forestry. To advise New Zealand on the "forest question," the India Office lent the services of an imperial forester for a one-year tour of duty on the Pacific periphery. This outside conservator—deprived of regulations, supervisors, forests, or revenues—failed to change minds and practices.[13]

Unlike India, New Zealand was a settler colony. An emigration boom in the 1870s included hundreds of newcomers from Nova Scotia, New Brunswick, and Scandinavia—people with cutting and milling experience. The Crown sold off kauri tracts or leased cutting rights for a steal. In a pattern of boom and bust, sawmills opened, then closed when the local resource ran out. No one penalized profligacy: the Crown collected royalties on cut timber only. Targeting the branchless boles, loggers abandoned the tangled crowns—epiphytic gardens—as slash, adding fuel for future fires. In the mills, up to 50 percent of the tree's remainder turned to sawdust, destined for dumping. Even with this extravagant waste, the decentralized industry produced excess supply for internal demand. Kauri, a premium durable, was absurdly cheap. As of the census of 1881, approximately 90 percent of houses in Auckland Province—the northern half of the North Island, Pākehā population ninety-nine thousand—were fashioned from old-growth kauri.

Excluded from a resource they formerly controlled, impoverished Māori in the "roadless north" turned to the informal economy. They gathered *kāpia*, or "kauri gum." Kāpia is neither gum nor rubber, but resin—like amber prior to fossilization. Europeans invented new uses for this natural rosin and Polynesian medicine. Like mastic in the Mediterranean, it became varnish and later a component of linoleum. Gum-bleeders tapped fresh "bush gum" from living trees prior to logging.

Bigger treasures lay underground. Using hooked tools, gumdiggers recovered chunks of subfossil copal from forest soils, and, more surprisingly, from swampy, treeless "gumlands" near the coast where kauri had grown in past epochs.

"Gumdigger" was used pejoratively. The word connoted marginal members of society, including Chinese veterans of the South Island gold rushes and, later, a sizable group of young male Croatians ("Dalmatians" or "Dallies"). British-born farmers also dug up "field gum" to get through lean years and off seasons. But domesticated settlers, people who lived in houses made of kauri, reserved opprobrium for tent-dwelling gumdiggers: vagrants, deadbeats, drunkards, aliens, trespassers. Theirs was a land of the lost. Backbreaking labor somehow evidenced laziness. Supposedly, the enterprise encouraged sexual promiscuity in Polynesian women.

Far northern Māori approached gumdigging as a modern means to a semitraditional way of life. They could be mobile, as they preferred. They could work as families on ancestral lands, as they preferred. They needed little equipment—spades, spears, sacks—and no capital. Denied access to the banking system, Māori made a rational, dignified choice to dig for copal. The global trade in varnish and linoleum sustained the Northland's iwis during the most challenging period in their history.

The poor's informal trade enriched capitalists. At each step in the process of turning raw gum into washed, scraped, sorted, graded, priced, and packed material from Auckland's wharves, Pākehā merchants made money. From the 1870s to the 1910s, when petroleum by-products destroyed the market, copal ranked among Auckland's leading exports. The resinous gift of paleo-kauris had greater exchange value than did old-growth kauris. The violent vanishment of the bush was a cultural imperative rather than an economic necessity. In the face of the rarest forests in creation, British settlers could only imagine a wealth of pastures.

By 1885, just enough parliamentarian representatives worried about the side effects of deforestation to pass a second Forests Act. This law enabled the creation of state forests on Crown lands newly acquired through the Native Land Court. For all of Auckland Province, the chief conservator, botanist Thomas Kirk, oversaw a handful of rangers. He warned that the kauri forest, "one of the grandest sights in the vegetable world," would be exhausted in two or three decades.[14] He guesstimated—a moral calculation, perhaps—that these endangered giants could live up to 4,000 years. Kirk's duration in office was pitiful by comparison. A newly elected government shuttered his office in 1888—a politically motivated austerity measure during New Zealand's long depression.

As luck would have it, drought accompanied misgovernment, setting the scene for disaster. The great bush fire of 1887–1888 devastated Kirk's crown jewel, Puhipuhi State Forest. The conflagration burned for weeks, consuming an estimated seventy-five million board feet of merchantable timber. The aftermath was "one of the most melancholy possible to the lover of sylvan scenes."[15] Officials in Auckland blamed locals, and locals pointed fingers at each other—Māori swiddeners, Māori and Pākehā gumdiggers, and would-be settlers hoping to force open government land through arson.

The second shutdown of state forestry coincided with the onset of industrial logging. In 1888, the Kauri Timber Company bought out distressed timber operations on the North Island and amalgamated the holdings. This Melbourne-based syndicate now had a monopoly on kauri forests not owned by the Crown. To satisfy shareholders, the company worked full speed, around the clock—with sawmills equipped with electric lights—to convert giant trees into joinery for homebuilders in Melbourne, site of a real estate bubble that rivaled the one in Los Angeles. The company paid no export duties because the state didn't bother to levy any.

In 1896—the year the Māori population fell to its nadir—New Zealand held a timber conference in the capital, Wellington, with noted Australian forester George Perrin as the featured guest. Like all the conferencegoers, Perrin accepted the inevitable extinction of the kauri forest. He viewed privatization and governmental neglect as water over the dam. The best available solution to the "kauri-forest question" was rational liquidation by the Kauri Timber Company, followed by proper imperial forestry. He looked forward to scientifically grown plantations of eucalypts, fast growers from Australia. Perrin held out hope that kauri might yet have an economic future in locales with marginal soil. Disputing Kirk's "great age" theory, he argued that big old kauris were scarcely older than Māori settlement. In the youthful colony, there was still room for native growth, just not slow growth.[16]

As the blazing nineteenth century reached its end, one of New Zealand's earliest Pākehā historians—a leading parliamentarian—wrote a poem, "The Passing of the Forest." In it, he eulogized arboreal kings, despoiled and discrowned.[17] His middling lament, later canonized in colonial literature, exemplified an affect seen throughout the Anglo settler world: colonists regretfully wallowing in their own violent improvements. The kind of people who recited this verse subscribed to *New Zealand Illustrated Magazine*, which in 1901 ran a photograph of gargantuan stumps in the bush used as platforms for portraits of middle-class parties—white people in white dresses, dark suits, and feathered hats.

The caption: THE DETHRONEMENT OF KING KAURI.[18]

CHILE / CALIFORNIA

Separated by more than five thousand miles of open ocean, the southern limit of kauri aligns latitudinally with the northern limit of South America's longest-lived and largest-growing conifer, a cypress species that can achieve 10 feet in diameter, 150 feet in height. In Spanish, this plant is called alerce.[19]

On a continent where gymnosperms are rare—in Amazonia, almost nonexistent—southern Chile constitutes an island of Gondwanan conifers. Endemic cypresses, podocarps, and the Western Hemisphere's sole surviving species of *Araucariaceae* (the family that includes kauri) here exist in temperate isolation, bounded by the Atacama, the Andes, and the Pacific.

Throughout the Pleistocene, including multiple ice ages, alerces persisted in Patagonia. Time and again, in extreme slow motion, they marched downhill as glaciers advanced and seas retreated. Unlike North American and European glaciers, which scraped vast areas clean of plants, Patagonian fjord-dozers stayed in their channels. To weather the frigid millennia, alerces hunkered down in the central depression between the Andean and coastal cordilleras—a refugium now largely underwater.

When humans arrived on the Chilean coast some 15,000 years BP, alerce hadn't yet begun its latest return to the slopes of the Andes.

In utter contrast to kauri, the oldest old-growth alerces—what few now remain—germinated in habitat denuded of megafauna and otherwise altered by humans. Hunters and toolmakers used Patagonian cypress for the entirety of the Holocene. After removing the fibrous bark from a felled or fallen alerce, Indigenous harvesters used wedges made of hardwood to split the tight-grained softwood into near-uniform pieces. The ease of production belied the phenomenal durability of the product. Alerce's resinous heartwood repelled rot in a rain-soaked region, making it ideal for canoes, utensils, and construction material. From bogs, Native woodworkers pulled up centuries-old windfall that appeared as new.

Beyond a moderate level of use, alerce is nonrenewable on a human timescale. In its mature phase, it grows more slowly than almost any plant. The species germinates in patchy cohorts after rare, localized disturbances such as fires, landslides, and eruptions. The post-Columbian period—particularly the nineteenth century—brought millennial-scale disturbances not in patches, but across whole landscapes. Change came like an inferno, with the overall effect of an ice age but without refugia.

When sixteenth-century conquistadors, drunk on blood and Incan silver, arrived in south-central Chile (Araucanía), they found a well-populated land. They enslaved Indigenous people and exported them to Peruvian mines, and coerced others to extract placer gold closer to home. Mapuches fought back, earning the fear and respect of Spanish soldiers, who compared them to Romans. During the episodic 350-year-long "War of Arauco," Native groups staged multiple coordinated revolts, repelled invaders northward, and maintained Río Biobío as *la frontera*. For as long as Lakotas and Apaches preserved their independence in North America, so did equestrian Mapuches in Chile's southern half.[20]

Beyond the frontier, a small number of colonial outposts remained. Spaniards retook and refortified Valdivia, a strategic location for imperial trade and defense. From the adjacent temperate rain forest, millers cut alerces into timbers for shipment to Lima. Farther south, on La Isla Grande de Chiloé, sat Castro, a provincial capital. Here, in the heart of the coastal alerce belt, colonists found cypresses so large that twenty people could link hands around their trunks. Here, local Mapuches (sometimes called Huilliches) paid tributes in the form of standard-sized alerce planks—two and one-half meters long. This small-scale commerce continued after the abolishment of the encomienda (labor-and-tribute) system. In spring and summer, Mapuches from Chiloé used alerce canoes to cross the inland sea to split alerces on the Andean side. The planks ended up in Peru.

At the time of Chilean independence in 1821, hardly any national capitals had less control over their hinterland than Santiago, located nearly seven hundred air miles from Chiloé Island, with rough country and armed Natives between. In the absence of state currency, cypress planks functioned as money in Castro. When the *Beagle* landed there in 1835, the crew from Britain had no way to pay for provisions. Darwin marveled: "A man wanting to buy a bottle of wine, carrys on his back an Alerce board!"[21]

In the second half of the century, the Chilean state came into its own by seizing the southland. It was a double effort: military "pacification" of Mapuches alongside settlement of "waste lands." By force and fraud, elites acquired title to vast expanses of expropriated Indigenous territory. They imagined forested estates as future wheat fields and cattle pastures. Colonization companies that promised to bring settlers to backwater regions received land grants from the state.

To modernize Valdivia and Chiloé, companies recruited several thousand Germans—the most desired immigrants in the Americas. Recruiters benefited from the upheaval in Europe following the revolutions of 1848. To set the stage for farmers and artisans from Hesse and Bavaria, and to hasten Patagonia's post-Indian future, boosters set fire to the forests around Lake Llanquihue during a dry spell in 1852. The conflagration blazed for months. The next year, town builders celebrated the birth of Puerto Montt, a new inland port named after the nation's president.

Between the port and the lake lay a ribbon of swampy forest, approximately twenty kilometers long. Hulking alerces rose from the mud and standing water, much like bald cypresses in the North American South. In 1863, during an intense La Niña drought that affected the entire Pacific Rim, local authorities ignited this temporarily unswamped habitat. Doing so, they committed mass botanical eldercide. In challenging environments, slow plants grow slower, and in this poorly drained, nutrient-poor area, multimillennial trees numbered in the hundreds or thousands. The fire burned so hot that the soil changed, rendering the land useless for crops as well as trees. By the turn of the century, locals referred to the weedy stumpscape as "the graveyard." Shingle makers mined the tombstones, except a few roadside specimens reserved sentimentally as ruins. A German gardener made a flower bed atop one. On the largest, named (perhaps ironically) *La Silla del Presidente*, groups of German colonists posed for photographs in suits, dresses, and fancy hats.[22]

Chilean forests were burned for Germans and sometimes by Germans, and when Chile developed a forestry ethic, German Chileans led the way. The key figures, Rodulfo Philippi and Federico Albert, hailed from Berlin.

Philippi, the brother of the leading organizer of Chile-bound Germans, emigrated to Valdivia in 1851, then moved to Santiago, where he rose to the top. He became a professor of zoology and botany at La Universidad de Chile and director of El Museo Nacional de Historia Natural. His career project was to inventory the endemic biodiversity of his adopted country. As early as 1866, Philippi comprehended the longevity of *Fitzroya* (a monotypic genus named for the *Beagle*'s captain by Joseph Hooker, friend of Darwin). The professor compared the tree to iconic species: baobab, ahuehuete, giant sequoia. From tree-ring counts, Philippi knew alerces could live 2,500 years or beyond.[23]

In the capital, Philippi raised awareness among his scientific network, but he had no power in the south. Even after the crime of forest fire entered Chile's penal code in 1874, local law enforcers turned a blind eye. Processes encouraged by the state—anti-Indian violence, settler colonialism, railway construction—created social and ecological feedback loops in the Southern Cone. One kind of fire made another burn hotter.

Federico Albert was two generations younger than Philippi, and came of age in a time more conducive to reform. After working in the natural history museum for a decade, he shifted his research to *el problema forestal*—the relationship between forest cover, soil erosion, and economy—an international discussion among scientists. In 1903, he pronounced *Fitzroya* largely "exhausted."[24] Thinking past the fiery era of settler colonialism, Albert wanted government agents to raise new commercial forests on classified public land using fast-growing Tasmanian blue gum and Monterey pine. When Santiago founded a forest department in 1910, Albert became first chief.

By technocratic consensus, "overmature" trees had no silvicultural value. "The forests of the Andes are too old," declared Bailey Willis, a US engineer hired by Argentina to study a trans-Patagonian railway route. "It may well be asked what is the value of the forests?" he continued. "Why not let them be burned and turned into pasture lands?" Experts like Willis were trained to have bigger thoughts and longer views than estate owners and campesinos, and they provided their own answer. A seemingly "worthless" forest protected water supply; water contributed to hydropower; and electricity created wealth for the nation.[25]

Willis's team included a geologist trained at Stanford, a university with a coast redwood on its seal. This Californian called alerce the "redwood of the Andes," because the endemic cypress reminded him of home. The biogeographic correspondence was latitudinally perfect. At the fortieth parallel, north as well as south, grew the tallest and densest forests in the Americas, with redwoods over 15 feet in diameter and 300 feet in height.

The redwood genus—another branch of the cypress family—existed throughout the North American West at the fiery dawn of the Cenozoic, the age of mammals. By the time humans made their way south from Beringia, coast redwood's range had contracted to the Oregon–California coast, including the Los Angeles Basin, where saber-toothed cats perched in conifers. The fog belt of northern California became the final refugium. Well into its second decamillennium with Amerindians, the species was anything but endangered, however. In mountainous mixed forests and riparian pure stands, coast redwood occupied something like half a million hectares at the time of European contact.

The conquest of the North Coast and its peoples began with the suddenness of an earthquake. Secluded in steep-walled valleys, many

Indigenous groups had escaped even indirect ails of Spanish colonialism. Natives here did not need to travel far for sustenance: Pacific salmon swam upstream to them. Watersheds defined annual cycles and linguistic boundaries. The lower Klamath River and a parallel stream, Redwood Creek, sheltered Yuroks (*Oohl*, the People) and the tallest of the towering trees.

Using elkhorn wedges, Yuroks converted fallen redwoods into planking for homes. For making a sweat lodge or a canoe—two kinds of ceremonial constructions—Yuroks required a living redwood, a sacred being. Before felling, the People propitiated the tree-person. Canoe makers further honored the tree's personhood by leaving a knob—a heart—in the bottom of the carved-out hull. Yurok paddlers spoke and sang to the hearted being that gave them access to river and sea.

At the onset of the Gold Rush, 1848, Yuroks numbered two or three thousand, compared to some ten thousand settlers in the entire territory of California. Two years later, California's approximately one hundred fifty thousand Natives met a greater onslaught of avaricious men. French and Anglo-American fortune-seekers came by boat, around the Southern Horn. At least eight thousand Chileans caught gold fever, too, and sailed to San Francisco Bay. Soon, no part of California remained a refuge: the Native population fell eighty percent in two decades.

North Coast peoples faced large-scale invasion without warrior societies, mounts, or guns—unlike Māori, who had time to master firearms, and Mapuches, who had time to master horses. In the 1850s and 1860s, Anglo-American volunteers waged exterminatory campaigns. Militiamen carried Colts and Winchesters acquired through federal programs, massacred innocents with the blessing of local and state governments, and later received reimbursements from Congress. Yuroks survived the genocide by hiding uphill and heading upstream.

The elimination of Yurok, Wiyot, and Tolowa villages from the coast presaged the clearance of coastal-access forests. Despite remoteness from markets, and low demand, logging commenced. Businessmen imagined future riches from 300-foot-tall straight-as-mast conifers composed of "clear all heart"—no holes, no knots, just vertical-grained heartwood.

By the normal fraud that characterized US land disposal, small firms accumulated large tracts of expropriated tribal land that had been intended for homesteaders. Thanks to mechanical "steam donkeys" and the indispensable labor of Chinese immigrants—repaid with xenophobic expulsions—a county named in honor of Alexander von Humboldt became the heart of redwood dismemberment.

As soon as sawyers and millers began handling the colossal logs, they saw that each contained five to twenty centuries of growth rings. In the US East, such ring counts would have generated astonishment. However, in comparison to the Big Tree of the Sierra—a related species estimated, to instant worldwide fame, to live three times longer—redwood longevity seemed less than stupendous. Sawyers with double-wide crosscut saws posed for photographs on trophy stumps. They betrayed no remorse cutting short the lives of ancient redwoods. The coastal species was as widespread as the mountain species was rare.

The transition from abundance to scarcity, like everything in Golden State history, happened at breakneck speed. "California will for centuries have virgin forests, perhaps to the end of Time!" predicted a German visitor in 1852, before announcing, with equal surety, "The doom of the red man is once and for all irrevocably sealed."[26] Less than three decades later, Anglo-Californians simply accepted that redwoods, identical to "poor Indians," were "doomed to fall before the advance of civilization."[27]

A business environment of oversupply, undercapitalization, and excess competition incentivized profligacy. Each small firm cut fast to avoid paying taxes on standing timber, and cut faster to pay off debts. The California State Board of Forestry, an advocacy commission created by the legislature in 1885, estimated a wastage rate of 70 percent for each felled redwood. Lacking regulatory power, commissioners could only lament that a resource once considered "exhaustless as the ocean" was "rapidly melting away."[28]

Humboldt County's cut-and-run era ended at the turn of the century, thanks to corporate consolidation. Increased efficiency didn't mean decreased extraction, though. When San Francisco needed to rebuild after the earthquake and firestorm of 1906, the industry rushed to provide.

With the opening of the Northwestern Pacific Railway and the Panama Canal, companies like Hammond and Palco looked beyond the Bay Area market. Production surged during World War I, with orders from the War Department.

Even advocates of corporate responsibility envisioned landscape conversion to pastures, orchards, or timber plantations. They approached grade A redwood as a one-shot crop. Many experts called for a nonnative eucalyptus belt to replace the liquidated old growth. Sustainable management of the native resource by the state was legally unviable. The executive authority to reserve forests (1891) and the organization of the US Forest Service (USFS; 1905) came too late. The entire habitat of coast redwood—including a Klamath River Reservation intended for the Yurok Tribe—was privatized by 1900.

Because seizures of land, then giveaways, had been so complete, redwood preservation required land purchases. In the post–World War I era, California pioneered nature protection as industrial philanthropy, with the Save-the-Redwoods League setting the example. The league resembled a pro-eugenics social club for nature-loving Republicans. Its founders worried about the future of America's "better" things—its megafauna, its megaflora, and its "native stock." They were nothing if not long-term thinkers. The Waspy people who organized to preserve the "oldest trees in the world" overlapped with dinosaur collectors, and they played up redwoods as relics of the age of reptiles.

The league raised millions—the Rockefellers pledged most—to purchase acreage at fair-market price. Efforts focused on a ribbon of extra-tall trees alongside Highway 101. As donated to the state of California, the reserve became drive-thru scenery—a screen to industrial logging—with pullouts at "groves" dedicated to benefactors and heroes, including Humboldt's pioneers. Through its network, the league helped to establish the California State Park Commission, which used public and private monies to acquire additional redwood properties.

Few Californians of the twenties had the luxury of steering touring cars down the "Avenue of Giants," but everyone in the fast-urbanizing state benefited from the resinous product marketed as "wood everlasting."

Like alerce, redwood defied decay, making it ideal for infrastructure—as long as prices compared favorably to concrete. For the building of California civilization, coast redwood "outranked all other natural resources," including gold, wrote Willis Jepson, the state's leading botanist, in 1923.[29]

To appreciate that evaluation, imagine a Jepson-era day trip from Los Angeles to Long Beach, beginning in a single-story house trimmed and floored with redwood. Before setting out, the homeowner uses a shower, sink, and toilet connected to water and sewer systems equipped with tongue-and-groove redwood stave pipes. Then he catches a trolley with redwood seats, and travels on tracks laid on redwood ties. The beachgoer glides past factories topped with redwood tanks, past the jungle of redwood derricks pumping oil from Signal Hill, finally arriving at a palm-lined beachfront equipped with redwood boardwalks and redwood piers, with a view of a port—soon to be the largest in the eastern Pacific—built atop redwood piling.

In the original Promised Land, the oldest and best conifers, the Cedars of Lebanon, had been reserved for the edifices of pharaohs, kings, emperors. Except as false-front movie sets, America's new land of promise lacked any equivalent to Solomon's Temple. But for one generation, at least, petit bourgeois Angelenos could afford their own private bungalows porched with "nature's lumber masterpiece." In the sunny southland, this durable expression of the California Dream was subsidized by stolen land and two thousand years of growth on the foggy North Coast.

JAPAN / TAIWAN / JAPAN

In contrast to California's extravagantly utilitarian use of redwood, Japan traditionally reserved its finest endemic timber trees, *hinoki* and *sugi*—two more species in the long-lived cypress family—for special things like shrines.

Widely distributed in the Northern Hemisphere at the beginning of the Cenozoic, sugi retreated, like redwood, to one coastal region—the Japanese Archipelago—by the Pleistocene. So-called Japanese cedar achieves gigantic proportions and contains durable wood that splits evenly and easily, similar to alerce. The heartwood of sugi is ideal for roofing in a rainy land. Long before Japan's industrial revolution, Japanese harvested and replanted sugi on an industrial scale. Today, the nation has many sugi habitats—mountain plantations, avenue plantings, shrine groves, urban parks—but no primeval forests. The qualified exception is Yakushima, a small, rugged, storm-battered isle in the Satsunan Islands, just beyond Kyushu, the southernmost of Japan's main islands. The rainiest locale in the nation, Yakushima receives four to ten meters of precipitation per year from monsoons and typhoons alike. Here, the oldest trees have lost their crowns in gales, and regrown them atop their massive trunks. Sugi wood from Yakushima, *yakusugi*, is exceptionally resinous and water-resistant, and therefore valuable.

In the Edo period (1603–1868), a clan from Kyushu gained monopolistic control over yakusugi. Local villagers paid tithing in timber. Like people in Chiloé, they used wedges to convert fallen trees into planks, which they carried down the mountain in wooden-frame backpacks. Local crews also leveled big standing sugis. Kyushu elites traded the prestigious heartwood to Honshu elites, who wanted yakusugi for edifices like Asukadera (Hōkōji), a Buddhist temple in Kyoto.

Throughout Japan, a building boom in shrines—and castles, and castle towns, all made of wood—occurred in the late sixteenth and early seventeenth centuries, after the consolidation of the samurai ruling class. The maintenance of all this wooden infrastructure—and the ritual reconstruction of major shrines on decadal schedules—encouraged long-term thinking about forest resources. No other civilization developed a richer material culture in wood, and no early modern state outside Germany developed a stronger institutional culture in forestry.

After the Meiji Restoration of 1868, authority over Yakushima's sugi transferred to Tokyo, the newly renamed capital. Imperial officials mapped, zoned, and regulated the remote island forest, but did not recognize the customary rights of islanders to collect wood for fuel. Decades of conflict ensued. By the time East Asian plant collector Ernest Henry Wilson, an employee of Harvard's arboretum, visited the "most interesting and remarkable forest in all Japan" in 1914, a détente had been reached: locals controlled the low-elevation coastal fringe; the state patrolled the upland interior.[30] In the 1920s, Meiji officials demarcated a small old-growth reserve—a Prussian-style natural monument—even as they financed a narrow-gauge railway to facilitate the extraction of big trees.

The university botanists who led the effort to create a national registry of natural monuments reacted against a different centralizing project: the "shrine merger program." In an ironic outcome of the nationalization of Shinto, the state closed tens of thousands of village-level shrines throughout Japan, relocated their *kami* (spirit beings), and deconsecrated their groves. State and private interests worked to liquidate surplus shrine trees, to the dismay of naturalists, who recognized that artificial forests had become natural habitats.[31] Meanwhile, at newly consolidated shrines, managers planned new sacred groves as sugi and hinoki plantations—pinetums of the Meiji state.

The contradictions between a forest-loving people and a forest-eating empire only deepened in the decades before World War II. How could Japan preserve its wood culture while modernizing its economy and equipping its military? The answer lay abroad. In the 1930s, as much as one-third of Japan's wood supply came as imports, mainly from its imperial rival, the United States. For all-purpose material, Japanese buyers favored Douglas fir from Washington State and, for special uses, Port Orford cedar, a species from Oregon with the look and smell of hinoki.

Japan also looked to its colonies for wood. At the same time that Europeans scrambled for Africa, at the same time that the United States seized the Philippines, Japan took Taiwan, Manchuria, and Korea through wars against China and Russia. In Korea, Japanese planted trees; in Manchuria, they plundered forests. In Taiwan, by contrast, they strategically harvested Asia's greatest conifers, hinoki-like species that surpassed even old-growth yakusugi in girth, height, and age.

Located at a tectonic collision zone, the island of Formosa (Taiwan) is one of the youngest mountain formations on Earth. The island's summit, Yushan, rose higher than Mount Fuji in the period that *Australopithecus* evolved into *Homo*. In a sense, humans have been there since creation.

Taiwan was the origin place of the Austronesian language group, from which Polynesian languages, including Māori, derive. Despite the relative closeness of mainland Asia and a history of Han migration, Formosa wasn't annexed by China until the seventeenth century—a response to competing Dutch claims issued from Fort Zeelandia.

From 1683 to 1895, imperial China struggled to incorporate Formosa. One island became two: a Han-dominated piedmont—a zone of camphor, bamboo, and tea—below the fog-shrouded, snow-capped, Indigenous-controlled highlands. Chinese officials marked the "savage boundary" with mounds and ditches, beyond which lived the "raw savages," respected and feared for their ritualized warfare. Han settlers were long forbidden to cross the frontier. After 1875, however, Qing policy shifted aggressively to "Open the Mountains, Pacify the Savages." China was heading for the conifers but lost control of Taiwan before reaching them—or even knowing they existed.

The new imperial power, Japan, had the resources, the expertise, and the drive to carry out a complete cartographic survey of Formosan forest resources. It wanted to show the "civilized world" it could succeed where China had failed. Japanese officials used language reminiscent of the white man's burden. They spoke confidently of plans to turn mountain bandits and tattooed cannibals into colonial subjects. "To subjugate Taiwan," said the governor-general, "we must conquer its forests."[32]

Within months of occupation, Japanese soldiers reported on forest giants in the Alishan Range at 6,000 to 9,000 feet.[33] Although located on the Tropic of Cancer, this mountainous area had a climate similar to California's North Coast or Chile's Valdivian coast. Here in the misty highlands, Japanese foresters were delighted to discover a population of hinoki and a new kind of super-hinoki, endemic to Taiwan, which they called *benihi*.[34] The Japanese, German, and US botanists who subsequently visited Alishan compared the big Formosan conifers to those of California. According to Ernest Henry Wilson, the oldest known (felled) tree had approximately 2,700 growth rings. He marveled: "The forests are easily the finest and the trees the largest I have ever seen."[35] It was like Yakushima, only better. Imperial officials called it a divine gift.

In 1902, the state commissioned Shitarō Kawai, the son of a samurai and a professor of forest utilization at the Imperial University of Tokyo, to write a development plan for Alishan.[36] Kawai had recently returned from a multiyear postdoc in Germany and Austria. After seeing Alishan in person, he decided this scenic wonderland deserved a world-class mountain railway like Austria's Semmeringbahn, with dozens of bridges and spiral tunnels. Such a railway would allow uphill transport of sightseers and hikers as well as downhill transport of whole logs. Parliament balked at the cost, forcing Kawai and other railway boosters to seek private financing. In the end, to controversy, the government bailed out the overbudget project.

To equip the railway, Kawai went to the United States. The professor wowed US executives with his cosmopolitanism and command of English. His envoy ordered logging cars and locomotives from Seattle, milling equipment from Milwaukee, and a cableway skidder from Brooklyn. As much as he desired US machines, Kawai deprecated the "extravagant American way" of cutting and running. For Alishan—a forest that "outclassed" anything on the West Coast—he promised to practice German forestry: removing overripe hinoki and benihi, and reforesting with sugi. Industrial experts in America called this "long-time management."[37]

By 1915, when the extravagantly overbuilt railway connected the sawmills in Kagi (Chiayi City) to the summit of the Alishan Range, Kawai had been marginalized. He became disillusioned with the project as politicians pressured managers to focus on the bottom line. Privately, he wrote an elegiac poem about millennial trees axed down. After his death in 1931, students and colleagues erected a memorial in the forest—a granite block from Japan engraved with characters that praised imperial forestry and Kawai's role in bringing tourists to the mountains and teachers to the savages. Nearby, a state monument honored the scores of Japanese workers who died in the construction of the railway.

Old-growth extraction at Alishan benefited the empire, not the public; Tokyo, not Taipei. Through the entire Japanese period, forest-rich Taiwan imported wood. Subsidized Alishan exports were destined for prestige projects—notably the Meiji Shrine (for the spirits of the late emperor and his consort) and the Yasukuni Shrine (for the spirits of military personnel who died in service of the empire). Two complete benihi trunks supported the main *torii* (gateway) at the Meiji Shrine. In 1920, a thin cross section from one of those trunks went on display at the Tokyo "Time Exhibition"—a government-sponsored event that encouraged Japanese citizens to further internalize habits of modern timekeeping.

The symbolic counterpart to the perfectly straight torii in Tokyo was the largest benihi on Alishan, a leaning behemoth over 170 feet tall and more than 60 feet around. As early as 1912, this cypress had been fenced off. Later, a Shinto priest consecrated the tree, girding it with a *shimenawa*, a ceremonial rice straw rope, twisted left, with downward-hanging paper inserts. As a *shinboku*, or divine tree, it sheltered kami. In a remarkable meeting of modernity and tradition, Japanese engineers laid the tracks of the Alishan railway directly atop its roots. Uniformed officials, soldiers, and schoolchildren assembled around the divine tree for photographs; trains belched coal smoke into its canopy. A sign estimated the tree's age at 3,000 years.

The consecration of Alishan's purported oldest tree belied the logging operation, which targeted ancient trees—overmature and overgrown, to use language of German-trained foresters. In terms of total area, the Japanese didn't cause widespread deforestation in Taiwan. They were not burning and privatizing the bush for agrarian settlers, as did New Zealanders and Chileans. They were not wastefully working their way through stages of frontier capitalism, like Californians. Instead, Japan trained the industrial might of the empire onto forestland seized by the state.

Regardless of economic context, violence against Indigenous peoples accompanied the onset of big-tree removal around the Pacific Rim. In Taiwan, construction of the Alishan railway coincided with a military campaign against resistant highlanders, notably Tayal language speakers. They would be "civilized or destroyed," predicted Shitarō Kawai.[38] After being resettled in the lowlands and forcibly educated, some Tayal men fought and died for the Imperial Japanese Army. By Meiji tradition, their spirits were housed at Yasukuni Shrine—surrounded by wood from their homeland.

After the military opened Alishan to tourists in 1920, the logging camp became accustomed to notable guests from Tokyo. Until the outbreak of the Pacific War, the logging railway doubled as a scenic railway, and a parade of literati, painters, scholars, and officials made the sinuous trip to Alishan, the base camp for hiking to Yushan, the empire's "New High Mount," higher than Fujisan. In a partial realization of Kawai's vision of regional development, Japan created a national park encompassing Yushan and parts of Alishan in 1937.

The Formosan highlands attracted elite foreign travelers, too, including Poultney Bigelow of Malden-on-Hudson, a figure made for an Edwardian melodrama: childhood-friend-for-life of Wilhelm II, with whom he played cowboys and Indians in Potsdam; canoeing partner of cowboy artist Frederic Remington; author of *White Man's Africa*. Bigelow considered Alishan a brilliant colonial triumph. Comparing Taiwan to the US West, he lauded the Japanese for educating its savages instead of killing them. Feted by his Japanese hosts in 1921, Bigelow participated in a tree-planting ceremony by the Shinto shrine in the logging camp. Amid Japanese cherry trees and the stumps of "monster cedars," Bigelow celebrated Alishan's conversion from a "wilderness of headhunters to a rich garden of modern civilization" and toasted the potential of Anglo-Japanese friendship.[39]

The future shocked both nations, and their forests. To fuel its war against the United States, Japan devastated its domestic sugi. In turn, the US military poured fire onto Tokyo, the world's greatest wooden city. In spring 1945, an empire's worth of millennial heartwood burned to bone-flecked ashes. From their B-29s—assembled in a Seattle factory constructed of millions of board feet of old-growth fir—US airmen targeted cultural institutions as well as civilians. The hinoki buildings at the heart of the Meiji Shrine suffered direct incendiary hits. To air-raid sirens and the screams of dying neighbors, quick-thinking residents ran to the shrine's pinetum, a refuge within a holocaust.

After the surrender of Japan, the history of Taiwan abruptly turned again. In the final stage of the Chinese Civil War, the island became the seat of Chiang Kai-shek's government in exile. In its nation-building phase, the Republic of China caused far worse deforestation than the reviled Japanese occupiers ever did. By the end of the twentieth century, though, Han-majority Taiwan reevaluated Alishan and its quaint railway as postcolonial heritage and flocked to see Yoshino cherry blossoms. The leftover giant cypresses, accessible by elevated walkways, are now a secondary attraction; people have forgotten the moment of publicity in 1973 when a professor at Chinese Culture University claimed to have found a 6,000-year-old benihi—the "world's oldest tree." The shinboku had died in the fifties but remained standing (minus its shimenawa) until 1998, at which point Taiwanese authorities designated a "new divine tree." A former state Shinto site was thus Sino-fied. A decade later, when Taiwan began receiving tourists from the People's Republic of China (PRC), locals complained about uncouth mainlanders who disrespected the island's (new) oldest living thing by smoking cigarettes in its hollow.

Benihi has bigger problems: The IUCN has listed the species as endangered since 1998, despite a logging moratorium a decade before that. The healthiest old-growth specimens grow in rugged, remote areas, making them easy targets of illegal loggers and also burl poachers—maligned as "mountain rats" in Taiwan—who sell burls to woodcarvers, who in turn make Buddhas and essential oils for tourists. Criminal gangs have lured Vietnamese migrant farmworkers to do the poaching.

As well as mutilators, there are protectors. Several Tayal communities that returned to mountain homelands after World War II have, since the end of martial law in 1987, asserted use rights by designating "divine trees" as attractions. For instance: Smangus, about thirty-five air miles from Taipei, was among the last communities in Taiwan to gain electricity (1979) and tarmac (1995). Once accessible—albeit on a winding, dead-end road—Smangus branded itself as a Xanadu and reorganized itself like a kibbutz. It began offering lodging to foreign hikers who wanted to be guided to a grove with a multitrunked giant purported to be 2,500 years old. The idea of using this "mother cypress" to revitalize the community came from ancestors, who spoke to the tribal leader in a dream. The old-growth relict, surrounded by cultivated forest, falls under the de facto supervision of Smangus, despite growing on government land within the proposed boundaries of a national park. In the early twenty-first century, responding to urban environmentalists, the Taiwanese state attempted to broker a large-scale reserve for highland cypresses, with a weak form of comanagement offered to mountain villagers. Many Tayal groups feared dispossession, and the proposal foundered; though, in a related legal case, Taiwan's High Court affirmed the principle of Native use rights. For the present, community-based ecotourism helps Smangus sustain itself as "God's tribe"—a Presbyterian Indigenous commune. The village converted to the Gospel in the 1950s, making its subsequent co-option of a vaguely Shinto tree tradition all the more "glocal."

Something analogous and strange has happened at Yakushima, Japan. During the postwar recovery, the state reopened the logging town in the sugi forest and introduced chainsaws. The island's population peaked in 1960 when hundreds of state-employed loggers clear-cut old-growth outside the protected area. A decade later, with the best trees gone, the state shuttered and demolished the town. Under new pressure from environmentalists, Tokyo reinstated a ban on extracting yakusugi and turned the better part of the island into a biological and recreational reserve. To stimulate Yakushima's postindustrial economy, the state programmatically remade Yakushima National Park into an ecotourist attraction.

This top-down governmental initiative benefited from two overlapping groundswells in late twentieth-century Japan—one from the left, one from the right. The former, the "spiritual world" (the Japanese equivalent of "New Age") emerged from global sixties youth culture. An early influencer was Sansei Yamao, cofounder of a commune called "the Tribe." The California beatnik Gary Snyder, who had come to Kyoto to study Zen, capped his education in Japan by living with the Tribe and marrying one of its members. Yamao went on to establish the "Hobbit Village" in Tokyo before going back to the land in 1977 at Yakushima, where he allied with local old-growth activists. Like his friend Snyder across the Pacific, Yamao wrote essays and poems on bioregionalism.

The second groundswell—more influential, more reactionary—was the so-called Jomon boom of the 1990s. After spectacular archaeological discoveries on Honshu, tourism boosters fetishized the once fringe theory that the ancient people(s) who defined the Jomon period (ca. 10,000–300 BCE) counted as the original Japanese. Against the standard narrative of proto-Japanese migrations from mainland Asia, Jomon enthusiasts embraced an alternative history of national indigeneity—even as the state denied legal recognition to Ainu and Ryukyuan peoples. On Honshu, heritage tourists visited "Jomon villages" and Jomon-themed festivals.

Far to the south, a powerful if ambiguous heritage site existed in the inner mountains of Yakushima. The island's largest sugi—the nation's largest plant—had been called "Jōmon Sugi" since 1967, not long after its disclosure. Yamao wrote a poem about this "holy old person" as a manifestation of a celestial buddha. He regularly walked to it, and around it, seven times, chanting *nembutsu* and leaving offerings to residing spirits. Yamao disapproved of sightseeing—he emphasized listening to the tree— but, through a convergence of bioregionalism and nationalism, Jōmon Sugi became a commodified symbol of the primeval "Japanese forest." In the 1990s, the emblematic plant appealed to cosmopolitan environmentalists as well as cultural nationalists who posited Shinto as an aboriginal forest-based religion. National language common readers instructed Japanese children to poetically reflect on the life of the 7,200-year-old (which, if true, would have made it the world's oldest tree by far).[40]

Visitation to Jōmon Sugi spiked after 1993, when UNESCO accepted Japan's nomination of Yakushima as a World Heritage Site. High-speed ferries and resorts began catering to ecotourists. The most influential visitor was Hayao Miyazaki, who came with animators from Studio Ghibli to do location research for *Princess Mononoke* (1997), an anime rich with animism. By the 2000s, the all-day hike to the last giant yakusugi had become an eclectic pilgrimage. On peak holidays, up to one thousand tourists congested at the tree, leaving litter, feces, and compacted soil— despite the addition of a viewing platform. Impacts on the "real-life enchanted forest" would have been worse were it not for the fact that the trail largely followed the old logging railway. At Yakushima, devotees of "forest bathing" and "power spots" now walk between iron rails from the twentieth century, past moss-covered stumps from the sixteenth century, to an ancient tree equipped with a security camera.

CALIFORNIA/CHILE

Yakushima followed California: The first protected area with big trees to earn World Heritage status was Redwood National and State Parks, listed by UNESCO in 1980. This recognition followed twenty years of political conflict in Sacramento and Washington over the value of the planet's tallest trees.

Proposals to add a federal buffer to the Golden State's patchy redwood parks came and went in the first half of the twentieth century. During the Depression, when numerous logging operations fell into receivership, the US Forest Service initiated a buyback program in redwood country. Insufficient funds prevented large-scale purchases. In the 1940s, the retired first chief of the USFS, Gifford Pinchot, urged Congress to purchase two million acres of northwestern California—a bold proposal that made it to House committee, no further.

When federal planners returned to redwoods in the 1960s, the desired outcome had changed from a silvicultural forest to a recreational park. Before oil and gas royalties (funneled through the Land and Water Conservation Fund) could be spent on redwoods, various legal parties had to agree on boundaries. Technically, Congress had to compel owners to sell or swap land through "legislative taking." To expedite land classification, the National Geographic Society conducted a land survey for the US National Park Service (NPS), an effort compensated by the society's exclusive announcement of its "discovery" of the world's latest "tallest tree," located in Humboldt County. The "Tall Tree Grove" on Redwood Creek, ancestral Yurok land, belonged to Arcata Redwood Company.

The upstart Sierra Club wanted to preserve the Redwood Creek watershed from rim to rim, though that meant going up against the "Big Three"—Arcata, Simpson, and Georgia-Pacific. The old-guard Save-the-Redwoods League focused on Mill Creek in Del Norte County, a smaller watershed owned by one small firm. The Sierra Club's leadership grabbed more attention: it published a book, *The Last Redwoods*; distributed a film, *Zero Hour in the Redwoods*; and took out full-page ads in the *New York Times* ("the last chance REALLY to save the redwoods").[41]

With his presidential pen, LBJ signed the park into being in 1968, a year of political fires around the world. Congress authorized $92 million; the transaction cost more than triple that amount after full appraisal. The nation's priciest park had no visitor services and no ecological coherence. Neither watershed had been protected in its entirety. Environmentalists portrayed the scrawny property at Redwood Creek as "the worm"—a "false-front" preserve that needed immediate enlargement. They revived the language of doom: "Time was running out" on the "deadline" to save the "last" of the "vanishing" forest.

In the 1960s, timber corporations had accepted a national park as inevitable and played defense. Now, in the 1970s, they attacked tree huggers as hysterical coercers. The tallest and oldest redwoods were already saved, they argued. Park expansion would drain public coffers, threaten private enterprise, and endanger the local economy. While Congress held hearings and debated bills, the Big Three ramped up operations—what environmentalists called "vendetta logging." By the time the national legislature authorized the additional taking of forty-eight thousand acres, four-fifths of the area had been reduced to stumps.

By signing the bill in 1978, President Jimmy Carter sided with environmentalists over another constituency, the AFL-CIO. Humboldt County had enough members of the Lumber and Sawmill Workers Union to organize large demonstrations in the Bay Area, and a big-rig convoy to Washington, DC, in 1977. "Save our jobs!" they had chanted from the Capitol steps. The hard-hatted demonstrators included two prominent Yuroks, Milton Marks and Walt Lara Sr. Although they cut redwoods for a living, they were better known in their tribal community as the first two chairmen of the Northwest Indian Cemetery Protective Association, which called attention to vulnerable sacred sites on state and private lands.

When the NPS conducted resource assessments of the enlarged park, it found archaeological evidence of Yurok villages and ecological evidence of Indigenous burning. The agency declined to consult tribal stakeholders about its management goal of restoring the forest to its "natural pre-European" condition. On the park's recently logged hillsides, rangers became de facto gardeners, planting redwood seedlings that would, they hoped, reach maturity in the twenty-second century.

The Yurok Tribe—once and still the most populous Indigenous group in the Golden State—was thinking far ahead, too. In the years before and after the turn of the millennium, it took coordinated steps to secure its long-term future. The tribe created a language curriculum; celebrated the construction of redwood canoes and redwood plank houses in traditional style; and, to stunning success, lobbied for the removal of salmon-blocking dams on the lower Klamath River. Today, with profits from the sale of carbon offsets, and assistance from philanthropies, the tribe buys back stolen land, parcel by parcel. Going forward, the tribal government wants to superintend both sides of the Klamath's mouth and to comanage sections of the park that include ancestral villages and sacred sites.

Seeds of indigenized management have sprouted. In 2019, the US Fish and Wildlife Service announced a partnership with Redwood National Park and the Yurok Tribe to reintroduce the California condor, another sacred species. The year before, in coordination with the UN Global Climate Action Summit in San Francisco, Yuroks hosted an event on Native forest management for a pan-Indigenous delegation. The gathering included a guided walk in Yurok Research Natural Area, an old-growth reserve owned by the USFS—at least for now. No community has a deeper, longer commitment to redwoods than do Yuroks, with untold centuries to come.

Of the coast redwood forest that existed in 1848, only about 5 percent exists now—all of it protected. But the overall area containing redwood is not dramatically lower. In other words, second growth dominates. The North Coast's hills and valleys are checkerboarded with redwood properties restocked with containerized seedlings cloned from pedigree stock. Although large firms vacated the area once the last stands of grade A trees were felled, reserved, or tightly regulated, a scaled-down industry continues. Coast redwood isn't endangered—partly because it regenerates from stumps, and mostly because California foresters never found a better tree for the North Coast than the endemic one.

The situation in Chile could not be more different, yet the outcome could scarcely be more Californian. Parallels between redwood and alerce became intersecting lines in postwar Pax Americana. As of mid-century, practically all alerce patches (*alerzales*) in the central depression had been extirpated, leaving two discontinuous mid-elevation habitats that alerce shared with southern beech. One was a temperate rain forest in the coastal hills, the other a slightly less mossy forest in the foothills of the Andes.

In the early 1960s, Seattle-based Simpson Timber Company—one of the Big Three operators in the redwood empire—determined that its remaining "virgin timber" would run out in a decade. Cushioned with cash, the company went looking for untouched forests. Simpson hired a timber cruiser to select the best property on the Pacific Rim, which led them to the Andean side of Los Lagos, northeast of Chiloé Island, where ten thousand hectares of old-growth alerce grew—an estimated one billion board feet of merchantable product. With the blessing of the Chilean state and with assistance from the US Export-Import Bank, Simpson oversaw the first industrial-scale logging of the "redwood of the south." It built a US-style company town, Contao, accessible by airstrip. Bulldozers and other heavy equipment arrived via Puerto Montt. Simpson hired Chilean managers and brought them to Arcata, California, for training.

Once milling commenced, Simpson discovered "disaster instead of bonanza." The old alerces had dry rot in their cores. One billion board feet shrank to seventy-five million. Looking back on the fiasco, Simpson's president, Bill Reed, regretted he hadn't invested in New Zealand or Taiwan. He admitted that the United States and Chile were overeager to demonstrate the superiority of Anglo-American industry. "We should have built [an operation] like the Araucanian Indians," Reed said, referring to his local employees.[42] Hand tools and pack animals would have generated more profit. When a made-in-the-USA Caterpillar broke down at Los Lagos, it took half a year to get replacement parts.

After less than a decade, Simpson cut its losses and left Chile. The decision was hastened by the rise of Salvador Allende, whose Popular Unity coalition called for the nationalization of the forest industry. A committed socialist, Allende hoped to redistribute estates to peasants and Indians while increasing production on state land. During the coalition's brief time in power, state foresters ramped up production at Contao and began new alerce operations on Chiloé.

Some tree huggers dissented. In 1970, in the revised edition of *The Survival of Chile*—the foundational text of Chilean environmentalism—Rafael Elizalde Mac-Clure called for a total ban on alerce logging. It's time to declare the species a national monument, the author declared. To kill the survivors would be a crime against nature. In a poem dedicated to alerce, written shortly before he committed suicide by self-immolation, Mac-Clure intoned: Your death is the death of life.[43]

After the CIA-supported coup of 1973, Chile reverted to form: disempowering Indigenous peoples, clearing native forests, and creating nonnative commercial forests. The Pinochet regime carried out conservation by gunpoint, executing the old plan of German-Chilean foresters on a new scale. By the 1980s, south-central Chile had more land covered in *Pinus radiata*—a fast-growing species once restricted to Monterey Bay, California—than anywhere on Earth, not counting New Zealand's North Island. Socially and ecologically, Chile's pine plantations were dead forests, fenced and guarded against peasants and Indians, sprayed with fungicides by airplane.

At the same time, the military dictatorship made a show of protecting Chile's endemic cypress, declaring *Fitzroya* a "monumento natural." This designation from 1976 categorically banned the cutting of "one of the longest lived species of the plant kingdom."[44] On closer inspection, though, the declaration permitted commercial logging of dead alerces, even whole forests of dead alerces, provided that loggers submitted government-approved work plans. This loophole incentivized illicit fires and the commercialization of salvage logging. The law burdened the one group of people with a record of sustainable harvesting—Mapuches (Huilliches) of the western Los Lagos Region.[45] Without proper paperwork, their traditional economy of plank-making was technically against the law.

The declaration—and a similar legal move in Argentina—got the attention of the international community. The United States defined alerce as a threatened species under the 1978 amendment of the Endangered Species Act and banned importation of alerce products. A decade later, the Convention on International Trade in Endangered Species (CITES) elevated *Fitzroya* to the class of prohibited trade items, comparing it to ivory. With financial assistance from the World Wildlife Fund, a Chilean NGO began monitoring illegal burning and logging.[46]

Reports of endangered, redwood-like trees traveled to the Golden State and possessed Rick Klein, a Humboldt County original. Klein had come to redwood and cannabis country with a commune from Minnesota. He had planned to continue to Canada, to dodge the draft, but in California he met a Chilean heiress recruiting hippies to gather on her estate. Klein hitchhiked down the Pan-American Highway and bummed around the Southern Cone for a couple of years. After Pinochet rose to power, Klein evacuated Chile, returned to the North Coast, and became a founding member of a grassroots self-taught legal organization that went on, improbably, to transform forest regulation in California.[47] Klein never lost his love for Chile and, in 1987, he began organizing trips to introduce fellow back-to-the-landers to the "lost forest of the Andes." He formed a new group, Ancient Forest International, with the goal of purchasing cathedral groves. With the moxie of a missionary, Klein secured a profile in the *New York Times* and a donation from Yvon Chouinard, founder of the California-based Patagonia, Inc.

Most importantly, Klein converted Douglas Tompkins, cofounder of North Face and Esprit. Recently divorced, and divested, Tompkins was in the market for a hideaway in Patagonia—a region he had visited as a young climber. He and Chouinard had established the "Californian Route" on Monte Fitz Roy, the peak depicted on the Patagonia brand label. Having profited from consumer capitalism, Tompkins now rejected it. In 1990, as partial amends, he created a foundation dedicated to "deep ecology," a philosophy popularized by mountaineers and redwood activists.[48] The same year, Klein accompanied Tompkins and photographer Galen Rowell on a wilderness adventure trip to Los Lagos. Arriving by Cessna, the California trio backpacked through the newly designated

El Parque Nacional Alerce Andino. Under the evergreen canopy, Klein pushed Tompkins to think bigger: Why settle for a ranch when he could afford his own private park?

The arrival of do-gooders from California irritated Antonio Lara, Chile's leading authority on alerce. A US-trained professor of ecology, Lara compiled the data that placed *Fitzroya* on the IUCN Red List. He had worked within regular political channels to reform logging regulations and practices before turning to activism. "I decided I had to get involved or see my research vanish," said Lara. "Here the only law is the law of the jungle."[49] To him and many compatriots, Ancient Forest International smacked of neocolonial grandstanding. Other Chileans appreciated Klein's freewheeling style, and founded allied activist groups.[50]

The verified evidence that experts had expected for decades—that alerce can indeed live beyond three millennia—came in 1993, when *Science* published an article coauthored by Lara on temperature signals recorded in tree rings. Lara's meticulous survey of stumps revealed that the oldest known *Fitzroya* had been chainsawed in 1975, the last year such logging was legal. The feller—almost certainly an economically distressed Mapuche—couldn't have known he'd killed a 3,613-year-old tree. The posthumous data made the news in California, where both redwood and sequoia were demoted on the list of longest-living plants.

By this point, Tompkins had sold his house on Lombard Street in San Francisco and moved permanently to Patagonia. Through land purchases, his "Pumalín Park" grew more than five times larger than Redwood National and State Parks—at a fraction the total cost paid by the US Treasury and the Save-the-Redwoods League. Rumors swirled around his wilderness grab: a nuclear base, a Jewish colony, a secret gold mine. As one of the largest landowners in Chile, Tompkins earned the ire of the Church, the military, and developers of timber and hydropower—especially once the Californian started criticizing Chilean environmental and economic policies.

During and after Pinochet, under the advisement of Chicago-trained economists, Chile became a laboratory for neoliberalism. The government privatized all it could, opening the country to foreign investors and buyers. One of the world's most isolated countries rapidly became a leading exporter of fish, wine, timber, pulp, and ever more copper. In fruit production, Chile became the Golden State of the Global South. During the Chilean "miracle"—so named by Milton Friedman—habitat destruction and habitat protection increased in tandem by the same rules. In a free market, multinationals and Chilean oligarchs could pursue extractivism, consolidating vast holdings, just as Esprit's founder could follow his ethos of "eco-localism" to protect hundreds of thousands of hectares.

Various NGOs, including the Nature Conservancy and the WWF, have directed resources to the Valdivian coastal rain forest, one of the planet's great carbon sinks. From San Francisco, the Save-the-Redwoods League has arranged a sister-park relationship between understaffed El Parque Nacional Alerce Costero and Redwood National Park. At "80 degrees of separation," both temperate rain forests are mixtures of relict stands, recent cuts, and second growth. Both face the law enforcement challenge of burl poaching. The NPS has sent rangers to Valdivia to consult; the Chilean national forest corporation (CONAF) has reciprocally sent personnel to Humboldt County.

Meanwhile, transnational teams who study forest canopies have drawn cis-Pacific connections between alerce and redwood. This type of aerial research—climbing meets ecology—was made famous at Humboldt State University in the 1990s and 2000s. North Coast botanists "discovered" and named the new tallest tree in the world (380 feet), climbed to the summits of reiterated upper trunks, and documented astonishing endemic ecosystems—terrariums in the sky. Similarly, in the ferny crowns of the Valdivian rain forest, scientists found critical habitat for the nocturnal marsupial called monito del monte (little mountain monkey), another Gondwanan relict.

On the ground, Indigenous leaders in Los Lagos Region began an initiative in the 2000s to establish a network of community-run parks called "Mapu Lahual" (land of the alerces). Like Yuroks, coastal Huilliches hope to continue harvesting cypress planks while hosting ecotourists.

However, Chile's Indigenous peoples lack the constitutional standing of Māori and US tribes—despite being proportionally more numerous—and thus the means to regain ancestral lands.

Most of the truly illegal alerce extraction takes place on private properties owned by the non-Indigenous. As suggested by criminal cases—prosecuted patchily—trafficking sometimes occurs with CONAF's complicity. "The corruption is tremendous, involving very important people," said Adriana Hoffmann in 2005, speaking as Chile's leading botanist and former director of its environmental protection agency. "There is always plenty of talk about saving the alerce, but nothing gets done and as a result, we are losing part of our patrimony."[51]

Ten years later, Hoffmann's friend and collaborator Doug Tompkins died of hypothermia after a kayaking accident. In the aftermath, his foundation—run by his second wife, a former CEO of Patagonia, Inc.—entrusted Pumalín to CONAF. The gift was publicized as the largest private land donation in world history. Compared to the redwood-saving Rockefeller family, Doug and Kris Tompkins were barely rich; compared to the Huilliches of Mapu Lahual, they were magnates of the first order. Chile's new national park, named in honor of a US American (granted honorary Chilean citizenship posthumously), contained one-quarter to one-third of all remaining alerzales.

Tompkins Conservation sponsors a restoration project called "Alerce 3000," the idea being that recovering from the nineteenth and twentieth centuries could take one millennium for the slowest big plants. In comparison, Humboldt County's second growth flourishes, yet fast-growing redwoods may yet grow out of time. On the scale of the next hundred years, how pacific will the climate remain? Will the summer fog that enables gigantic growth still roll in each morning? Will the El Niño-Southern Oscillation that connects California and Chile conform to the cycles recorded in tree rings?

The endurance of settler polities is another open question. In 2019, the year El Parque Pumalín Douglas Tompkins came into being, urban Chilenos took to the streets in massive demonstrations. They protested systemic inequality—something Mapuches have experienced since the birth of the nation. Beyond a certain threshold, everything burns, and the fires of the nineteenth century still smolder.

AOTEAROA

In small and large ways, New Zealand remains an outlier. Of Pacific Rim polities that destroyed forests of big old conifers in the nineteenth century, it's the only one that didn't establish a state park devoted to them in the twentieth. The absence of a "Kauri National Park" betrays Pākehā ambivalence and Māori resolve.

The year New Zealand graduated from colonial status, 1907, coincided with peak native timber production. The new Dominion of New Zealand permitted the Kauri Timber Company and other private interests to purchase properties directly from Māori, while continuing to lease Crown land for clearance. With the advent of refrigerated shipping to Britain, New Zealand recommitted to an economy of grass and mammals, meat and dairy. At the same time, the Dominion finally began regulating kauri logging and making plans for afforestation. A 1913 royal commission made recommendations on land classification—a watershed in the history of Aotearoa. To replace once-extensive and biodiverse "natural forests," the commission laid the groundwork for intensive "exotic forests" composed exclusively of Monterey pine, the "Great Timber Tree of the Future." In the settled opinion of New Zealand's homegrown experts, kauri took a millennium to mature, making it useless for forestry. The future belonged to a conifer from the other side of the Pacific.[52]

Land classification included a scenic category, and commissioners recommended the reservation of the last, best, biggest, oldest kauris. They focused on Waipoua, a rugged area south of the mouth of Hokianga Harbour. They envisioned a postage-stamp park—a "forest museum"—buffered by a pine plantation, and patrolled by forest rangers alert to Māori "trespassers" and Dalmatian "gum thieves."

The outbreak of the Great War delayed the founding of a forestry agency. In the interim, an eminent imperial forester came out of retirement, moved to New Zealand, and attempted—and failed—to convince commissioners they were all wrong about kauri. David Hutchins, a veteran of India and Cape Colony, with training in France, was insufferably didactic, reiterating his points as rhyming mnemonics. He didn't make friends by calling kauri clearance an anarchical holocaust, a national scandal, a dark blot on Anglo-Saxon civilization. He compared New Zealand expertise to witchcraft and mused that kauri would have fared better in a France Australe. The war provided an opportunity for a reset, he argued, if only New Zealanders appreciated that kauri grew faster than forestry species in Europe. In one century—not one millennium—the state, staffed with retrained veterans, could supervise an ecological transition from overmature kauri forests to all-age forests. Looking past giant relics of a doomed species, Hutchins trained his mind's eye on a standard-sized "kauri tree of the future."[53]

Hutchins died in 1920, the same year New Zealand organized its State Forest Service. Rangers wore badges that featured—like Hutchins's headstone—a stately kauri. Ironically, the agency's goal of permanent forests provided a rationale for accelerated clearance of the natural endowment. State foresters deemed *all* native conifer species deficiently slow for sustainable forestry. Whether kauri grew faster than Scots pine in France hardly mattered when Monterey pine grew faster in New Zealand than any conifer ever measured. It was a California miracle tree. By 1930, the North Island's pine production equaled that of kauri.

The hyper-rationalization of forestry took place against shifting attitudes among the Dominion's urban middle class. Landscape painters now romanticized the bush; bird-loving women formed the country's first nature protection club; and a succession of historians, poets, and novelists portrayed settler culture as rapine and recklessness. In *The Story of a New Zealand River* (1920), Jane Mander narrated an unhappy marriage between a cultured woman and a type A timber boss, "a coming man" whose guiding star is the biggest kauri in the best bit of bush, thousands of years old. "Nothing to beat it outside of California," he says. "She's coming!" he announces, as his quarry falls to earth. After witnessing the sacrilege, his wife feels the fire of marital revolt.

Contradictions within Pākehā culture were apparent in 1935, when the Northland hosted Prince Henry, Duke of Gloucester. Henry loved to drive touring cars and to make tour movies with his portable cinematograph. For his royal pleasure, a giant umbrella-type kauri, 20 feet in girth, reputedly one millennium old, came crashing down on cue. After filming the arboricide, the prince visited Waipoua to marvel at the scenery. The big-tree forest was newly accessible by a controversial road. Conservationists in Auckland feared the scenic track would become a pathway for logging.

If anything, the road saved the forest, because the country's largest surviving tree—girth, 50 feet—was discovered during construction. In the twenties and thirties, visitors called the roadside attraction "Big Tree," "Big Kauri," "Captain Ellis" (after the director of forests), and ultimately "Tanemahuta" (after Tāne, Māori god of forests and birds, though often mistranslated as "King of the Forest"). Whether the appellation originated with local Māori is unclear. In any case, a mighty tree named for a Polynesian deity appealed to Pākehā in a time of neonative eco-nationalism. Kiwis began calling Tāne Mahuta the oldest thing in the Dominion, even older, perhaps, than California's Big Trees. By the time of the national centennial in 1940, the giant had appeared on commemorative stamps.

The symbolic adoption of Tāne Mahuta coincided with the high tide of the "great New Zealand myth." Schoolchildren of the interwar years—including Māori forced to speak English—learned a just-so narrative, assembled by white pseudo-anthropologists, about the Polynesian discovery of Aotearoa by Kupe in AD 950, followed by the "Great Fleet" of AD 1350.

At Waipoua, the contradictory messaging regarding priceless old trees within a valueless old forest led to the Dominion's first major environmental controversy. During World War II, the Forest Service logged at the edges of Waipoua for timber for minesweepers; after the war, the agency framed kauri as essential material for returning soldiers in need of bungalows. Conservationists, led by Auckland University zoologist William Roy McGregor, drew a line around Waipoua. The "last virgin" kauri forest was an interdependent, harmonious organism. If left intact, it would regenerate everlastingly. McGregor wanted an inviolate national park to atone for the wanton recklessness of British settlers. Unlike David Hutchins, he had Pākehā support for his ecological moralism. We have, he wrote, torn a whole chapter from the Bible of Nature, and hold one last crumpled page in our hands. If we throw it away, the future holds the certainty of remorse.[54]

The Forest Service retorted: The settler period might have been wasteful, but nature can be wasteful, too. Kauris will last longer in a managed forest, for man improves on nature. The government will prevent fire through forest thinning and salvage logging, whereas nature enthusiasts would lock up decadent and dead trees, and let the forest burn in future drought. Their plant museum is really a tree cemetery. Waipoua's natural existence for thousands of years hardly guarantees continued existence. Scientific forestry is the true measure of perpetuity.[55]

The outcome of the conflict pleased no one. Waipoua Forest Sanctuary, established in 1952, was neither a production forest nor a national park. To advise the Forest Service on the administration of the reserve, the state appointed an advisory committee. It contained no Indigenous stakeholders, even though the Northland remained majority Māori. Members of the local iwi, Te Roroa, had complained for decades that the Dominion, in its efforts to consolidate Waipoua, had coerced land sales without just compensation.

Over the twentieth century, as they recovered their population, Māori engaged in political activism and cultural revitalization. For example, Te Puea Hērangi (Waikato), known in the English press as "Princess Te Puea" for her family's connection to the Māori King Movement, dreamed of a new Great Fleet. Her organizing coalesced with the construction of a ceremonial war canoe (*waka taua*) for the 1940 centennial. With permits received from Wellington, and offerings made to Tāne, Māori artisans felled three large kauris and performed funerary rites. Then they hewed and hollowed, joined and carved. Their sacred work, performed in traditional manner, went on for three years. The red-painted waka, named *Ngātokimatawhaorua* after Kupe's legendary vessel, held 150 men and their *mana* (spiritual prestige). Smaller canoes were constructed for the supposed sexcentenary of the Great Fleet in 1950. Queen Elizabeth II watched the ceremonial war canoe's second launch, in 1974, the year Waitangi Treaty Day became a public holiday—the result of long political struggle. Each successive February 6, more and more canoes joined the fleet, a modern indigenous tradition.

Grassroots activism eventually produced structural changes in law during the Labour government of the 1980s. Māori became the nation's second official language. Acceding to decades-old demands, parliament allowed iwis to seek redress from the state based on the Māori-language version of the Treaty of Waitangi. A state-supported tribunal processed thousands of claims, and parliament approved scores of settlements. Iwis received Crown lands as well as payouts in compensation for their losses. In particular cases, iwis accumulated enough capital to buy back portions of their homelands, one way of keeping the ancestral fires burning (*ahikāroa*)—a central metaphor in place-based Māori identities.

Simultaneously, the Realm of New Zealand restructured its own territory and its relationship to native and nonnative species. In 1987, the government bifurcated protection lands from production lands, and divested from the latter. State forestry died. When the Realm privatized its "exotic forests"—avian deserts in a country famous for birds—some of the plantations were purchased by iwis with reparation monies, others by foreign corporations. In contrast, state lands with native trees became nature reserves under management of a new Department of Conservation. Only about one-quarter of the nation's landmass still consisted of "natural forest," and the DOC tried to make them more natural. Once a poster colony for introducing species, New Zealand became an international leader in mitigating invasive mammals and weeds. As for kauri, Wellington made it all but illegal to fell the species, even on private land.

The dissolution of the New Zealand Forest Service provided an opportunity to revisit the idea of a national park at Waipoua. Te Roroa opposed any change of land status before the outcome of the Waitangi Tribunal. The local iwi had historical grievances against the Crown and the Dominion—for dispossessing them of forest and farmland in the settlement era, and for stigmatizing them as gumdigging trespassers and fire-starting menaces in the conservation era. In 1992, when the iwi presented its claim to the tribunal, it began with an invocation sung by an elder, who invoked the authority of megaflora and elderflora: "Stand tall, Tāne Mahuta, stand tall as you have done for aeons."[56]

The Te Roroa settlement, approved by parliament in 2008, provided an apology, $9.5 million (NZD), and legal assurances. The state recognized Waipoua as a tangible and intangible treasure (*taonga*) of Te Roroa, and placed a statutory overlay (*te tārehu*, literally, "the mist") over the reserve. Parliament acknowledged ceremonial sites within the forest as sacred and off-limits, and identified Te Roroa as the guardian of Waipoua and its *mauri* (life force). However, day-to-day management of the reserve remained with the DOC.

After the settlement, national park advocates pushed again and reached a new sticking point: Te Roroa's demand for co-governance. Much like the Yurok Tribe in California, the iwi conceived nature tourism in terms of heritage. It wanted visitors to appreciate Waipoua in relation to Hokianga Harbour, site of embarkation for their legendary ancestor Kupe. For economic inspiration, the iwi looked to Yakushima, Japan, and its World Heritage sugis. To grab the attention of Japanese tourists, Te Roroa invited the mayor of Yakushima to Northland in 2009 for a signing ceremony that connected Tāne Mahuta and Jōmon Sugi in a "Union of Ancient Trees."

As fate would have it, 2009 was also the year scientists identified an emergent forest disease caused by a novel species in the genus *Phytophthora* (plant destroyer). Later named "kauri dieback," the pathogen used zoospores to swim through soggy soils from tree to tree, blocking sap flow, resulting in lesioned trunks and bleached crowns. Feral pigs increased the speed and scale of dieback. Despite an extensive (and controversial) poisoning campaign, New Zealand's native forests were, ecologically speaking, overrun with four-legged creatures. Bipeds, too. In a wretched irony, people who loved kauris unwittingly infected them. Tree huggers became vectors.

At Tāne Mahuta, tourists now pass through a shoe-disinfecting station before stopping at a viewing platform. Decked in ferns, shrouded in mist, the giant reduces many New Zealanders to tears. This kauri, unknown and unnamed until the 1920s, has become a bicultural emblem of belonging for modern Māori and postcolonial Pākehā. More than any other settler society—admittedly, not the highest bar—white New Zealanders have made a collective attempt to make amends for the related colonial legacies of racism and extractivism. This can be seen at the Waitangi Treaty Grounds, the Realm's key heritage site. Here, Māori guides have been entrusted with the role of narrating the country's genesis. By the end of the tour—including musical and dance performances—guides have repeatedly exhorted visitors to drive to Northland's western coast to see the nation's oldest tree.

In 2016, amid kauri dieback, Te Roroa issued a vision document for a future "Waipoua Kauri National Park." The report stressed that forest mortality was a post-settlement issue with social and spiritual ramifications, especially regarding Tāne Mahuta. The emblematic tree had enormous mana; Te Roroa enhanced its own mana by serving as guardian. Likewise, the brother-tree relationship with Jōmon Sugi brought "international mana-enhancing opportunities." As much as Te Roroa wanted recognition as sovereign comanagers of the oldest living being in Aotearoa, it worried about a devastating loss of mana if it became known unjustly as the iwi that lost Tāne Mahuta. That outcome would have "acute intergenerational consequences."[57]

In its conclusion, the iwi quoted John Muir, big-tree disciple of California: "When we try to pick out anything by itself, we find it hitched to everything else in the universe."[58] Muir had spoken cosmically, but his insight applies historically to the universe of the Pacific Rim, to the realms of kauri, alerce, redwood, sugi, and hinoki. Ancient populations of giant conifers, isolated for hundreds of millennia, became hitched together in the long nineteenth century through colonialism and fossil fuel capitalism. A pan-Pacific outlook reveals parallel forces liquidating the planet's superlative forests and, ultimately, convergent forces shoring up compensatory reserves.

These sanctuaries are not refugia where time stands still and change comes slowly. Except in imagination, such islands no longer exist.

V.

CIRCLES AND LINES

SYLVAN SPECTACLE — HEALTHY EXTINCTION — TALKATIVE TREE RINGS — UNSTABLE CHRONOLOGIES — AMERICAN MORTALITY

The biggest old trees and the oldest big trees—the planet's superflora—are restricted to the Sierra Nevada. Upon its exposure in the 1850s, giant sequoia became a dualistic symbol of antiquity and health, endangerment and doom. US Americans exploited, preserved, and conceptualized the "Mammoth Tree" in original and outrageous ways. Living sequoias became mental objects for thinking about geological time, while eviscerated sequoias—material objects with annual rings—served historical time. Sections of logged stumps from just beyond the boundaries of Sequoia National Park went to a lab in Tucson, where they aided the establishment of dendrochronology, a discipline founded by A. E. Douglass. Contemporaneously, in science museums in Manhattan and London, racist educators—including eugenicists—inscribed civilizational timelines on cross sections from California. It became a twentieth-century custom for white people to convert concentric layers into linear, progressive narratives. Ecologically, the same tree rings could be interpreted as chronicles of episodic droughts and fires. Forest managers now desperately try to save sequoias from the fire-suppressing policies of their predecessors. These past-correcting, future-minded efforts occur against a backdrop of megadrought and the shifting baselines of climate change.

SYLVAN SPECTACLE

Giant sequoia is nonpareil: no equal in mass, almost no equal in age. From the moment that modernity hit the Sierra—amid the Gold Rush and the genocidal US conquest of Alta California—moderns approached sequoias as antiquities. They immediately perceived the "Big Tree" as venerable, with none of the delay that attended alerce in Chile, kauri in New Zealand, and, for that matter, coast redwood in northwest California. Then, as now, a threefold combination staggered sensibilities: size plus age plus rarity. In its Holocene habitat, *Sequoiadendron giganteum* exists only in the Sierra Nevada, and only there in about seventy discrete populations. A widely spaced group of mature sequoias—cinnamon trunks leading to billowed green under a vault of azure—is the opposite of a gloomy forest. "Where are such columns of sunshine, tangible, accessible, terrestrialized?" wrote John Muir in a moment of pantheistic rapture.[1]

No plant went from general obscurity to global celebrity so quickly. Frontier California was an international meeting ground, and within months of the first newspaper reports of the "great cedar tree" of Calaveras County, European collectors had departed for the botanical El Dorado. One of the first to arrive was an employee of an English nursery. He forwarded Sierran samples to Britain's preeminent botanist, John Lindley, who in December 1853 gave the "vegetable monster" its first scientific name: *Wellingtonia gigantea*.

Without setting foot in California, Lindley made statements on arboreal age that would be echoed interminably: These giants, 3,000 years old, were saplings when Sampson slew the Philistines, when Paris absconded with Helen, when Aeneas carried Anchises. And so on. Almost any celebrated figure or pivotal event from the mytho-historical past could and would be compared to the life of a *Wellingtonia*—or, as US chauvinists countered, *Washingtonia californica*. Exposed to sequoia temporality, moderns immediately composed mental timelines: the Story of Western Civilization from Troy and Athens to Yorktown and Waterloo.

Euro-Americans who came of age in the antebellum period knew the classics and the Bible, and they had developed a habit of comparing US scenery to European ruins. At the same time, they had developed a taste for novelty, pulp, and con artistry—a popular culture of sensational juvenility. These two cultural streams came together in the Sierra Nevada, issuing weird and contradictory outcomes.

Many people initially assumed that the "Mammoth Tree" was a frontier hoax. California settlers bragged only half in jest about monstrous cherries that could fill a dinner plate, petite pumpkins that could squash a house, half-acre tree stumps that could accommodate railroad tunnels. In the proto-photographic era, the public desired physical proof, and entrepreneurs worked to satisfy. If they couldn't bring the world to Calaveras, they would bring novelties of forest antiquity to the city.

Precedent existed for their brazenness. In the 1820s, US Americans gawked at the "Big Black Walnut Tree" of Lake Erie: first carved into a public house, then shipped to Buffalo, where it served as a grocery, then on to Rochester, where a promoter sent it via the newly completed Erie Canal to Manhattan. Gratified by ticket receipts from this "Prodigy of Vegetation," the promoter arranged for further showings in Philadelphia and finally London. The floor of the walnut drawing room, ornamented with medallion rug, accommodated thirty standing guests, who could admire genteel art that blanketed the walls. The exterior trunk of the "Monarch of the Forest" carried this quote from Psalm 104: "O LORD, how manifold are Thy works! In wisdom hast Thou made them all: the earth is full of Thy riches."[2]

A supersized western version of the tree salon originated in July 1853, just weeks after the public unveiling of sequoia. To the best of Euro-American knowledge, the grove in Calaveras County was singular in existence. Two fast-thinking brothers took out a preemption claim, a legal process by which a citizen-squatter could take possession of 160 acres of public land—in this case, unceded Miwok land—and obtain title at rock-bottom price even before the government surveyed it. Straightaway, the brothers sold the "original Big Tree" to a friend, William Hanford, who hatched a plan. He employed local miners to remove 40 vertical feet of the spongy outer bark in sections marked for reassembly. After girdling this wonder of the world, they attacked it with mining augers, bringing the whole thing down after two weeks of drilling.

The shell of the sequoia went by wagon and steamer to San Francisco. On Bush Street, workers reassembled the bark to form a cozy parlor with wall-to-wall carpet, a piano, and enough seating for "all the wives of Brigham Young and all the husbands of Lola Montes."[3] For three dollars, couples could attend cotillion parties. To ensure customer satisfaction and to preempt complaints of imposture, the proprietor displayed a radial section from the felled tree that contained an uncountably large number of narrow growth bands.

After a month in residence by the Bay, the first "Mammoth Tree" traveled via clipper around Cape Horn, arriving in Manhattan in early 1854. Hanford negotiated with P. T. Barnum to show the sequoia at the New York Crystal Palace, an exposition hall that had, the previous year, hosted America's first world's fair. Dissatisfied with Barnum's terms, Hanford rented his own space on Broadway and prepared advertisements for his 3,000-year-old. He moved too slowly. Before the *Gigantea americanum* could open, Barnum began selling tickets to his own "California Cedar Tree." Flummoxed and incensed, Hanford tried to prove that Barnum was a faker—which he was—but Barnum outplayed him at the confidence game. The impresario's exhibit, though much smaller, seemed believable. Hanford's improbably large tree-room looked counterfeit in comparison. His "Vegetable Monster" flopped. Before it could be shown again, the bark turned to ashes in a warehouse fire.[4]

Meanwhile, the owners of the Calaveras Grove sold another giant, the so-called Mother of the Forest, to frontier hucksters. The buyers approached their vandalism professionally, erecting scaffolding, then excising the outer bark—up to two feet thick—to a height of 116 feet. Marked for reassembly and shipped to Manhattan, the husk found a home at the newly renovated Crystal Palace. The glass-domed exhibition space opened on Independence Day 1855 with the vegetable wonder at its center. The paying public gaped: It looked like a genuine Big Tree, not another gimmicky tree-room. Internal scaffolding held up the hollow trunk. After a yearlong run, the proprietors sent the exhibit across the Atlantic to England's bigger, better Crystal Palace, the pet project of Prince Albert. Potted cycads and eight small sphinxes held court beneath the *Wellingtonia*, which stood opposite replicas of Ramses II from Abu Simbel. Antiquities of Egypt and California shared physical and psychic space in the pleasure dome until 1866, when a fire consumed the Victorian menagerie.

Not everyone approved of sequoia sideshows. The commercialization of Big Trees prompted the earliest calls for state preservation of megafloral populations. Moralistic commentators from New England interpreted tree parlors not as reflections of US culture but as distortions, even perversions. By luring men away from families and communities, the Gold Rush had corrupted traditional values. Only lucre-crazed men polluted by gold would have the cruelty and the inhumanity to destroy plants that would have been deified in heathen ages, critics opined.

In fact, one of the main vulgarians endeavored to be a model Yankee. In lengthy letters to kinfolk in Massachusetts, Ephraim Cutting called himself the "runaway son" and the "stayaway uncle"—attempts at humor that read as shame. Please remember me to mother and father, he instructed his brother from the camp called Murphys. Just give me one more year: "This gold digging business is the most depressing uncertain & lottery like occupation that ever was thought or heard of." Please convince our parents to sit for a daguerreotype, he asked, shortly before receiving the melancholy intelligence of his mother's death in the hallowed family home. Let me at least donate some of my gold dust for a marble headstone, he wrote. Domestic thoughts gave him "blue fits"; he craved reminders of Boston. When he asked his brother to send books from his trunk, he told him to be careful with the wooden chips inside—relics of the original monument on Bunker Hill.[5]

The Calaveras Grove, located a day's walk from Murphys, presented a novel business opportunity. Cutting met the holders of the original Big Tree, and "they appear to be very fine men—but it looks like a crazy speculation—10 or 12 thousand dollars invested in the bark of a single tree." The more he thought about it, the less crazy it seemed. If and when P. T. Barnum purchased the multimillennial giant—a complement to General Tom Thumb, no humbug—the sellers would take home big profits. Cutting convinced his mining partner to trade their water stock for cash, and soon they signed papers making themselves half-owners of a flayed sequoia in Manhattan. After the affair ended in flames, Cutting felt a damned fool. Nonetheless, he held out hope that his wild goose chase would end, that his day in the diggings would come, and that he would return to New England some Thanksgiving as a nonprodigal son.

It never happened. Cutting entered the timber trade, planted chestnuts in the Sierra, and grew a garden. He and other bachelors-for-life formed a community of exiles: "We eat and drink and fuss and fret—pay taxes and grow old." They stayed long enough to see their share of Bostonians: each summer after the Civil War, tour groups passed through Murphys on their way to the Big Trees. Talking to these New Englanders on holiday, Cutting felt like an outsider.

The touristy grove contained approximately one hundred living monuments—minus two. Each giant earned a name and a collectible stereographic card. Private managers turned the downed trunk of the "original Big Tree" into a bowling alley and erected a dance gazebo on its stump. Their on-site hotel sold pincushions made from the bark of the original, and, for a higher price, wooden cups and candlesticks. A cross section—"an astonishing chart of tree time"—had been sliced off and propped up so that visitors could have deep thoughts.[6] "They [the sequoias] began with our Modern Civilization," wrote a renowned eastern journalist who traveled west with the Speaker of the House to mark the end of the Civil War and the forthcoming completion of the transcontinental railroad. "They were just sprouting when the Star of Bethlehem rose and stood for a sign of its origin; they have been ripening in beauty and power through these Nineteen Centuries; and they stand forth now, a type of the Majesty and Grace of Him in whose life they are coeval."[7]

Postbellum tourists to Calaveras continued by stagecoach to Yosemite Valley and the Mariposa Grove of Big Trees. This grove had been "discovered" in 1857 about twenty miles south of the glacial gorge. Whereas private interests acquired Calaveras, Congress protected Mariposa as part of the 1864 Yosemite Grant, signed by Abraham Lincoln. This marked the first time in world history that a settler state reserved land from private entry for the public enjoyment of nature—though only after a volunteer state militia, the public's excluders, had burned the villages and food caches of local Miwoks.

Mariposa's first state land manager, the Yosemite Commission, did not by today's standards practice much stewardship. In 1881, it allowed a stagecoach company to punch a passageway through the basal fire scar of one of the grove's giants. The result, the Wawona Tunnel Tree, became one of the world's most recognizable plants and one of the leading attractions in California. Tourists paid to pass through the bole, and bought trinkets such as penknives fashioned from its former core.

In the 1880s, as wagon roads penetrated the Sierra Nevada south of the Kings River, California settlers identified broader sequoia habitat and even bigger trees. Once an outpost of the General Land Office opened for business, a brief land rush ensued. To preserve the Big Tree named in honor of Ulysses S. Grant, warrior-savior of the Union, the US general surveyor for California withdrew the land around it, despite lacking authorization. The rest of the giants went up for grabs. Three enterprises— two capitalist, one socialist—worked to consolidate landholdings in *Sequoiadendron's* prime habitat.

"It would be nothing short of vandalism to indiscriminately destroy these sentinels of past centuries, as has been done in several parts of California, by ruthless ravagers of the Competitive system," wrote Burnette Haskell, leader of the Kaweah Co-operative Commonwealth.[8] An anarchist turned utopian socialist, Haskell wanted to create a perfect medium of exchange based on a unit of work time, not gold or silver. When Kaweahans paid each other, they used scrip called "time-checks." In a show of cross-species respect, the socialists named the Sierra's largest tree—presumably the one that had labored longest on growing—after Marx.

In 1890, as the socialists at long last completed their timber access road, US president Benjamin Harrison signed a bill that created small-scale General Grant National Park and large-scale Sequoia National Park—the world's earliest public reserves of elderflora, and the first protected areas for a single species. The big park enveloped the commonwealth's land claims. Because the Land Office had technically never issued title to the Kaweahans, the government could now cite them for illegal timbering. The new guardians of the "Giant Forest," US Army Rangers, cleared out the "trespassers" and renamed the Karl Marx after their supreme commander, General Sherman, hero of wars against Confederates and independent Indians.

America's dualistic response to Big Trees—to exploit and to preserve, to venerate and to cheapen—continued through the end of the nineteenth century. Shortly after the establishment of Sequoia National Park, federal officials contracted with a logging corporation, the beneficiary of land fraud, to eviscerate a giant sequoia just outside General Grant National Park. This multimillennial plant became yet another parlor. It adorned the main rotunda of the US Government Building at the 1892–1893 exposition in Chicago, the biggest world's fair so far, marking the 400th anniversary of Columbus. Curators named the sacrificial tree the General Noble in honor of the late John Noble, Civil War veteran and former secretary of the Interior Department—the agency in charge of public land. After its sojourn in the White City, the General Noble came to rest on the National Mall, where it became the Smithsonian's toolshed.

The company that sold the General Noble was the Kings River Lumber Company, incorporated in San Francisco in 1888. By legal and illegal means, it acquired the densest, grandest expanse of *Sequoiadendron* ever known, a montane bowl called Converse Basin. With venture capital, the company built a flume that fell 4,200 feet over fifty-four miles—a giant waterslide perched on redwood trestles on the edge of steep canyon walls, the forerunner of amusement park log rides. Giant pieces of raw sequoia floated from the Sierra to the Central Valley in half a day. A technological wonder and a financial disaster, the flume needed constant repair.

In 1895, in the wake of failed reorganization, impatient debt holders—primarily a Canadian bank—took over the firm and renamed it Sanger Lumber. Ownership gambled on full-scale production, knowing full well that sequoia wood was brittle compared to redwood. Workers extended the narrow-gauge railroad from the top of the flume into the heart of the basin, where they erected a new sawmill. During Sanger's short life, the company converted some eight thousand titanic sequoias into grape stakes.

Only one of Converse's elders survived. At the northernmost tip of the mega-grove, at the edge of a slope overlooking the jagged mouth of Kings Canyon, stood a stupendous plant with a fire scar and a tag, BOOLE—the surname of Sanger's foreman. The tree's designator, a Fresno doctor who worked in the timber camps, said, "Frank Boole was the squarest man I ever knew."[9] In 1901, when word got out that Boole planned to cut the "largest tree in the world," Sierra Club members from Berkeley and the newspaper editor in Fresno raised a fuss. The foreman issued a terse statement of reassurance: "I am of the opinion that the big tree will protect itself, as I believe it could not be cut down with any profit to the company."[10] He didn't yet have the tackle to handle the hillside job. Under Boole's watch, output peaked in 1903, when some seven hundred workers produced 191 million board feet. To increase the speed of tree felling, Sanger turned to dynamite. Perversely, this bank-controlled business detonated the planet's most supernal stand of trees without turning a profit. Hundreds of blasted sequoias never even went to mill.

The Boole Tree—one of the top ten largest ever measured—was ultimately saved by insolvency, but storytellers prefer simpler tales. After the fact, two versions of a preservation legend arose; both still circulate. In one, Frank Boole shamelessly spares the tree just to perpetuate his name. In the other, the foreman selflessly spares the tree—decrees its protection, pleads for its life—and workers name it in recognition of his clemency.

Of Sanger's common laborers, neither legends nor histories were told. Obituaries from Converse Basin give anti-utopian glimpses of dismemberment and death: *Caught by a grab hook and dragged; killed while switching cars; thrown under the wheels of a slowly moving truck; crushed in a falling pile; killed by a heavy steel wire cable which broke loose; blown to atoms thawing out six "frozen" sticks of dynamite.* In 1903, a day worker died on day one after timbers slipped from a truck. Lingering for hours in agony, he refused to state his identity. "Smith is as good a name as any to be buried with," he told the camp doctor.[11]

The following year, Converse Basin's namesake, the engineer C. P. Converse, waded into San Francisco Bay, his pockets laden with railcar coupling pins. His body washed up weeks later. Compared to powder monkeys in the Sierra, Converse lived a long life, 87 years—long enough to have lost a small fortune from the Gold Rush. After a previous suicide attempt, this pioneer of Fresno explained matter-of-factly that he was "getting old and was in everybody's way."[12]

For so many of his cohort—the white men who made the Golden State bleed—way out west was no place to be a senior.

HEALTHY EXTINCTION

In the first half century of US California, as the population grew from 100,000 to 1.5 million, far more people read about Big Trees than saw them, and most of the writing issued from travelers who came in and out. They expressed conflicting ideas about the monster plants. Christian-minded visitors anticipated that living sequoias might yet witness an evangelized Earth or, less cheerfully, that they might crackle in the last conflagration of the planet. Science-minded visitors spoke of biological longevity and evolutionary doom in the same breath.

Mammoth, a word associated with North America's original iconic extinct species, evoked immense size and temporal immensity. In the early republic, novelty seekers could visit the "Mammoth Room" in Charles Peale's famous Philadelphia Museum and see a chauvinistic skeleton— ANTIQUE WONDER, LARGEST OF *Terrestrial Beings*, NINTH WONDER OF THE WORLD!!![13] By the Gold Rush, Mammoth-mania had waned; US Americans were ready for a new national symbol of organic antiquity. By likening sequoias to mammoths, they exhumed a cluster of ideas: American uniqueness and bigness, biblical gigantism, antediluvian time, and total extinction. As the cognate of an ice age species, the "sylvan mastodon" seemed practically pre-extinct.

Nineteenth-century geologists and biblicists often arrived at the same conclusion: the olden Earth had harbored prodigious organisms that mostly died out. In this context, mammoth trees counted as anachronisms. When entering a Big Tree grove, many visitors, including Emerson, quoted Genesis on life before the Flood: "There were giants in the earth in those days."[14] Mixing old religion with new science—and updating giant mammals to giant reptiles—California's state botanist described rear-guard "post-diluvian monarchs" as "coeval with the mammoth saurians and other monsters of the coal period."[15]

The era's most important treatise on sequoia temporality came from botanist Asa Gray, Darwin's chief defender in America. In 1872, Gray gave his presidential speech to the American Association for the Advancement of Science (AAAS). He'd just been to California—his first trip to the Far West—where he goggled at the botanical "wonder of the world" at Calaveras and Mariposa, and enjoyed a private tour of Yosemite with John Muir. Gray returned by train (including a stop in Colorado to climb a 14,000-foot peak named in his honor) and used his time in Pullman Palace cars to write his speech, "Sequoia and Its History." The professor refrained from trite remarks on size and longevity. He was far more interested in the isolation of the species; it appeared to be sui generis.

Gray laid out three hypotheses in the form of questions: (1) Are the trees veritable Melchizedeks, without pedigree, and fated to be without descent? (2) Are they now coming upon the stage—or, rather, *were* they coming before man's interference—to play a part in the future? (3) Are they remnants, sole and scanty survivors of a race that has played a grander part in the past, now verging to extinction?

Speaking from the pulpit of the congregational church—the closest thing to a convention center in Dubuque, Iowa—Gray praised the Darwinian revolution for allowing humankind to investigate the changing conditions of life over millions of years. Gray presented sequoia as further proof of Darwin's refutation of special creation—the idea that God had created organisms perfectly suited to their time and place. When we see how eucalypts thrive in California, and how sequoias flourish in

England, said Gray, we must abandon the notion of primordial adaptation of plants and animals to their present habitats.

Gray argued that noble sequoia, far from being perfectly adapted, was no longer adaptable: a slight drying of the climate would "precipitate its doom." Gray built his argument on recent fossil discoveries of leaves, shoots, and cones. Paleobotanists had dug up forerunners of the iconic "American" conifer all over the Northern Hemisphere—from subtropical, temperate, continental, and Mediterranean zones (Texas, Montana, Silesia, Tuscany), to subarctic (Iceland, Sakhalin), to polar (Ellesmere Island, Spitsbergen). These widely distributed fossils demonstrated that California redwoods came from "ancient stock" that evolved in warmer times. The record of the *Sequoioideae* subfamily went back to the late Cretaceous period, roughly 70 million years ago. Only with the onset of the ice ages, in the Pliocene (more new) epoch of the Quaternary period, did redwood species disappear from Asia and Europe. As Gray told the story, planetary changes left sequoia behind; new, more competitive life-forms evolved; an epic genealogy approached a dead end.

This classic speech on evolution reached a wide audience as part of Gray's book *Darwiniana*.[16] Biology resembled a river, he explained, not an ocean. The energy of evolution flowed incessantly and circuitously; the stream didn't follow a straight course called progress. To Gray, the process of extinction was more inspiring than the notion of Creation, which was inflexible and changeless. After acknowledging that Darwinism remained repugnant to some, he bore testimony to its rightness. Just as religion had survived "the notion of the fixity of earth," it should "equally outlast the notion of the fixity of the species which inhabit it." Gray echoed Charles Lyell, a believing Christian, who had argued in *Principles of Geology*: "The successive destruction of species must now be part of the regular and constant order of Nature."[17] Gray's reformulation: faith in an order, which is the basis of science, cannot be dissevered from faith in an Ordainer, the basis of religion.

John Muir followed this discussion and contributed to it with a paper read at the AAAS conference in 1876. At this juncture, the Scottish immigrant and vagabond naturalist knew more about the geobiology of the

Sierra Nevada than anyone. Muir argued that the island-like distribution of sequoia groves could be explained by the past location of glaciers. The sequoia-less gaps in the foothills generally corresponded to the pathways of ice rivers in the Pleistocene (most new) epoch. Sequoias on "lofty protective spurs" had escaped the ice, Muir inferred. In the current Holocene (wholly new) epoch, the species had reclaimed much of its former range in the southern Sierra by seeding onto newly deposited moraine soil in a manner similar to Mount Lebanon's cedars. In the northern Sierra, sugar pine outcompeted giant sequoia.

According to Muir, then, sequoia existed in a moment of expansion within a larger period of decline. Because he couldn't find a fallen specimen outside the boundaries of any extant grove, he deduced that habitat size had remained stable for many centuries. While admitting uncertainty in the details, he believed that lawless mystery had given way to harmonious science: the relationship between glacial history and biological history was factual. Muir marveled at the "great radical fact" that "all the present forests of the Sierra are young."

Using long-term thinking, Muir candidly addressed the question that had haunted Big Tree enthusiasts since the 1850s: *Is the species verging to extinction?* No species, including humankind, lasts forever, he wrote. Sequoia is closer to the temporal abyss, "but the verge of a period beginning in cretaceous times, may have a breadth of tens of thousands of years." Responding to casual pronouncements that the Sierran redwood had lost their ability to reproduce, as indicated by the dearth of seedlings, Muir offered conditional reassurance. North of the Kings River, he conceded, the childless and companionless patriarchs seemed "doomed to a speedy extinction, as being nothing more than an expiring remnant, vanquished in the so-called struggle for life." In the south, however, young and old flourished side by side. Unless destroyed by man, the decadent mastodon of the vegetable kingdom should last until at least AD 15000.[18]

Muir and most Euro-American observers struggled to reconcile species-level rarity with population-level vitality. Again and again, writers averred that the thick-barked Big Tree had unrivaled resistance to fire, astonishing powers of regeneration, and total immunity from insects, fungi, and diseases. The species possessed perfect health, claimed explorer-scientist Clarence King; nothing indicated degeneracy.[19] No one ever found a sequoia dying of old age, echoed others. Paradoxically, sequoias seemed to succumb solely to their vigor—"tumbling to ruin in mid career." The California State Board of Forestry called such gravity-powered deaths the "only examples of suicides in the great class of exogens, or outside growers."[20]

Sequoia health claims contained a kernel of botanical truth, but the salient context was Golden State boosterism and health tourism. In nineteenth-century medical bulletins and travel guides, paid experts and privileged invalids vouched for the life-preserving qualities of California's salubrious climate. Supposedly, the disease-free atmosphere perpetuated animal as well as vegetable tissue; youthful energy pervaded the state, inducing low mortality rates and high life expectancies. Anecdotally, surviving Natives of California—those who hadn't been murdered—lived to extraordinary ages, decades longer than other Amerindians. One eastern tourist who visited the Mammoth Tree Grove of Calaveras County also paid a visit to "the oldest woman in the world," a 140-year-old mestiza in Los Angeles.[21]

Verifying the exact age of any old organism is notoriously difficult. At annual resolution, it can only be done with woody plants—exogenous growers. From the outside, wood grows with the addition of cambium. As each growing season winds down, the cells in the outer cambium layer decrease in size until cell production ceases. The contrast between the densely packed, thin-walled cells of the "latewood" and the following season's "earlywood" is visible as a "ring." Each tree ring marks a stoppage of growth—not necessarily the passage of one year. Temperate rings are more consistently legible than tropical rings because of the regular interruption of winter. By the late eighteenth century, most European naturalists accepted that lignophytes on the Continent produced one ring per year. After the invention of increment borers in nineteenth-century Germany, cambial growth became the metric of scientific forestry.

In the United States, skepticism continued. "The discussion on the age of the Sequoias has brought up the question of the truthful record of age as indicated by so-called annual rings," wrote Bernhard Fernow, the first chief forester of the United States, in 1888.[22] This was hardly a theoretical issue, given property law. In the US East, frontier surveyors had typically hacked "blazes" on "witness trees" to mark boundaries. Multiple surveys or overlapping surveys could produce dueling claims to title. When disputes went to court, it became necessary to determine which survey had priority—that is, who hacked first. By "blocking" a witness tree—by cutting out the section of the trunk containing the blaze—one could count the layers of cambium over the surveyor's scar, thus producing a relative date. Many litigants dismissed the scientific evidence as bogus. Two cases traveled all the way to the Supreme Court.

Sequoias were far too large for block cutting or increment boring. To enumerate their annual growth layers required a clean and complete radial section of the trunk—in short, arboricide—and a patient and meticulous observer. In 1859, botanist John Torrey made the first trustworthy count from the "original Big Tree," a surprising 1,120 rings: "The facts show that the tree lacks almost three centuries of being half as old as it was said to be!"[23] Rather than disappointment, Torrey felt amazement at sequoia's capacity for growth. A few years later, California state geologist J. D. Whitney made a follow-up count. "The Big Tree is not that wonderfully exceptional thing which popular writers have almost always described it as being," he echoed. Whitney estimated that the "original" had lived 1,300 years—"not so great as that assigned, by the highest authorities, to some of the English yews."[24]

Even after Torrey and Whitney, the life span of sequoias lengthened in the popular imagination. "The age of the Sequoias is the one point most hopelessly befogged to the ordinary tourist," complained a professor at the University of California in 1886.[25] Wild talk of three or four millennia must be relegated to the realm of absurdity and impossibility, he wrote. He blamed the "vaporings" of newspapers, though, in truth, misinformation derived from reputable sources, too. John Muir claimed to have counted 4,000 rings; a member of the California Academy of Sciences raised the limit to 6,126; and the biologist David Starr Jordan, first president of Stanford University, expressed confidence that sequoias could live 8,000 years.

The most thoughtful—though not the most influential—statement on sequoia longevity came from botanist William Russel Dudley, professor of botany at Stanford. In 1905, Dudley spoke to a group of Columbia University alumni in California and described his fieldwork in Converse Basin, where he had made ring counts on logged stumps. He had disproven the relationship between sequoia girth and age, and dismissed all reports of 4,000 years and up. At the same time, Dudley documented 2,425 tiny rings on a single stump—the oldest (dead) life in the world yet known.

In his talk, released in print after his untimely death, Dudley focused on a slightly younger specimen and its recuperative powers over 2,171 years. After providing an analysis of cambium healing, the botanist became a nonhuman historian, compiling a chronicle of arboreal existence. After germination, the first major event in the tree's life occurred at age 516, when fire swept through the basin, leaving a fire scar. Over the next 105 years, the tree overlaid that injury. Afterward, this no-name sequoia experienced 1,196 years of uninterrupted growth, before another fire, another period of wound covering, and so on, until a near-fatal conflagration at age 2,068. Dudley conjectured: had the US government held on to Converse Basin, instead of privatizing the land, this plant would have closed its latest wound around AD 2250.

If we save constitutions and other priceless parchments, he concluded, why not the witnesses of nature, too? Sequoias documented fires, droughts, and periods of rain in their tree rings. Every last Big Tree should be preserved so that this irreplaceable data could someday "be read and recorded by skilled hands and interpreted by the best intelligence."[26]

Unbeknownst to Dudley, that dawn was already approaching. Lamentably, most early interpreters of sequoia tree rings wouldn't share his sensitive vision.

TALKATIVE TREE RINGS

In the early twentieth century, in overlapping ways, pedagogues and dendrochronologists turned sequoia tree rings into instruments. Not by accident, the first samples for science museums and research labs came from the same devastated habitat that supplied the final samples for sideshows and fairgrounds.

In 1890, railroad monopolist Collis P. Huntington agreed to do a friendly favor for Morris Ketchum Jesup, fellow Manhattanite and director of the American Museum of Natural History. Jesup had a pet project: the Jesup Collection of American Woods, consisting of pieces of every native species. He intended the pièce de résistance to be a sequoia cross section. Jesup, a former banker, was rich, but not that rich. To obtain a Mammoth Tree, he needed a robber baron.

The following summer, the museum dispatched a timber expert, S. D. Dill, to San Francisco, where he obtained a letter of reference from the office of the Southern Pacific Railroad. Dill then traveled by stagecoach to Converse Basin. The camp manager showed Dill a fine tree that had been felled for the museum ahead of time. Unimpressed, Dill went hunting on horseback. Just outside the boundaries of General Grant National Park, he found his quarry. It bore a nametag: "Mark Twain." The author's namesake towered almost 300 feet above the forest floor, where its flared base measured nearly 90 feet in circumference. The lower half was all trunk, no limbs. Although the General Grant loomed slightly larger, Dill felt sure he claimed the better specimen—"the handsomest tree I have seen." He got there just in time. "This and a few others is all that remains of a once magnificent grove of Sequoias," Dill wrote.[27]

When the Mark Twain came crashing down in October, most workers watched from a distance; three fearless ones stood on the stump. After the thunderous impact, a posse of mustachioed men posed triumphantly atop the trophy. For the photograph, they brought accoutrements of manliness and civility: rifles, axes, pipes, fob watches. The man in the highest position held in his lap a well-worn book—perhaps a Twain. The denominator of the Gilded Age would have appreciated the grand absurdity of destroying a 1,300-year-old tree for a museum display. Workers dynamited the bulk of the bole into chunks for milling into posts, ties, and shingles—but they handled the lower section with care. With a superblade (two crosscut saws brazed together) they removed two pairs of four-foot-thick cross sections. These matching slabs, sixteen feet in diameter, would never fit in a freight car, so the lumberjacks used irons to split them into twelve pieces: one center circle and eleven equal wedges, numbered for reassembly.

After long-distance transport by horse-drawn carts, trains, and steamer—all charged to Huntington—the first set of twelve arrived in New York City. Museum staff glued the segments together, filled in the cracks with wood scraps, and polished the cross section to a smooth finish. By the end of 1892, visitors to Darwin Hall could admire the twenty-four-ton slab, propped on its side, awaiting completion of the Jesup Collection. "For the sake of science was this monster dismembered," reported the papers, "a part of him brought across the continent into a city built by a race which did not exist when he first began to stretch upward."[28]

The second sequoia kit traveled farther—around Cape Horn, across the Atlantic to Liverpool, then London. The trustees of the Natural History Museum had wanted a sequoia for years, and they piggybacked on Jesup's initiative. In 1893, the second slab made its debut in the High Victorian Gothic edifice in Kensington, causing a sensation. The corresponding displays at Central Park West and Cromwell Road would, in the long run, make the Mark Twain the world's most visited sequoia—eclipsing the murdered Mother of the Forest previously installed in the same two metropoles.

To cut down a tree that had lived more than a millennium seems "almost as heartless as to shoot a human centenarian," wrote a commentator in 1903 regarding the scientific showpiece in Manhattan. "But there are moments when the fell act is pardonable." A tree that otherwise would have met a "humdrum fate" had instead become a "fully-installed professor of history." The investiture ceremony had occurred a few months earlier when curators mounted labels directly on the tree rings to show historical events. Thus "given speech" by the museum, the sequoia "gave daily lectures" on the past. No one had ever seen such a thing.[29]

Curators grasped for a pedagogy to match their teaching tool. In addition to a conventional history timeline, they pinned to the wood a history of science timeline broken into five subunits: philosophy of biology, general biology, comparative anatomy, paleontology, and embryology. The result was a messy mélange, a confusion of temporal trajectories. To help guests interpret the display, the AMNH produced a twenty-eight-page leaflet. The museum director praised how the tree had become a "potent educational agent," whereas a slab by itself would have the same teaching value as a circus giant.[30]

When British curators copied the idea a couple of years later, they opted for pedagogical clarity. On the Kensington slab they painted terms and dates that comprised a straightforward history of England, the United Kingdom, and the British Empire, with the usual highlights and sporadic lowlights: "Great Fire of London," "Collapse of the South Sea Bubble."

On a master timeline, the invention of timelines would appear *after* those disasters of 1666 and 1720. In medieval Europe into the early modern period, temporal visualizations had taken varied forms, with rivers, trees, animals, bodies, hands, wheels, chains, or columns representing the structured passage of time. Then, in 1765, to acclaim, the English chemist Joseph Priestly simplified things. He published "A Chart of Biography," a visually innovative diagram that used parallel lines above a temporal axis to represent the overlapping lives of two thousand individuals (all men) from 1200 BC to AD 1750. Priestly's clean, precise layout became the template for most subsequent timelines.

A timeline—secular, linear, universal—was modern time, idealized. It complemented ideas of individualism, historicism, and historical progress. Globally synchronous linear progressive time became a precondition of Western modernity. Colonialists and capitalists propped up linearity and uniformity, and attempted to bring down cyclical cosmologies and solar time. State-sanctioned violence accompanied the regime of clocks and time zones. A timeline made these upsets look orderly, even natural. Imposed across the stunning organic materiality of one thousand concentric tree rings, a sequoia timeline became modernity's ultimate naturalization. It may seem strange that Westerners took so long to co-opt nature's circles for history's arrow, but three preconditions had to happen: the public acceptance of temperate tree rings as annual markers;

the mechanical felling and transport of cross-sectioned megaflora from the Pacific Rim; and the professionalization of museums. By 1900, everything had fallen into place.

In a synchronous development befitting a timeline, a university scientist gave tree rings technical speech at the same time that museum curators gave them colloquial speech. The brilliant astronomer A. E. Douglass, originally from New England, spent his career under the clear dark skies of the Sonoran Desert. He became a towering figure at the University of Arizona. Without any training in biology, he founded and named the discipline of dendrochronology, laying out its rudiments in a 1909 article for the *Monthly Weather Review*.

Here's a simplified and idealized description of Douglass's technique: You, the scientist, extract a core sample from a fallen forest tree you call A-1. You don't know when this tree died. From the outer end of the sample, mounted and sanded for enhanced viewing, you discern a distinctive multiyear pattern of rings—say, three narrow bands, then a thick one, then an extremely thin one. Your goal now is to find that same distinctive sequence—an arboreal code—in younger and older specimens of the same species from nearby locales. You go back to the field. After strenuous hand cranking and tedious observations with a magnifier, you recognize in your fifty-third tree the same pattern from the inner end of the sample. Using graph paper, you create two representations—"skeleton plots"—of the series of ring widths from A-1 and A-53. By overlapping the two plots at the matching sequence, you create a cross-dated record of longer duration than either individual series.

Now you repeat the process, looking for new patterns, again and again, overlapping backward and also forward until you reach a living tree or a tree with a known death date. With that temporal anchor, you can now assign absolute dates to all the rings in the master chronology. You may have a separate "floating chronology"—overlapped plots, decades or centuries long, lacking calendrical assignation, just a relationship to a relative date. With enough perseverance and luck, you eventually find a sample that "bridges the gap" to the chronology you previously compiled and linked to Gregorian time.[31]

Clear in the abstract, tree-ring science becomes knotty in practice. As Douglass learned, it doesn't work with all woody plants. An ideal specimen fulfills four requirements. First, the tree produces one growth ring per year. Second, the growth of the tree is dominantly controlled by a single limiting factor—preferably a climatic signal like precipitation or temperature. Third, the width of the growth rings corresponds to the availability of that input, meaning the rings can serve as proxy data. Dendrochronologists draw a distinction between "complacent" and "sensitive" tree rings. A lucky tree growing in a habitat with perennial water tends to be complacent: its rings are regular, and thus hard to read for signals. Trees in stressful locales and arid environments tend to be sensitive. Conifers tend to be more sensitive than angiosperms, and certain conifer species are especially sensitive; they react in an exaggerated way to feast-or-famine conditions. Such species serve the needs of lab work provided they meet the fourth requirement: trees must occur over a wide enough area—with the climate signal recorded in their rings being equally widespread—to allow for unbiased sampling.

Douglass began with ponderosa pine in Arizona. Almost accidentally, this place- and species-specific method revolutionized Southwestern archaeology. By cross-dating samples from pine beams in ancestral Puebloan ruins at Chaco Canyon, New Mexico, Douglass determined their construction dates. In December 1929, *National Geographic* announced with fanfare Douglass's discovery of the sample that bridged the gap in the Puebloan chronology: "The Secret of the Southwest Solved by Talkative Tree Rings." In subsequent statements, the professor likened tree rings to annual reports, yearbooks, calendars, and diaries. Although archaeology made Douglass renowned and influential, and provided essential lab funding, it never rose to the top of his intellectual interests. He had invented dendrochronology to investigate sunspots and their possible climatic effects. For this work, Douglass needed a longer chronology than ponderosa pine could provide. Thus he turned to giant sequoia.

On his first day in Converse Basin in 1915, Douglass experienced a stroke of "good fortune"—a dynamited tree. Acting as coroner,

Douglass precisely dated the monumental plant he called "D-5." After just a few weeks in the demolition zone, Douglass had secured a cross-dated chronology of 2,200 years.[32]

He returned in 1918, hoping to extend the chronology to three millennia. Overwhelmed by the number of stumps, he tried to limit the selection by girth but determined that largest did not mean oldest. Finally, he found a butchered group of super-elders in the "World Fair District," the grove that had supplied exhibits to Philadelphia in 1876 and Chicago in 1893. Douglass took samples from those stumps—still decorated with the calling cards of tourists—and the recently named "Moving Picture Tree," a sequoia blasted for the benefit of a film crew and left to waste. It had lived nearly 2,500 years. Nearby, Douglass found his desired stump, on which he carved the identifier "D-21." It had lived 3,220 years—the new oldest (dead) tree.

Douglass had come to Converse Basin on advice—and with accompanying reference letter—from Ellsworth Huntington. In the interwar years, this Connecticut Yankee became the most famous, influential, and controversial geographer in the United States. His career revolved around the "geographic theory of history," sometimes simply called the "Huntington theory." Like his contemporary Oswald Spengler (author of *The Decline of the West*), Huntington looked for fundamental causes and deep patterns in human history over the *longue durée*.

Huntington argued that climatic change—and corresponding social change—occurred in "pulsatory fashion." A clement and seemingly stable climate could abruptly shift to prolonged drought, and vice versa. As far back as classical Greece, thinkers had postulated the existence of cyclical rainy periods, what climatologists now call pluvials. Whereas Aristotle assumed such events were geographically restricted, Huntington believed that pulsations like jet stream oscillations affected the planet in synchrony, albeit with regional differences. By decoding planetary and cosmogenic pulses, scientists would, Huntington predicted, reveal the "pulse of progress"—the controlling factors behind the evolution of races, the dawn and twilight of civilizations. By understanding circles (Earth's past), he would understand lines (human history).

In his enthusiasm, Huntington succumbed to teleology: he developed his Big Idea first, then went looking for evidence. He traveled to Turkey, Palestine, and the US Southwest to measure lacustrine terraces—the bathtub rings of dried-up paleo-lakes. And he went "prospecting for large stumps" in the Sierra Nevada in 1911 and 1912. "It may seem a pity that trees thousands of years of age should be cut for fence posts and 'shakes,' but it is fortunate for our present investigation," he wrote.[33] He and his assistants worked facedown on 451 fresh stumps, doing their best to ignore biting ants.

In *Harper's* magazine—and in a leaflet for the NPS published by the government—Huntington made the usual frothy commentary: this very tree had been a sturdy sapling in the days of Exodus, mature in the time of Marathon, unchanged through the fall of Rome and the coming of the Dark Ages. Then the geographer added his twist: Whereas others could only rhetorically compare California flora to antiquity, he could scientifically explain ancient history with sequoia. The secret of the Big Trees clears the way for the solution of one of the most profound and far-reaching of the problems of history, he wrote. Inverting Edward Gibbon, he placed climate at the top of the list of factors that contributed to the decline of Rome. He likewise ascribed the collapse of Mesopotamia to drought. Climate explained the Black Death, the spread of Islam, the Mongol invasions—all these and more. His tree-ring theory stood "firm" after "rigid mathematical test."[34]

Astonishingly, the pontificator from New Haven never cross-dated his 451 specimen trees. Although inspired by dendrochronology, he didn't practice it. In a report published by the Carnegie Institution, Huntington meticulously explained his dubious methodology.[35] He started by measuring the width of rings in units of ten, often making multiple transects per stump. Later, he averaged these decadal measurements and graphed them as a composite curve, an undulating line that supposedly illuminated climatic cycles. He smoothed the curve mathematically to correct for observational error, the jaggedness of stumps, and the basal flaring that distorted the width of growth rings. The greatest corrections affected the small sample of trees from pre-Christian times, the period that interested him most. To shore up the left end of his graph, Huntington decided to incorporate climate data from before the Common Era that he had collected in Central Asia. Rather than testing a hypothesis that sequoia could be used to deduce past climates in distant places, he reified his assumption.

For Huntington and his donors, understanding the climatic past meant protecting the racial future. Ungovernable cycles had shaped the development of civilizations and races, they believed, and now white people had the power, through science-based governance, to safeguard the outcome of the latest pulse of progress. Aligning politics with science, Huntington in the 1930s served as treasurer and president of the American Eugenics Society. Prior to that, in 1921, he prepared a fifteen-foot-long version of his climate curve for the Second International Eugenics Congress, proudly hosted by the American Museum of Natural History.

The original Mark Twain timeline had lasted a mere decade. Curators disassembled the cross section in 1912 and moved it to the Forestry Hall for the grand opening of the long-delayed and instantly old-fashioned Jesup Collection of North American Woods. The refurbished cross section now featured gold-paint century markers, but no events besides "Began growing 550 A.D." and "Cut in 1891." In keeping with the desires of the late Jesup, the descriptive label in front of the sequoia presented information on economic botany.

For the Eugenics Congress, curators jumped at the chance to remake the fusty Forestry Hall. New displays included "Approaching Extinction of *Mayflower* Descendants," "Eugenical Sterilization in the United States," and "Difference Between White and Negro Fetuses." To make room for these and other pedagogical objects—including the brains of fifty "insane criminals"—the museum banished all the trees except the Mark Twain. It sat by the entrance/exit, where it reclaimed its role as professor of history. Curators strung white threads between dates on Huntington's chart and corresponding tree rings to show "Changes in Climate for 3,000 Years—Their Effect on Natural Selection and the Mixture of Races, Through Food Supply, Health, Migration." Long after the jarred brains and plaster-cast fetuses had been carted away, Huntington's chart remained by the tree. It was still there in 1941, the year the United States declared war on the Third Reich.[36]

While sharing some of Huntington's politics, A. E. Douglass showed more circumspection in his public speech and more fastidiousness in his sequoia research. He had transported some fifty radial samples—large wedges—to his lab at the University of Arizona. There he shaved off thin slices for mounting. When Douglass determined that Huntington's quick-and-dirty counts had produced errors of decades—in one case, centuries—he noted diplomatically that his friend had been interested in hypothetical centuries-long cycles, which allowed for a margin of error. Douglass did not have any margin because the sunspot cycle operated on a short time scale—eleven years. The astronomer returned to California a third time to figure out a single layer, 1580 CE, that appeared in some trees, not all. After resolving the "doubtful ring," Douglass built ingenious machines with lenses and mirrors—the "periodograph" and the "cycloscope"—to obsessively compare skeleton plots.

In the end, he failed to find definitive evidence that sunspots affected sequoia growth. Perversely, his machines suggested the existence of dynamic cycles—"cyclics"—that appeared, disappeared, and repeated without regularity. Douglass couldn't explain these patterns. As an astronomer, he favored the predictability of planetary revolutions, lunar tides, and heliocentric orbits. Earth's climate was chaotic by comparison. Nonetheless, he remained optimistic that the "complex" or "mosaic" of signals in the sequoia record would someday be correlated with proxy data from other trees in other parts of the North American West.[37]

This project would fall to his advisee, Edmund Schulman—anything but a *Mayflower* descendant—who would go on to document California conifers far more sensitive than giant sequoia, and surprisingly older.

UNSTABLE CHRONOLOGIES

The Jesup Collection—an artifact of the late nineteenth century, when national literacy encompassed economic botany—never returned to view in Manhattan. But the Mark Twain stayed. In 1928, curators renovated the sequoia slab a third time, aiming for increased legibility. Following the British model, they affixed a single historical timeline. Relatively few events took up space—war and diplomacy gaining precedence over science and invention—and they did so evenly, next to century markers at each hundredth ring. This "space-for-time" approach became standard practice. Many curators in other cities patterned their cross sections after this one in Manhattan.

The NPS likewise facilitated standardization. More than twenty timelines derived from a single fallen tree in Sequoia National Park. Starting in 1923, the superintendent supplied slabs to schools and museums free on request, as long as recipients paid shipping. With each 1.5-ton piece, the NPS sent interpretive instructions, including a chronological list from which events could be selected. Inevitably, each section of trunk contained a different number of rings (1,500 ±200); to the frustration of curators—though to the ignorance of museumgoers—the NPS could not say with exactitude which rings belonged to which years. Sequoia time kits traveled to Oslo and Stockholm and a few more locations overseas. Mainly they moved within the United States, including unlikely destinations such as the parking lot of the Automobile Club of Southern California on Figueroa Street in downtown Los Angeles. In tidewater Virginia, Colonial National Historical Park erected a big-tree timeline in 1931 for the sesquicentennial of Yorktown, the last major battle of the Revolutionary War.

Sequoia's superintendent reserved the largest slice for the University of Arizona—a favor to A. E. Douglass. In return, Douglass went to the national park in 1935, took some shallow borings from the General Sherman, and provided an uncharacteristic piece of puffery: an age estimate of 3,500 ± 500 years. The professor revived a long-standing rivalry between Fresno County, home of the General Grant, and Tulare County, home of the Sherman. In 1931, a committee of six engineers, in concert with the State Chamber of Commerce, had announced a verdict: Sherman was the world's largest living thing; Grant, the world's oldest.

Perhaps Douglass realized that his drawn-out work with *Sequoiadendron* had more promotional than theoretical value. Tucson's mammoth display—a local children's favorite—explained the significance of tree-ring science to Southwestern archaeology, not astronomy. Gushing about nature's Rosetta stone, the local paper predicted that dendrochronology would go down as one of the great advancements of the century; national leaders would seek out tree-ring scientists for long-term predications. Immodestly, Douglass approved a museum label for the 1902 sequoia growth ring, inches away from Columbus and Marco Polo: DR. DOUGLASS BEGAN TREE RING WORK.[38]

Compared to published and graphical timelines, which included hundreds or thousands of names and dates, timelines on tree rings looked radically simple. Curators handpicked salient events—ones with high mnemonic value—revealing their own prejudices. The most variable tag was the penultimate one, for which curators turned to whatever appeared to be the last pivotal event before the death of the specimen. If not Dr. Douglass, then "Marconi Wireless" or "World War Armistice."

The earlier tags repeated themselves. I know this because I obsessively tracked down twenty-five cross sections from California—mostly giant sequoias, and a few coast redwoods—installed by US Americans in the first half of the twentieth century, including one gift to France. The following dated events, clustered by theme, appeared with greatest frequency:

American Revolution/Declaration of Independence	22
Discovery of America by Columbus	21
Pilgrims/*Mayflower*	14
Battle of Hastings/Norman Conquest/William the Conqueror	13
Magna Carta	12
First Crusade/Second Crusade/Last Crusade	12
Charlemagne crowned	11
Civil War begins/Civil War ends/Lincoln assassinated	10
Leif Erikson/Norse/Vikings in America	10
Mohammed born	10
Fall/pillage/burning/sacking/destruction/dividing/capture/ invasion of Rome	9

These timelines—artifacts of white supremacy—encapsulate a totalizing meta-narrative. The divinely ordained course of empire moves westward from the Old World to the New World, from Christian Rome to Reformation England to the twice-born US republic, with America's empire of liberty serving as the endpoint of a trajectory, the conclusion of civilizational progress, the final stage in Christian time. A commenter on the original AMNH display summarized the narrative this way: "The rapid growth of inventive genius and the increasing freedom of thought."[39] In collective memory, famous men, primarily explorers and generals, carry the banner of advancement. The prominence of Vikings reflected the eugenical valorization of the "Nordic race," as well as the efforts of Scandinavian immigrants to claim American heritage. Mohammed's salience is also explicable. Waspy curators would have preferred a marker at Year Zero—"Jesus Christ born"—and, indeed, some of them went ahead and added the Nativity. Honest ones knew that sequoia slabs available to museums were not quite coeval with Christ, and they accepted the Prophet as an Abrahamic placeholder.

None of these settler timelines referenced California's precolonial past. The landing of Sir Francis Drake—an analogue to Leif Erikson for the mythmakers of Anglo-Saxon California—occasionally merited mention, as did the founding of Spanish missions, but the Golden State before the US conquest was literally timeless. The privileging of fixed points in linear time facilitated the (further) erasure of the Native histories, irreducible to dated or datable events. The only Indigenous peoples who sometimes appeared on tree rings were Mayas and Aztecs—calendrical peoples who comported with Western ideas of civilization.

Although chauvinistic and triumphalist, timelines could not close off ambiguity. All the kingdoms and empires inscribed pedagogically upon tree rings formed a perfect record of civilizational impermanence. Signifying both immortality and death, a sequoia display could alienate as well as educate. In a letter to Ezra Pound from 1935, E. E. Cummings described the "miracle" of the natural history museum in Manhattan, where nothing was natural—just stuffed animals and sectioned trees. At the Mark Twain, Cummings read the labels from "the centre or birth of that tree to its circumference or murder. Of course if that tree hadn't been murdered,& murdered crosswise,that tree would have remained a mute inglorious milton."[40]

During World War II, the sectioned meaning of sequoia time became more precarious. In the bleak midwinter of 1941, barely two weeks after the bombing of Pearl Harbor, the superintendent of Kings Canyon National Park gave a Yuletide speech at the General Grant—officially the "Nation's Christmas Tree." As was customary, a crowd of hundreds sat with thermoses and blankets on folding chairs for the noontime program: flag and wreath ceremony, brass and vocal solo, patriotic sing-alongs, carols, prayers, poems, and written messages from the president of the United States and the governor of California—all broadcast on the radio.

The superintendent turned somber as he invoked the "haunting specter of a world in convulsion, of a humanity which seems engaged in a struggle designed for its own destruction." In this time of catastrophe, he said, when America's enemies worked to exterminate every ideal of the Prince of Peace, the noble sequoia offered true inspiration, a model of resoluteness. Great civilizations have "come from savagery to wealth, refinement, and culture, and have been destroyed; while only a few feet of growth have been added to its trunk." The unwavering tree was a "living presence of the Creator" to which the nation could turn for a Christmas message. If we but listen, he continued, a voice will come from the tree— from God himself—saying that the hope for humanity, the hope against darkness and slavery, rests with Americans. The outcome was not foreordained. "If you fail, then on a cross-section from my trunk, perhaps by a generation centuries from now, a marker will be placed at the growth ring for 1941, indicating that here the United States of America died and with it the Christian Civilization."[41]

During the Cold War, the imposed consensus about timelines, like so many things in America, fractured. In a 1965 novel concerning a draft-dodger in Paris, the main character visits the Jardin des Plantes, where he sees the giant sequoia cross section—a gift from the American Legion. After reading the labels for the birth of Christ and the landing of the Pilgrims, the expat laughs bitterly and suggests an update: "February 7,

1965: Uncle Sam bombs small helpless nation." His friend adds another: "Start of World War III."[42]

Later in the century, curators on both sides of the Atlantic revised sequoia timelines to make them politically correct, though no less androcentric. At the Natural History Museum in London, the Mark Twain moved to a new location—a landing in the main hall—to befit its updated signage. Global synchronicity replaced national trajectory: the Ghana empire and the Mongol dynasty sandwiched the Magna Carta; Gandhi sat inches from Newton and Shakespeare. By excising once memorable events like "Athelstan's Great Victory over the Danes" and the "Battle of Bannockburn," the timeline became more inclusive yet more fragmented.

Similar revisionism occurred at Muir Woods National Monument, a coast redwood reserve not far from San Francisco. The timeline here rivals the two Mark Twain slabs as the most visited in the world. The original 1931 display included "Battle of Hastings," "Magna Carta," "Discovery of America," and "Tree Cut Down." In the multicultural 1980s, Aztecs and Anasazi supplanted English kings and barons, and "Tree Falls" replaced the overly candid terminus. After further revision in the 2000s, Columbus completed his three-step downgrade from discovering the New World to landing in it to merely sailing toward it.

Many tourists recognize the Muir Woods cross section from *Vertigo*, Alfred Hitchcock's 1958 masterpiece, a film that circles around time and temporal bewilderment. In a haunting scene, the character of Madeleine, played by Kim Novak, becomes agitated by the thought of "all the people who have been born and have died while the trees went on living." She seems to go into a trance as the camera pans over the timeline. Her black-gloved finger points to a spot between the chalked 1776 ring and the edge of the trunk. In the ghostly voice of Carlotta, she speaks simultaneously to Scottie, the dead tree, and the living forest: "Somewhere in here I was born . . . and there I died. It was only a moment to you. You took no notice."

Vertigo informed another classic film, Chris Marker's experimental short *La jetée* (1962). Through black-and-white photographic stills and a voiceover, Marker told an end-of-the-world time travel narrative. After World War III, in a bunker beneath the radioactive ruins of Paris, government scientists forcibly transmit a prisoner to the prewar world to summon aid. During his painful forays into the past, the protagonist meets a déjà vu woman; they fall in love. At one point they go to the Jardin des Plantes and stop at the sequoia slab. The protagonist points his finger beyond the timeline on the tree. I come from *there*, he says.

Taken as a whole, the genre of the big-tree timeline—a material attempt to permanentize a pedagogy of historical thinking—proved to be unstable. Over the twentieth century, the same display in the same national park could signify national greatness and technological progress, or postcolonial regret and civilizational decline, or apocalyptic anxiety. Now, in the "post-truth" period of extreme partisanship in the United States, one can just as easily imagine an executive order to erase multicultural revisions from tree rings, or the opposite: an agency directive to inscribe replacements for the *Mayflower* and the Gold Rush—arrival of the first enslaved Africans in Virginia (1619), onset of the California genocide (1846). The governor apologized for the latter in 2019.

Out of view of tourists and cameras, the other kind of linear timekeeping with sequoia has become more stable. In the 1980s and 1990s, teams of dendrochronologists—many of them employed at the lab in Tucson founded by Douglass—went back to Converse Basin and the Giant Forest and narrated fire histories on sequoia terms, belatedly realizing Dudley's phytocentric insight. Astoundingly, many old sequoias had survived 100+ fires. To figure out which year a blaze occurred, scientists needed to locate the exact point where fire scar met growth ring. An increment borer couldn't do the job, so they took chainsaws to felled and fallen trees, and extracted wedges, a form of data mining conducive to unbiased sampling. The end result was the world's finest annual-resolution chronology of fire.

Scientists also cored outer rings from living sequoias, which could now be correlated with local weather readings. Douglass had not attempted such correlations because the instrumental data only went back to 1895. As of the 1980s, the data was deep enough for researchers to deduce an important relationship. In mild or moderate drought years, sequoia tree rings exhibited low sensitivity. But for the three most intense drought years then on record—1924, 1976, 1977—every tree from every site exhibited high sensitivity. A common signal existed.

Dendrochronologists went back to their samples and looked for this climate signature across the centuries. They calculated a mean of 4.5 extreme droughts in central California per one hundred years. These famine years were not evenly spread. For example, between 699 and 823 CE, sequoias endured fourteen years of restricted growth. By contrast, the period from 1850 to 1950—California's first century as a US state— exhibited the lowest frequency of drought for any hundred-year period in the last two millennia. In other words, Euro-American colonizers reaped the rewards of the dumbest luck.

Enlarging their scale from California to the North American West, the tree-ring scientists in Tucson vindicated Ellsworth Huntington's emphasis—though not his methods nor analysis nor politics—on transformational drought. Huntington's reputation had fallen through the floor in the postwar period, as geographers rejected his work as crude and racist environmental determinism. For their part, midcentury historians downplayed sudden swings in climate as causative factors. By the dawn of the new millennium, climate change, and anxieties about it, prompted reevaluations.

Using sequoia and other sensitive conifers from across the region, scientists assembled a gridded data set of tree-ring chronologies. Their networked data showed numerous dry periods in western North America's past more severe and more sustained than anything in the instrumental record. Seven megadroughts occurred between 860 and 1600 CE, including a thirty-year period leading up to 1300 CE, by which point Chaco Canyon's social and ritual life had collapsed. Within this multi-century phase of far-reaching aridity—a subcontinental effect of the planet-wide "Medieval Climate Anomaly" (aka Medieval Warm Period)—one desiccating year, 1580 CE, stands out. Many trees survived by shutting down growth, as evidenced by absent rings.

Sequoia chronologies of drought and fire aren't equivalent to national or civilizational timelines. The "tree story" contains no plot or meta-narrative, just a record of change over time. Drought rings and fire scars can be keyed to the Christian calendar, or any other calendrical system, or none at all. The tree-ring archive defamiliarizes the past—a benefit to long-term thinking, but a detriment to temporal feeling. A cambial data set doesn't feel like a piece of Americana like a sequoia slab. Indeed, no artifact can embody a sequoia chronology, which is a graphical representation of statistically smoothed data extracted from scores of samples. Through dendrochronology, sequoias speak collectively as populations, and speak honestly. However, their individual arborescence—their personhood—gets lost in the process.

In contrast, the Mark Twain is personally affecting. In Central London, the cross section fills an appropriately grand space. When I first paid my respects, I overheard an awestruck grade-schooler ask his parents in a sorrowful tone: "Why would they cut this down?" When I returned in 2017, the timeline on the varnished slab had been revised yet again. The scale remained global, but the emphasis had gone back to science and technology. At each hundredth ring, curators now listed the human population—from 345 million in 1000 CE to one billion in 1800 CE.

In Manhattan, the other Mark Twain sits in a dim, unrenovated warren. Compared to the museum's airy dinosaur halls—renovated with a gift from David Koch, a key benefactor of climate denialism—the Hall of North American Forests is a time capsule. It evokes the era when the privileged and the powerful considered it good form to cut down a millennial tree for educational purposes.

The future of sequoia timelines—and mammoth trees themselves—may be virtual. It would be simple for a coder to create a touch-screen *Sequoiadendron* to represent a 3,000-year-old tree in the Giant Forest. The graphical interface could have three hundred tabs, one for each century, showing the global average level of atmospheric carbon. The first 298 labels would be an education in repetition: 280 ppm. Then, in 1900, an uptick to 296 ppm. And then, in 2000, an anomalous and alarming number: 370 ppm.

Future students of US history may learn a narrative by numbers—280, 296, 350, 400, 500—the way past ones memorized the ships and voyages of Columbus. In retrospect, the coal revolution, which conferred disproportionate power to Britain and the United States, was far more consequential than any individual, including Wellington and Washington.

AMERICAN MORTALITY

Sequoia history in US time swung between extremes—from sideshow tree parlors to protected groves, from land giveaways to capitalists to land withdrawals for the public. After Congress and the White House, respectively, created Sequoia National Park (1890) and Sequoia National Forest (1893), US conservationists congratulated themselves on protecting the best Big Tree habitat for the forever future. Small-scale buybacks rounded out the effort. For fifteen dollars per acre, the ghostly Converse Basin returned to federal ownership in 1935.

Saved trees could still be imperiled. In the initial view of the NPS (founded in 1916), giant sequoia faced two main threats: tourists and fires. It did something to manage the former, and everything to eliminate the latter. From an ecological historian's—and a sequoia's—point of view, rangers got their priorities fatefully wrong.

Tourists had behaved badly in the groves for decades, removing strips of bark, nailing signs to trunks, and climbing onto fragile burls to pose for photographs. The epicenter of concern was the Grizzly Giant, a beastly tree with incomparably gnarled and bulging arms, a leading contender for the title "world's oldest living thing." Footfall in Mariposa Grove flattened the tree's understory to a bare, hard surface, prompting the NPS to erect fences, and to commission a study from a plant pathologist. In his 1927 report, the tree doctor warned about "gradual killing" by "unconsciously predatory" enthusiasts. At the Grizzly Giant, he saw "no possibility of recovery."[43] This alarmist conclusion was seemingly validated by the fall of the Massachusetts Tree, the first giant to tip over since the 1870s.

In the 1930s, rangers responded to the perceived crisis. They constructed fire hydrants in Mariposa, suspended car traffic at the grove's *other* drive-thru tree, and rerouted the scenic road away from the Grizzly Giant, now protected by a triple-barbed-wire entanglement. The superintendent of Yosemite National Park, a veteran of World War I, had been influenced by his experience on the Western Front. He adorned his "low German type" barrier with plantings of azalea, ceanothus, and dogwood. With a sign at the grove's entrance, the superintendent admonished visitors to achieve a finer integrity of soul by cultivating humility and contemplating their mortality in the presence of a species from the age of reptiles, and specimens standing tall since Jesus walked the shores of Galilee.

To the consternation of managers, giants continued to fall in Mariposa. The Stable Tree crashed to the forest floor in 1934, followed the next winter by the Michigan Tree and the Utah Tree. The second sequoia named after Mark Twain—supposedly the tallest in the Sierra—collapsed in 1943. After another study, the NPS eliminated camping from the grove, built a rock wall to buttress the Wawona Tunnel Tree, and spread limbs around its trunk to discourage people from walking on its roots. Tourists simply hauled the wood away.

The dual mission of the NPS—to preserve nature for people's enjoyment—guaranteed contradictions. Rangers never stopped automobilists from driving through the Wawona; they encouraged it. This attraction brought endless positive attention to Yosemite, and the NPS hated to alienate its car-loving constituency. After providing enjoyment for decades, the structurally compromised tree buckled and fell in the snowy El Niño winter of 1968–1969. It persisted in memory as an FAQ: *Where is the tree you can drive through?*

Two more of Mariposa's tall sequoias keeled over in 2003, and speculation immediately turned to tourism. "We may be loving them to death," commented one ranger.[44] A decade later, Yosemite's superintendent called the grove's condition an embarrassment, and closed access for the duration of a wholesale renovation paid in half by philanthropic donors. Crews tore out the asphalt loop, replacing it with trails and elevated boardwalks. Mariposa reopened in 2018. Tourists—up to a million per year—leave their cars in a giant parking lot a few miles away and board shuttle buses. After disgorging, they walk to the lifeless Wawona, renamed the Fallen Tunnel Tree.

At Sequoia National Park, tourism likewise concentrates in one extralarge grove, the Giant Forest. Before the 1930s, when the park gained a paved access road, this sequoia population was barely visited compared to Mariposa. To drum up publicity, Sequoia's superintendent presided over naming ceremonies. President Warren Harding's unexpected demise in 1923 provided such an opportunity. Superintendent John White wrapped a Big Tree—hailed as the second largest in the world with a purported 5,000 years of growth—in an oversized Old Glory. "Neither pharaoh nor emperor has ever had as enduring a monument as the plain American citizen who was our 29th President," White wrote. "When the pyramids are crumbling to dust the Warren Harding sequoia will be but swelling to a fuller growth and majesty."[45]

By 1950, Sequoia National Park had congested roads and packed parking lots—postwar signs of success. To protect enjoyers at Giant Forest Village, managers issued a death warrant for a prime-of-life sequoia that leaned ten degrees in the direction of a cabin. A pair of contractors felled the 2,222-year-old in six hours. The concessionaire who had pressured the park to take it down got emotional at the sight: "Everyone seemed to speak in low tones as if gathered at the bedside of a dear friend who was about to depart this world." In 1967, the NPS authorized removal of another old-growth liability—this time over the objections of its employees.

The agency's prevailing culture shifted in the 1960s and 1970s from aesthetic to ecological management. After interminable bureaucratic planning, the Giant Forest got a makeover in 1997–2005. The NPS relocated the tourist village, demolished the remaining buildings, rehabilitated roads and parking lots, and instituted a shuttle bus system for the top attraction, the General Sherman. Paradoxically, changes meant to encourage contemplative visitation discouraged long visits.[46]

Restoration efforts also included fire, a feature of sequoia ecology that went missing from 1890 to 1970. The diminishment began earlier, in the Gold Rush era, with the violent exclusion of Miwoks and Paiutes, groups that met and mixed in the High Sierra. Natives had used fire as a seasonal tool in their mobile subsistence economy. Annual small-patch low-level blazes improved hunting and gathering without destroying sequoias. Just the opposite: anthropogenic burning promoted seedling recruitment by opening the serotinous (flame-released) cones, creating space for light, and adding nutrients to the soil. All of America's ancient sequoias had germinated in landscapes modified by people.

Revealing their racism, many early tourists at Calaveras and Mariposa speculated that fire-stained sequoias must have been carelessly used as fireplaces or backlogs—or maliciously targeted. Natives loved vandalism, supposedly. The limited range of the Big Tree was charged to the "destructive propensities of the Indians."[47]

Once federal foresters arrived on the scene in the early twentieth century, they perceived a dearth of sequoia seedlings, and responded with a policy of fire suppression. "Fires and young trees cannot exist together," wrote a leading US forester in 1911. "We must, therefore, attempt to keep fire out absolutely." He dismissed proponents of light burning as misguided followers of the "savage" ways of the "redman."[48] The city engineer of San Francisco likewise blamed the "Digger Indian System of Forestry" for the threatened state of the species.[49] Aldo Leopold—future author of a canonical book on land stewardship—contrasted "Piute Forestry" with forest fire prevention, like Prussians contrasted "Polish management" with German forestry.[50] Total suppression became such a shibboleth that Congress, when appropriating funds to Sequoia National Park for firefighting equipment in 1922, stipulated that none of the monies could be used for "precautionary fires," later known as prescribed or controlled burning.

Ironically, pro-natal policies increased seedling failure. Young sequoias do best in nutrient-rich soil with plenty of sun—the kind of conditions that occur after moderately hot ground fires. In the new fireless regime, shade-tolerant white fir outcompeted giant sequoia. To give sequoia saplings a fighting chance, land managers thinned the competition in the famous groves. This policy, described as "vista clearing," seemed to have added benefits: reducing fuel load and improving scenery. Desperate for new sequoia growth, managers resorted to gardening. During the

Depression, CCC boys in Sequoia National Park hand-planted thousands of seedlings from an on-site nursery.

In the 1960s, the NPS questioned whether its efforts to preserve charismatic megafauna and emblematic megaflora had done more harm than good. The 1963 "Leopold Report," named in honor of Aldo Leopold's son, the lead author, was a clarion call for ecological management throughout the park system. The report argued that overprotection had turned the park-like forests of the Sierra Nevada into dog-hair thickets. It called for restoration of biotic associations that existed "when first visited by the white man."[51]

Even as officials in Washington, DC, put the finishing touches on new fire policies, a lightning strike blew the top off the California Tree in the General Grant Grove. After the crown kept burning, out of reach, for several days, the superintendent made an emergency call to his top crewman, Charlie Castro, then on assignment in Montana. Castro numbered among a small fraternity of high-climbing sequoia firefighters, all of them local Indians. A Miwok-Paiute born in Yosemite Valley, Castro toured as a jazz drummer in the off-season. With a rope around his chest and a cord on his belt, Castro fearlessly climbed a tall fir not far from the branchless bole of the burning sequoia. Nearly 200 feet up, he anchored his rope, then swung down, like Spiderman, to the sequoia's lower crown. Embers falling on his head, Castro continued upward to 250 feet, and took a fortified position in the smoky canopy. He lowered his cord to on-the-ground firefighters, who attached a high-pressure hose. After pulling it up, Castro spent hours in finger-numbing cold and hair-singeing heat, singlehandedly saving the California Tree—and, by extension, the Nation's Christmas Tree—becoming a legend in the process. "It was like being the first person up El Capitan," he recalled proudly.[52]

In 1967, the year Castro won an award—his second—for firefighting heroics, the NPS began to reverse its fire suppression policy, and designated jointly managed Sequoia & Kings Canyon National Parks as one of its flagship experiments. Soon, the agency established a "natural fire zone" in the backcountry where lightning-caused fires would be allowed to burn. In the Giant Forest, rangers began an annual regimen of prescribed burning. They hoped to lightly burn the park over many years, one small piece at a time. Despite the best efforts of managers to educate the public, these low-intensity fires ignited controversies. Tourists disliked seeing the blemishing of iconic named giants—the ones rangers called "monarchs." After years of complaints about "government vandalism" and "barbecuing," the NPS adjusted its fire management plan to preserve the aesthetics of the monarchs as much as possible.

Science rather than politics prompted a second round of adjustments. By 2000, dendrochronologists in the Sierra Nevada had proven that the anti-Indian era of fire suppression—an absence of scars in the rings called the "Smokey Bear effect"—was unprecedented in the last three millennia. For fire-adapted *Sequoiadendron*, the first half of the twentieth century represented a lacuna: a failed generation of seedling recruitment. For the post-glacial period at Giant Forest, scientists documented a wide range of fires, patchy to uniform, low to high intensity. Rather than shielding sequoia from further disturbance, fire ecologists wanted to facilitate a disturbance regime that would produce a "mosaic" forest. They no longer spoke of "restoring" groves to "primitive" or "original" conditions. The modified goal: a resilient forest within the range of past variation.

The efficacy of prescribed burning became obvious with the appearance of sequoia seedlings, then saplings, in the Giant Forest. But the slowness of the process—years of planning for each ignition, followed by waits for weather windows and political windows—prevented the NPS from burning the whole park once, even burning all the major groves once, before a megadrought began in 2000. A historical fire deficit remained. The red tape frustrated ecologists, who, taking the long view, considered immature sequoias as important as the biggest old ones—the "trees of special interest," to use managerial language. By the categorization of the NPS, roughly one hundred giants in Sequoia National Park count as special, and of those, a handful, including Grant and Sherman, have been designated "culturally significant."

The two generals have grown younger with time. In the 1990s, Nate Stephenson, a forest ecologist with the US Geological Survey, created a new-and-improved mathematical formula for estimating sequoia age. He tested his formula on hundreds of stumps in Converse Basin, disproving for good the old assumption that biggest means oldest. The Nation's Christmas Tree did not witness the Star of Bethlehem, after all. By Stephenson's estimation, the General Sherman was only 2,150 years old and the Grizzly Giant a shocking 1,790 years young. The most senior Big Tree, wherever it was, probably lacked significance, culturally speaking, because of its relative smallness. The diameter of the oldest known logged sequoia (3,266 years) was barely half that of Sherman. "Even the largest sequoias are middle-aged at most," said Stephenson, "and they're still growing fast like teenagers."[53]

Stephenson came to Sequoia straight out of college as a volunteer, and hung around long enough to become a seasonal ranger. He continued that summer work—all for the love of backpacking—after earning a doctorate. His big break came in 1988, when James Hansen of NASA testified to Congress about climate change. The federal government responded with the US Global Change Research Program. Stephenson submitted a proposal for a research station at Sequoia, won a grant, and thereby earned, for the first time, full-time employment in the forest he had adopted as home.

Stephenson became the go-to expert on sequoia in the first two decades of the new millennium, when near-continuous drought gripped California. Using tree rings as proxies for snowpack and summer soil moisture, scientists determined that this twenty-plus-year period, when the Golden State's population approached forty million, was the driest in the Sierra Nevada since 800 CE. They attributed nearly half the intensity to anthropogenic forcing.[54] In other words, heat-trapping emissions turned a "normal" drought into a "thousand-year drought"—assuming the idea of normality retained relevance as environmental conditions became untethered from historical baselines.

Because of aridification, Stephenson's research shifted to forest dieback. More than one hundred million conifers in the Sierra Nevada died standing in the 2010s, with the death rate approaching 90 percent for ponderosa pine. For that same time period, the estimated mortality for sequoia—less than 1 percent—seems phenomenally low, proof of the superlativeness of the species. However, the drought did produce widespread foliage dieback on sequoias, something never documented before. When I visited Stephenson in 2017, he showed me a harbinger of the future: a sequoia branch riddled with the tracks of bark beetles. "Beautiful, isn't it?" he quipped. The record dry year 2014 showed that *Sequoiadendron* isn't invulnerable to predation, as long asserted. Under enough stress, even superflora has its kryptonite.

What will happen when dryness like 1580 CE returns—as it surely will—in combination with a heat trap of 500+ ppm? That goes beyond the known range of past variation. The tree-ring record no longer serves as an operational manual for a low-to-no-snow Sierra. "It's a matter of triage, buying time, and hedging bets," Stephenson told me. He chose his words carefully. His research station had survived the administration of George W. Bush, and the transition to Donald J. Trump. In terms of sequoia policy, he's a pragmatist, having worked through his own climate grief. The NPS can actively manage "maybe 1 or 2 percent of the Giant Forest," he said. "We can't garden a whole national park." Stephenson prompted me to imagine a scenario: in future hotter droughts, rangers could water the monarchs (triage), apply pesticides generally (buying time), and facilitate the outmigration of the species (hedging bets).

In its current prime habitat—island-like flanks of the southern Sierra—sequoia has no place to go altitudinally. They need to go north. In the late nineteenth century, multiple authors anticipated this move; in 1906, *Sunset*, a lifestyle magazine published by the Southern Pacific Railroad, called for a Sequoia Week to match Arbor Day. Sequoias needn't be fenced-in veterans on exhibit as remnants of a vanishing race, *Sunset* said. Forget fatalism, think optimistically like true Americans: man can conceive a larger forest of Big Trees.[55] The USFS soon thereafter experimented with sequoia seeding near Lake Tahoe. A century later, history came full circle: A for-profit timber company took the lead in sequoia conservation. Sierra Pacific Industries, a family-owned business and the largest landowner in the northern Sierra, began creating living gene banks using nursery-grown trees germinated from seeds harvested from cones collected from Sequoia National Park with NPS permission. The company's "outplantings" occur at a variety of altitudes and latitudes in the Golden State's northern half.

Is assisted migration more than sentimental? "Sequoia could go extinct and the ecosystem would barely notice," Stephenson told me for shock value. In the functional language of ecology, *Sequoiadendron* counts neither as a dominant species nor a keystone species nor an indicator species. Its total carbon storage—negligible. The only "ecosystem service" a mature sequoia provides that other Sierra conifers cannot is nesting habitat for reintroduced condors. However, these rare plants provide temporal services for modern people. "Visitors come here because the big old trees ground them in a world of worry and crisis," Stephenson continued. "Sequoias give them peace, give them comfort—even if it's an illusion—that some things never change." But now the NPS cannot keep things unchanged. "How do we communicate this to the public?" he wondered aloud. "What will it do to the psyche? People get angry."

When I spoke with Stephenson, on a hot September day, fire burned in two groves south of the park. Three years later, in 2020, multiple megafires visible from space ravaged the Sierra. Californians began to measure the sizes of "fire complexes" in relation to New England states. On a carbon calculator, the Golden State's reductions in greenhouse gas emissions from 2001 and 2020 were neutralized by the estimated 440 million metric tons of California forest carbon that entered the atmosphere during the same period.[56] Bigger fires burned hotter and higher, producing pyrocumulonimbus clouds reminiscent of Hiroshima. Extreme heat from excess fuel—a consequence of past fire suppression—allowed the SQF Complex to spread vertically to the upper canopies of sequoias, torching thousands of healthy millennials. When preliminary estimates came in, experts communicated shock and dismay: up to 14 percent of large sequoias in their natural habitat may have perished in a single event. Technicians wept in the field. "The apocalyptic chickens are coming home to roost, way sooner than we thought," said the chief resource manager of Sequoia National Park. "If we don't get prescribed fire in those groves that have not had any for 100 years, we lose 2,000-year-old monarchs in a fire."[57]

There wasn't time: the dreaded moment arrived the next year, 2021, when the lightning-caused KNP Complex burned through the heart of Sequoia National Park. Before evacuating, firefighters installed sprinklers at the General Grant and wrapped the General Sherman in aluminum blankets—acts of desperation and public relations. The relatively flat Giant Forest escaped the worst as a result of topography and past prescribed burning. In the immediate wake of this positive news, the California State Legislature passed a landmark bill meant to encourage controlled burns throughout the state, following the historic example of Native peoples. Then came the devastating news. Running crown fires had "nuked" whole groves of ancients in sections of the park with steeper slopes and greater fire deficits. Crestfallen rangers anticipated they would need to plant seedlings where severe heat had overbaked the serotinous cones.

Tagged on a timeline, 2020–2021 could thus be summarized: ONE IN FIVE ELDERS OF SUPREMELY FIRE-RESISTANT SPECIES GOES UP IN FLAMES.

Ever since the Mammoth Tree became public knowledge, white Americans have fixated on its demise in national time and, inversely, the mortality of the racialized nation in one sequoia's lifetime. All the bravado about the linear (Western progress) and the sempiternal (a nation for all time) barely conceals anxieties about the cyclical (the rise and inevitable fall of civilizations). To remain the nation of futurity has required forever war. The militaristic personal names of the two "oldest" sequoias—evocations of the Civil War and the overlapping "Indian Wars"—suggest as much. It remains to be seen: Can a majority of Americans reform national patterns of long-term thinking, and respond to the mortality of the nation's most emblematic megaflora, without invoking race or the Apocalypse? Ingrained habits die hard. Ancient sequoias are for now living in US time; more accurately, they're dying from settler time. In the short history of Big Trees in the United States, there have been humble moments of cross-species temporality and shameless displays of temporal business as usual. The troubled early history of dendrochronology—more Eurocentric than phytocentric—shows the potency of settler time, and

its poverty. Ennobled by the awesome materiality of nonhuman timefulness, leading observers of standing and sectioned sequoias fixated on thin
white lines.

In May 2001—in retrospect, the final spring of the American
Century—recently inaugurated George W. Bush visited the Giant Forest.
White House speechwriters contacted Nate Stephenson ahead of time to
confirm the age estimate for the tree beside the ad hoc presidential lec-
tern. As if reading from a timeline, the Commander in Chief intoned:
When the *Mayflower* arrived, this sequoia was already here. When the
seal was fixed on the Magna Carta, it was here; when the Roman Empire
fell, and earlier, when it rose, it was here. "And had Christ himself stood
on this spot, He would have been in the shade of this very tree." We can't
peer into the centuries ahead, said the born-again 43rd president of the
United States, but it will be to "our lasting credit if these works of God
are still standing 1,000 years from now."[58]

In synchrony, the White House abandoned the Kyoto Protocol.

VI.
OLDEST KNOWN

BROOKLYN TO TUCSON — LONGEVITY UNDER ADVERSITY — ANCIENT BRISTLECONES — SACRALIZATION AND CALIBRATION — PROXY DATA

In the late 1950s, Great Basin bristlecone pine—a species with scant prior cultural significance—became instantly world famous as the "oldest known living thing." Tree-ring dating proved that stunted pines could surpass giant sequoias in age. These "Methuselah trees" were discoveries of Edmund Schulman, a Jewish-American scientist whose quest for arboreal longevity mirrored his academic precarity. Against tribulation and prejudice, Schulman advanced dendroclimatology, or the reconstruction of climates with cambial data. By insisting on precise dates, and by focusing on "dwarfs," he made a historical departure from the age estimations and megafloral bias of nineteenth-century foresters, tourists, and Humboldtian naturalists. Thanks to Schulman and his successors at the Laboratory of Tree-Ring Research, ancient bristlecone pines from the White Mountains of eastern California became model specimens for computational analysis. Radiocarbon daters and climate modelers embraced their utility. Living lab specimens became objects of touristic veneration, despite their unmonumental appearance and their high-altitude habitat above a little-loved desert. In the end, the abstract science of tree rings recapitulated the material history of sacred trees.

BROOKLYN TO TUCSON

Edmund Schulman begins his diary in neat cursive on the day he turns nineteen—just another workday in Brooklyn, 1927. He's a "convict of circumstance," having punched the clock each birthday since age twelve. The onset of adulthood feels like midlife crisis. In a different setting, a young man of Edmund's intellect would be studying Talmud or finishing college. Instead, this eldest son of a baker-turned-cabinetmaker works as the second breadwinner and the family accountant. His parents hail from the Russian partition of the Polish-Lithuanian Commonwealth. Adrift in a city of millions, they and their five children relocate every few months—South Williamsburg to Long Island City to Brownsville to Woodside to Bedford to East Williamsburg. Mom has a history of making security deposits on cheaper apartments without consulting the others. Now she's splurged on their first-ever living room set—something they deserve, and cannot afford. Edmund lies awake at night, brooding on money and fate.[1]

Some of the family decampments are precipitated by the youngest, Sonny. He suffers from a mysterious neurodegenerative disease. As a child, he could speak with a lisp; now, at puberty, he's lost all speech, becoming an "idiot-animal-primitive" who soils himself and breaks everything. His fits last for hours. On one of many occasions when Sonny runs away, the NYPD picks him up and dumps him off at a hospital. After locating his little brother in the insane ward, Edmund slips banknotes to a doctor to release him.

The Schulmans have a dream of escape: Zion. Sight unseen, they buy rights to purchase property within the "Switzerland of Palestine"—a planned health resort on Mount Canaan, overlooking the Sea of Galilee. The Holy Land will be reclaimed like the American West, says a Yiddish-English investment brochure that invokes Elwood Mead, commissioner of the US Bureau of Reclamation. A Lower East Side company that promotes Zionism through real estate speculation handles the money. Each time Edmund deposits cash toward the once-and-future Land of Milk and Honey, he worries his family is being swindled.

He works at United Naval Stores Company in Brooklyn. From his start as an office boy, he's hustled his way to senior sales. The company buys pine tar—and by-products from Standard Oil—at wholesale, and sells them, drum by drum, at retail to roofing contractors. Edmund recently completed his 800th sales call, and still he can't get a raise. His horrible boss, who sexually harasses the switchboard operators, will "never die of an expanded heart."

For escapism, Edmund reads the latest on boxing, college football, and aviation, and goes to movie palaces with live orchestras. The silver screen—including features like the original *Fast and Furious*—calls out to him, but he hates himself for enjoying entertainments meant for the gullible masses. He loathes "ridiculousity." Using his library card, he balances his media diet with a canonical list of highbrow books in multiple languages. From the main library, it's a short walk to the lake in Prospect Park, where he paddles rental boats in an attempt to make himself husky and to improve his rotten constitution. Desperate to fix his pimply face, he buys an exposé on the evils of processed food.

Wishfully thinking he can complete a degree while working full-time, Schulman chooses New York University (costly) over City College (free) for the reason that NYU offers business degrees. He wants to get rich. Instantly, he regrets he can't take subjects like art history. In economics, he finally makes a friend, who asks for help with homework. Edmund feels confused and hurt when his erstwhile buddy drops him: Is it because I'm poor? Because I'm Jewish? For Rosh Hashanah, Year 5688, he gets a haircut and sees another film, flouting Jewish law. He disbelieves in God and savors ham and bacon. "It is all antiquated barbarism," he writes.

Days later, he's coughing up blood. The old devil TB brings up existential thoughts: Why is there so much sorrow? Why is there life? Why am I at all? He has three breakdowns in nine months. Something is terribly wrong with his body. A teenager shouldn't have heart palpitations, chest pains, and digestive complaints. His teeth need inlays. In his condition, he can't continue night classes. He must find a competent doctor, once he has time—and money.

In August 1928, Edmund takes a short vacation. At kosher camp in the Catskills, an evening speaker makes an impression. Oppression over millennia has stimulated the brains of Jews, making them superior, the man says. In America, Jews have enjoyed life without accustomed persecution—and also without accustomed accomplishment. That will change through the stimulation of freedom. American Jewry will throw off petty inhibitions and archaic prohibitions, embrace scientific thinking, and usher in a new golden era.

On New Year's Day, Edmund goes to Manhattan for a special event at the American Museum of Natural History. The director of Harvard's Blue Hill Observatory dazzles the audience with a slide-illustrated presentation on galactic history and Einsteinian physics. To Edmund, the darkened room feels more like a temple than a movie palace. He makes a resolution: I will become an astronomer. The next day, he starts studying constellations and reviewing advanced algebra. He needs a quiet place to study, away from the discord and slovenliness of home. He must move uptown, near City College—the school of choice for so many bright young Jews, now that Harvard has a quota. But he can't afford his own rent. Although the Palestine property now belongs to the family outright, the value of their would-be resort investment has plummeted with the global economy.

After new bouts of blood-spitting and rectal bleeding, Edmund uses his banked Christmas bonus to take a proper summer vacation. In the Catskills, he meets a girl, who raises his hopes, then breaks his heart: "I am just a coordinated piece of animated ephemerality, caught in a finite wisp of eternal time." His love interest does a lasting favor, however. She refers him to a doctor, a pulmonary specialist, who finds tubercles in his lung. The diagnosis provides existential clarity. Just like that, Edmund tells his boss he's quitting; tells his family he's heading out west, where the stars shine bright through clean, dry air.

He decides on mile-high Denver, home of the National Jewish Hospital for Treatment of Consumptives and the Jewish Consumptives' Relief Society. He mails a trunk stuffed with books, and follows by train. He deposits his savings ($200) in a bank, then heads to the public library with a note of sponsorship from his Gentile landlady. He plans to complete an autodidactic curriculum in calculus, physics, astronomy, and geology. To keep himself well-rounded, he checks out *The Golden Bough* and self-help books on dietetics and psychology. "Do I have blue streaks because I'm egotistical?" he wonders. He buys cod liver oil—the first medicine he's taken in his life.

"My chances of getting a part-time laboratory job, (or even a full-time position) and being a Jew are practically nil," he acknowledges. He reads the new memoir by Rebekah Kohut, a leading figure in Jewish social service, which renews his determination. While Jews thrive on persecution, they also thrive on opportunity, Kohut says. In America, a transformed Jewry, with the freedom and lightness of classical Greeks, has come into its own after two thousand years. Those who succeed must expect from Gentiles the tribute of misunderstanding, envy, derogation, and senseless criticism.[2]

The Depression adds insult to Edmund's prospects, but he's acquired the doggedness of an American salesman. He writes to Lick Observatory in California and Lowell Observatory in Arizona. Dr. Vesto Slipher at Lowell writes back and refers him to a physics professor, a Jew, at Denver University. The professor in turn gives Edmund advice on approaching the director of Denver's observatory. Schulman makes the approach, to no avail.

Running out of cash, he moves someplace cheaper and drier: Albuquerque. He keeps writing letters to Slipher, director of the premier astronomical facility in the United States. Edmund eventually receives a frustrating reply: There's no work for you at the lab; you would be a better candidate if you had a lab background. When his mother wires him money to return east, he continues westward to Flagstaff. Drawing on his experience with cold calls, he goes to Slipher's office a half-dozen times, at long last catching the astronomer. The persuasion produces nothing. Edmund leaves the observatory sobbing: Goddamned world!

His diary changes to shorthand—more evidence of relentless self-training. Somehow, he manages to take piano lessons. He enrolls in Arizona State Teachers College with tuition money from his aunt's husband, a bootlegger. Living in the dorms in Flagstaff, practically the only Jew in a town of 4,000, he becomes the butt of pranks involving putrid liquids and telephone calls. Edmund becomes fatalistic about his chances in Arizona, the 48th state, with its cowboy culture. Maybe I'll just change my name to Ben-Yehuda (son of Judah), he quips.

Giving it one last go, Schulman sends exploratory letters to the University of Arizona (UA) in Tucson, where an observatory has recently opened. The director, Andrew Ellicott Douglass, is an institution builder—department chair, interim president, college dean. Self-consciously, Douglass honors his Yankee pedigree: his namesake great-grandfather surveyed for George Washington; his father, and both his grandfathers, served as New England college presidents. His doctorate is an honorary one, from his alma mater, an Episcopalian college. A big fish in a small desert pond, "Doctor D." has placed Arizona's fledgling research university on the scientific map, as evidenced by *National Geographic*'s coverage of his cross-dated tree-ring discoveries.

Schulman's diary goes blank for over one year. When entries resume in January 1932, he has transferred to UA. He loves the aridity and the mountain vistas with giant saguaros. As Purim approaches, Edmund gets a letter from his mother, Sophie, reminding him to eat a hamantasch. "Where can I get one in this town, I wonder." He's accustomed to loneliness, and the financial Sword of Damocles. Now that he's used his savings on the first tuition payment, he must scrounge up $450 to finish the semester.

Declaring astronomy as his major, he goes to the observatory to get his course plan approved by Douglass, his assigned adviser. Having learned from hard experience, Edmund is determined not to make a first impression as a nuisance. He doesn't (yet) ask for a job. Instead, he asks for permission to take Douglass's new course on cambial interpretation. "He was inclined to pat himself on the back a bit about his tree ring work," observes Edmund. "I suppose he is quite justified in doing so, at that." After Douglass mentions offhand that spherical trigonometry is useful for dendrochronology, Schulman immediately begins a comprehensive review of the subject.

When lectures begin, Edmund is surprised by the degree of difficulty and by the paucity of published work that explains the tree-ring technique. Mostly, he's dismayed by the required cost of a professional-grade magnifier. He wears down Douglass with eight requests for work. The professor concedes there could be a job if Schulman were an expert on tree rings. Douglass regularly receives requests from Southwestern archaeologists to date wood samples, and he could use an assistant who met his standards. Edmund convinces him he can, and, just like that, he has keys to the lab. A graduate student in astronomy warns him that Doctor D. is "a hard man to work for. He is a good fellow and all that but a terrible stickler for principle, for order." A 33rd-degree Mason, Douglass is the kind of person who views Rotary International as the culmination of Western Civilization. He always wears a necktie.

By the end of the semester, Doctor D. is so taken by Schulman's brilliance that he adds the undergraduate to his research grant from the Carnegie Institution. Schulman begins plotting data from multimillennial sequoias, developing tables of standardized tree growth, making thousands of calculations with a slide rule. The soft money is a lifesaver, though Schulman knows he deserves at least double his wage. "I can't complain, however. Science is not ordinarily a money-making proposition."

If only it were: With $10,000, Mom and Pop and his siblings could, he fantasizes, move to Palestine and thrive in the healthy, citrus-scented atmosphere, while Sonny receives care in a Hadassah house. Brooklyn is killing the Schulmans. Now they want to move out west, too. The eldest, Doris, plans to be Edmund's secretary once he becomes a professor.

In spring 1933, at age twenty-five, Schulman completes his Bachelor of Science and immediately starts a master's program. He attends meetings of the recently founded Maimonidean Society, UA's first club for Jewish student life. The faculty sponsor is Alsie French Raffman, full-time adjunct instructor of freshman English and a member of the Daughters of the American Revolution. She's also the wife of a Jewish mail carrier from the Bronx. Through marriage—the result of courtship at a sanatorium—this Presbyterian from Indiana has become active in Temple Emanu-El, a Reform synagogue with services in English. Fewer than one hundred Jewish families live in Tucson, so the congregation cannot afford to be exclusive. Raffman joins the choir and later becomes the founding president of Hadassah's local chapter.[3]

The year 1935, when Edmund attains his first advanced degree, marks the 800th birthday of Moses ben Maimon and the tenth anniversary of the Hebrew University in Jerusalem. As president of the Maimonidean Society, and head of its Zionist committee, Edmund organizes an event in honor of a pioneer university 7,500 miles away that provides safe harbor to German refugees. Alsie Raffman is invited to the event, though she's caring for her ailing husband. Within weeks, she's a TB widow.

Gratifyingly for Edmund, employment follows graduation. Thanks to a new multiyear grant from the Carnegie Institution on weather prediction, Doctor D. hires him as a tree-ring researcher. For his first business cards and his first letterhead, he names his purview: "Climatological Research." After the grant runs out, Schulman returns east to get a second master's degree—this one from Harvard's Blue Hill Observatory.

In December 1937, the regents of the University of Arizona approve the creation of the Laboratory of Tree-Ring Research (LTRR), hoping it will become the "world-center for this work." For now, it's the only center. An independent unit of UA, the lab is the fiefdom of Douglass, age 70. Still the preeminent scholar on campus, he reports directly to the president. Doctor D. invites Schulman back to Tucson not on soft money but on salary, albeit a tiny one. Schulman has every reason to believe he will be promoted to director someday, once Douglass retires. He's in debt to his mentor, literally. To keep the Brooklyn family afloat, Edmund has repeatedly borrowed cash from Douglass. He owes so much, the young man lists the old man as the beneficiary of his life insurance policy. Tragically, just as Schulman begins his salaried career in academia, his sister Doris dies at age 31.[4]

His titles are Instructor in Dendrochronology (for the Department of Anthropology) and Assistant in the LTRR. He's the lab's sole employee—the first person in the world hired as a tree-ring scientist. Assuming the editorship of the *Tree-Ring Bulletin*, he takes responsibility for systemizing his adviser's methods into an academic discipline. Edmund announces his career goal to Doctor D.: A "world-wide tree-ring map of past climates and the cycles therein." Douglass supports this agenda, especially the last two words. The astronomer can't let go of sunspot cycles. As Schulman matches, then surpasses his mentor's tree-ring skills, he becomes more interested in the geocentric—including newly theorized Milankovitch (orbital) cycles—than the heliocentric. But the dutiful advisee masters the use and maintenance of the cycloscope, Douglass's optical contraption. On his own, he keeps adding disciplines to his purview. In spring 1939, he goes to Berkeley to train in plant biology.

As Germany and Russia prepare to reinvade his parents' homeland, Schulman conducts his first season in the field. Whereas Douglass stopped with sequoia, his assistant makes long-term plans for a systematic investigation of western conifers, looking for the oldest specimens with the most sensitive rings. Besides his increment borer, Edmund's key research tool is his "trusty Hudson." During the teaching months, he lets his younger brother Harry, now a machinist in LA, borrow the coupe. Pulling rank, the first son writes a stern note on university letterhead, chiding his sibling for not keeping him informed of radiator maintenance. In a summer of solitude and science, Schulman cores old conifers on the sides of mountain roads in Plumas County, California, a world away from the smoke and din of the Brooklyn Navy Yard.

LONGEVITY UNDER ADVERSITY

In 1940, when Schulman submits his draft card, he's 5'9", 180 pounds, and under stress. Amid dreadful reports from eastern Europe, he participates in a statewide conference of Arizona's Jews to discuss the catastrophic urgency of the Palestine question. UA's Maimonidean Society, under the co-leadership of Edmund and Alsie, joins the national Hillel. The yearbook notes the accomplishment, while mangling "Hillel" as "Hifflin," and omitting the *c* in his surname. The extended Schulman family is counting on Edmund's success at the university. Sonny has mercifully passed away, but the surviving sister has her own adversities. An aunt's jewelry business is on the ropes; an uncle's furniture outfit is closing. The clan is bound for southern California, with Edmund's rental house as their waystation. Although he gets a raise for being a "brilliant research man," his salary remains meager.

In 1941, a soft money lifeline comes from Los Angeles: the city wants to understand historical streamflow in the Colorado River basin. The drought years 1934 and 1940 have clipped Commissioner Elwood Mead's rosy assumptions about the water available for future growth. More pressingly, the US Bureau of Reclamation and Los Angeles need to know if they can operate Hoover Dam for maximum hydropower for war production without drawing down the reservoir, Lake Mead, to critically low levels. They desire a regional climate forecast for the 1940s. Because instrumental data for rainfall and streamflow barely extend into the nineteenth century, the best guide to the past—and future—is tree rings. Schulman accepts the challenge of locating statistical samples of trees whose cambial growth corresponds to hydrology across the seven-state basin.

Using his funding, he takes a short leave on half pay from UA to begin PhD coursework in climatology at Harvard. Through Blue Hill Observatory, he obtains two deferments. He thinks of his research as his war effort. In addition to his bread-and-butter course—Anthro 160: Introduction to Dendrochronology—he teaches an overtime course for servicemen on weather prediction. "Things are hopping in everything but finances," Schulman tells his family in October 1942.[5]

In spring of '43, he presents a preliminary report to the Los Angeles Department of Water and Power, and in spring of '44, he submits a longer version to Harvard as his dissertation. Doctorate in hand, he gets a miniscule raise that lifts his salary on a par with Douglass's half-time sinecure. Unbeknownst to Dr. Schulman, Doctor D. adjusts his own salary so that he still gets more. Edmund can hardly complain about his life in worldwide context. Come summer, while millions die abroad, he's back on the road—his unrationed tires patched again and again—in a grand circuit of California, Oregon, Washington, British Columbia, Alberta, Montana, Wyoming, Colorado, New Mexico. His hydrological research has expanded to all the major watersheds of western North America.

Through his field research, Schulman ascertains the widespread western occurrence of "overage conifers"—trees that have lived beyond their "normal" life expectancy. Excluding sequoia and redwood, Schulman initially considers exceptional anything over 500 years. He focuses on Douglas fir and piñon pines. Site analyses indicate that slow growers in adverse environments live longest. Their climatic signals are complicated by the frequency of double (false) rings, locally absent rings, and simply hard-to-read rings. Schulman drowns out the noise with more and more data—trunkloads of timewood in labeled envelopes.

In 1945, Schulman publishes a progress report on "runoff histories in tree rings." Here, as in past publications, Edmund makes prefatory remarks about Douglass, archaeological dating, and sunspot cycles, but the language now sounds rote, less deferential than obligatory. Downplaying archaeology, he forecasts a two-part research agenda for the lab: "mapping" past climates around the planet; creating probability models for long-term regional weather. He refers to dendrochronology "as a branch of physical science." Using uncharacteristic hyperbole, he conceptualizes western conifers as "natural gauges" of precipitation and runoff, and Arctic conifers as "living thermographs."[6]

He continues working with LA Water and Power after the war, extracting and measuring tens of thousands of tree rings from California's southern coastal ranges. In the lab, he identifies several "severity periods"—later known as megadroughts—including a sixteenth-century episode twice as severe as anything in the instrumental record. Without naming names, Schulman debates a rival, Waldo Glock, a geologist turned dendrochronologist who got his start in Tucson. Glock disparages Schulman's understanding of plant growth. Cross-dating only works locally, Glock claims, because each site is different, and trees respond to multiple inputs. He lampoons Schulman's emphasis on abnormal trees at abnormal sites. Schulman counters that sensitive trees at sensitive sites offer the best, longest data sets—provided that ring widths can be correlated to a single limiting input like rainfall or temperature.

In 1945, Schulman contemplates the next decade of his career project. He finally has the luxury to think ahead: at age thirty-six, he receives a tenure-track position as assistant professor of dendrochronology. That summer, as the Pacific War comes to its calamitous end, he begins his quest for the oldest overage conifers. He returns home worn out and sick. "I may have to start thinking about my state of health for the first time," he admits.

He could use help in the lab. In 1946, Douglass wins approval for a staff line. The newly hired assistant, whose highest degree is a Bachelor of Arts from UA, instantly earns almost as much as Schulman. He's a Navy veteran with prior experience as a naturalist at Mesa Verde National Park and as a police officer in Tucson. He's a people person. Originally from Kansas, his surname seems chosen by central casting: Smiley. Soon after making the hire, Doctor D., now 79, begins making private "Notes about Edmund." They've had a "sharp talk." Douglass thinks his protégé has turned against his cycloscope.

Schulman is riding high, having just addressed the Colorado River Basin States Committee and submitted his final report. He acknowledges Alsie, his faculty friend, for her editing. The report contains growth curves for fifty-two field stations, with cambial data correlated against streamflow and rainfall records. It comprises a 658-year index of Colorado River runoff. His main conclusions: on average, one year in fifty brings severe drought; consecutive severe years are uncommon, but not rare; Hoover Dam has yet to be tested by such extremity.[7]

In 1947, Schulman is promoted to associate professor with tenure, Douglass praising his "unusually technical work." Edmund repays the favor by organizing the birthday celebration for Douglass's 80th at a faux-fancy Mexican-themed restaurant, El Merendero, in the foothills of the Santa Catalina Mountains. Schulman has earned the right to think of himself as a UA director in waiting.

If he could peer into the future, he would see himself struggling for breath in California's White Mountains. They rise to 14,000 feet in the immediate rain shadow of the Sierra Nevada. High, dry, cold, and austere, the Whites are emblematic of the Great Basin, a vast inland area with no outlet to the sea. Beyond the Whites rise mountains beyond mountains—up and down, up and down, all the way to Utah—with sagebrush at the bases and bristlecones on the summits. This is not a species with deep-set human connections. Numic peoples of the Great Basin—Paiutes, Western Shoshones, Goshutes—collect edible seeds and tell sacred stories about mid-elevation single-leaf piñon pine. High-altitude bristlecones, neither mythic nor merchantable, receive less attention from Natives and also settlers, though miners cut them for timbers and shepherds burn them in campfires. The USDA has described the species in purely utilitarian terms: it forms a "protection forest" on dry southern slopes, preventing erosion and conserving moisture.

The vegetative cover of the upper Whites isn't exactly forested—bristlecones and associated limber pines are widely spaced, short, and contorted—but it's called a forest because it falls within Inyo National Forest, reserved by President Theodore Roosevelt for aquifer protection. The range falls within the watershed of the Owens River, a landlocked stream that Los Angeles has siphoned away.

Of the hundreds of mountain ranges in the Great Basin, the Whites are where Schulman lands, for a reason: the Cold War. In 1948, with a special-use permit from the Forest Service, the Navy builds a Quonset hut research station at Crooked Creek, elevation 10,150 feet, to test infrared-seeking missile technology. The military gets permission to construct a thirty-mile dirt road, steep as can be, from Owens Valley. A Navy scientist turned professor at UC Berkeley uses Crooked Creek for high-altitude physiology tests.

Come 1950, the Navy transfers White Mountain Research Station to the University of California. The Office of Naval Research, the Rockefeller Foundation, and the brand-new National Science Foundation (NSF) all contribute funding to the facility. The local forest ranger, Al Noren, is pleased to have improved access to an extraordinarily stout bristlecone he recently located. He nominates the specimen to the American Forestry Association's "Social Register of Big Trees" as the champion of the species.

All of this is yet unknown to Edmund when he gets into a quarrel with Doctor D. in late 1948. The immediate cause is Smiley, who Schulman judges unfit for research and teaching. Having a better candidate in mind, he wants to dismiss the newcomer. Douglass is appalled. He plans to promote the war veteran even before he finishes his master's degree. Feeling disrespected, Schulman demands greater authority in the lab. I'm not an undergrad anymore, he complains; I'm 40 years old. He types an on-the-record letter to Douglass. The director disapproves of the message, the tone, and particularly the date: November 25. The letter gives proof that Schulman treats Thanksgiving as a workday. The Brooklynite doesn't anticipate how his pro forma accuracy will offend the New Englander, who pays utmost attention to dates.

One year later, Edmund weds Alsie. Their nuptials appear in the local Jewish newspaper, for the community loves Alsie "like a daughter unto Israel."[8] She's fifteen years Edmund's senior, with a longer career at UA. Unfortunately for her, the Board of Regents has a rule against the employment of married couples. Alsie Schulman gets a one-year waiver while appealing her case. Andrew Ellicott Douglass writes a personal letter to the president. He vouches that Alsie knows Latin, and Latin provides the best education. She shows dedication in teaching; moreover, she keeps a refined and charming house. The president coolly defends the rule as anti-nepotistic.

Instead of a honeymoon, the newly married Dr. Schulman embarks on a four-month solo research trip to Patagonia thanks to a grant from the American Academy of Arts and Sciences. In his absence, Alsie—her career in doubt—sends Christmas greetings to Douglass on holiday stationery inside a card showing the Virgin Mary and the Divine Infant. In January, the president reverses his position, allowing Alsie to remain employed, albeit on annual contracts.

In Schulman's absence, Douglass broods on the future of the lab, his baby. He has no biological progeny. Doctor D. thinks his star student needs reeducation. "Notes on Edmund" becomes a manifesto with many working titles: "Directorship," "The American Code," "The American Humanity Code," "American Group Structure," "The American Way," "Apostle's Creed," "Ability Needs Character," "The Right to Lead."

Douglass is losing his grip, but his philosophy hardened years ago, as demonstrated by a 1925 speech to Phi Kappa Phi on American values and long-term thinking. We need to cultivate "time-consciousness" in young adults so they will take care of "super-human organisms" that can last longer than any individual. The most important of these organisms is the nation or race; churches and universities are vital, too. "Your University is a living thing; if you work for it, it will work for you; if you fight it, it will fight you." Unlike Nietzsche's Superman, Douglass's collegiate "Super-person" realizes that caring for humanity is working for God. The long-term problem in the United States, following mass immigration, is "race versus education." Proper education will wear down racial divisions and bind diverse populations with Christian love.[9]

A quarter century later, Douglass's ideas have become personalized around his lab. He decides he doesn't know his mentee, despite having mentored him for seventeen years. By working against Smiley—and, by extension, the lab itself—Schulman has revealed his incomplete Americanization. Growing up in Brooklyn, he lacked the opportunity to acquire values exemplified by collegiate sports. A director must behave sportingly. Such a gentleman derives standards of decency from Christian and Anglo traditions. Immigrants from despotic nations lack such humanity. The Jews have been most oppressed and thus deserve greatest consideration, but in return they must strive to adopt the American humanity code. The difference between a director and a dictator is capital C Character. Following the example of Christ, a director treats subordinates with faith and cooperation, not suspicion and competition. Brainiacs do not have the right to give orders just because they're brainy. Advancement by force is the Russian method of totalitarianism. Although immigration and war have eroded the American ideal, it can be rebuilt. As the Good Book says, Israel can be restored by a remnant.

It's 1949, the height of the postwar Red Scare; the year the Soviets test their first atomic bomb. The popular linkage between communism and Judaism holds strong in the early Cold War, and Tucson becomes more anti-Semitic. The town's WASP elites have formed a new, exclusive country club—a response to the Jewish population having grown by 1,000 percent (to roughly four thousand people) during the war. At UA, the admissions office still enforces its "New York" quota. The arch-conservative editor of the *Arizona Daily Star*, the morning paper, attracts the attention of the Anti-Defamation League. If we recognize the Jewish right of reconquest in Israel, he argues, how will we stop future Mexicans from reclaiming Arizona from our great-grandchildren?[10]

In February 1950, Douglass prepares the latest version of his screed and mails it to Schulman via the embassy in Chile. Schulman is coring alerces and araucarias in Patagonia, riding horseback in gaucho clothing, and having the time of his life. He poses for a photo on a boat in Puerto Montt, develops the film in a local shop, and titles the portrait "The Boss." Upon his return to Santiago, he opens the package from Doctor D. The manuscript inside bears the title "Studies in American Character."

Via Rio and New York on Pan Am, Schulman returns to an uncertain workplace in Tucson. Whatever he thinks of Douglass's diatribe, it doesn't stop him from appearing as the main speaker at a twenty-fifth-anniversary jubilee event for the Hebrew University. Constant research provides constant distraction. He corresponds with chemist Willard Libby at the University of Chicago, who recently announced the discovery of carbon 14 (radiocarbon) dating, a method validated with wood of known age, including samples from the LTRR. Edmund looks forward to extending his tree-ring database with a new three-year grant from the Office of Naval Research.

Instead of the code of the gentleman scholar, Schulman follows the code of the principal investigator. Every year, he's hustling for soft money, out of ambition and also necessity. Edmund's salary is $200 more than that of his wife, an adjunct instructor in a humanities department with the university's lowest salaries. He applies for a Guggenheim. His mentor misses the recommendation deadline and hastily posts a perfunctory note to the foundation, praising Schulman's intellect: "He is Jewish and has the high ability that occurs in that race."

In 1952, Douglass begins planning his retirement and sends his thoughts on succession to the president. The founding director describes a lab of three domains: archaeological (Smiley), hydrological (Schulman), and astronomical (Douglass). Mendaciously, he asserts that his mentee lacks astronomical expertise. If Schulman's promotion to directorship occurs too soon, warns Douglass, the lab could devolve. Doctor D. knows that dendroclimatology will in the long run benefit Arizona, but he believes that the "everyman" Arizonan—including the cowboy legislature—disdains that kind of research. By contrast, the link between tree rings and Southwestern ruins is political gold, something Schulman would understand if he were educable.

This summer, more than ever, Edmund needs to escape to the field, though it means leaving Alsie behind, again. By this point, he has measured hundreds of thousands of rings in the inland North American West—a semiarid bioregion he now understands as uniquely suited to dendroclimatology. Nowhere else are there so many ancient conifers with sensitive growth rings on adverse sites, with specimens distributed across latitude, longitude, and altitude, permitting wide and selective sampling. With data covering 850 years, he can identify multiple megadroughts, including an epic dry period in the late sixteenth century. Hoover Dam could not operate in such an event, he warns. He feels confident he will someday be able to test planetary theories of climatic change over the last millennium, and calculate statistical probabilities of regional droughts and floods. He just needs the right old trees.

On the steep southern slope of a glacial valley above Ketchum, Idaho, Schulman locates his first pine beyond the 1,000-year barrier—a limber pine. The gnarled tree he calls KET-3966 is a revelation. Previously, he's focused his attention on the dry lower borders of western forests. Now, he shifts to the upper tree line. An article in the magazine of the Office of Naval Research catches his eye. The founders of the White Mountain Research Station promote "California's laboratory above the sky" with a photograph of Crooked Creek Station framed by pines "believed by some to be older than the redwoods."[11] Schulman reaches out for information, which in turn leads him to the main believer, Ranger Al Noren, who is busy writing memos in support of a nature reserve around the champion tree he has come to call the Patriarch.

Schulman returns to Ketchum in September 1953 to fell his revelator pine—1,650 years old. Or, as he will later write in the passive voice:

"As a type, it was cut down . . . for detailed analysis."[12] This is the first time in history that a human purposefully kills a millennial tree knowing absolutely its age beforehand. Schulman's accomplice is Frits Went, a Dutch plant biologist at the California Institute of Technology. To his Caltech buddy, Schulman goes by "Ed." Huffing, they struggle up the scree. The only stable footings are roots, which seem to hold up the mountain. From their quarry's last stand, Ed and Frits see the ski runs of Sun Valley in the distance. On the long drive back to Pasadena, they detour to the White Mountains and core the Patriarch and other nearby bristlecones and limbers. These samples belong to the same age class as the cross section in the car trunk, but with more sensitive rings. Excited, Schulman returns to Crooked Creek in October, before the station shuts down for the winter.

For the new academic year, Schulman stays in California as a visiting professor at Caltech, where he mingles with some of the world's greatest scientific minds as well as a who's who of Jewish émigrés. The invitation comes from a lab that won an Atomic Energy Commission grant to study isotopes, including isotopic carbon in wood. Ed works alongside geochemists Harrison Brown and Samuel Epstein, two researchers who understand already that industrial emissions from coal and oil will at length result in planetary change. Evidence that tree rings are recording the anthropogenic forcing of carbon 12 in the atmosphere will be offered in 1955 by chemist Hans Suess, in a research letter on "modern wood."[13] Soon thereafter, Suess joins the Scripps Institution of Oceanography at UC San Diego, where his social network overlaps Schulman's. Suess coauthors an article—little noted at the time, later canonical—that refers to the fossil fuel economy as a "large scale geophysical experiment."[14]

In the short term, Suess's discovery of the changing ratio of C-12 (common), C-13 (rare), and C-14 (trace) necessitates a calibration of the radiocarbon dating method. After Scripps, the two universities to build C-14 labs are, not coincidentally, centers of archaeology: the University of Pennsylvania and the University of Arizona. Schulman knows his ultra-old pines will eventually be important in this endeavor. But he keeps his research notes close to his vest. First things first: he must select the best limber-bristlecone site in the Great Basin, then build a cross-dated chronology for that site.

In March 1954, he previews his quest in a research letter for *Science*. Brilliantly, he captures a botanical insight in a pithy title: "Longevity under Adversity."[15] In the journal of record for the US scientific community, he provides the first documentation of pines older than one millennium. He predicts that 2,000-year-old limbers and bristlecones will be found. The takeaway: little old trees deserve more scientific attention than big old trees. Schulman looks ahead to a comparative study of xerophytic conifers in the North American West, Patagonia, and Central Asia. In private, he begins to refer to *over*-overage conifers—his favorite data sources—as "Methuselah trees." They take him beyond the "BC barrier." Consciously or not, his biological maxim echoes a phrase used to describe the children of Israel: strength under adversity.

A couple of months later, working with lab assistant Wes Ferguson, Schulman cores a 3,100-year-old bristlecone in the eastern Great Basin—the second-oldest plant yet dated, second only to the "D-21" sequoia of his monumental mentor. But there are fewer Methuselahs overall in White Pine County, Nevada, and tree rings are less sensitive there than in Inyo County, California. Besides, Crooked Creek has bunkhouses, electricity, and an access road maintained by UC. A "siege of ill health" prevents a return visit to the Whites, unfortunately. Ed commiserates with Sam Epstein on the matter of ulcers. "As for me," he writes, "I've been trying out that old Chinese proverb: There are years in relaxing the vigilance."

Come fall, he's back in Tucson, teaching, counting rings, cross-dating, graphing, running statistical analyses. The tree-ring lab sits under the bleachers of the football stadium, facing west, with no air conditioning, so the heat can be unbearable, except at night. Smiley is now a permanent presence, having been appointed to the faculty—without a PhD—during Schulman's leave in Pasadena. Edmund keeps his head down, focused on his research. In January 1955, he asks UA's president to sponsor his NSF proposal to study millennium-long tree-ring histories.

While waiting on his grant application, Schulman begins to deal with the fallout of finding trees as old as sequoias. He receives letters asking for age dates and locations. He replies vaguely. Separately, he writes to Inyo National Forest, expressing concerns about tourists and souvenir collectors. When, in 1955, features on multimillennial bristlecones appear in the *Los Angeles Times*, *Sunset*, *Ford Times*, and *Natural History Magazine*, Schulman hedges again, saying he must analyze a mountain of detailed evidence before he can make a definitive announcement.

In January 1956, the NSF awards Schulman a three-year grant to further his investigation into longevity under adversity. At last, he has funding to hire his own full-time assistant. He sees the summit in his mind: a definitive tree-ring survey of his model species in its model location. He types a letter to *National Geographic*, pitching an article on "Methuselah pines." My study is comparable in scientific importance to Douglass's 1929 article on the dating of Puebloan ruins, he states matter-of-factly. An editor responds: The magazine might be interested, but only if your work could be presented in a way that appeals to 2.1 million laymen. If not, the magazine could make room for a "filler" piece called "The World's Oldest Trees" that covers sequoia, too. Readers like big old trees.

In June, Edmund embarks upon his life's most important fieldwork. He's tempting fate at high altitude: his doctor has ordered him to stay home. Alsie, who suffers from atrial fibrillation, chooses not to come along. I'll be out of touch for weeks, he writes his brother, after scolding him for keeping him in the dark about the latest family crisis. In an emergency, he says, you can probably reach me through the Navy station. "Big things are afoot."

ANCIENT BRISTLECONES

Schulman doesn't get carried away like Humboldt. He knows the Patriarch and its largish neighbors aren't that old—relatively speaking—just lucky. Their habitat, though harsh, is slightly better than that of others in the Whites. To find the oldest bristlecones, Schulman seeks out steeper, looser, drier slopes. He finds strange pines whose trunks grow as slabs instead of rounds. Sustained by a single strip of bark, such a tree grows horizontally more than vertically, and ever so slowly.

Its growth rings are infinitesimal. It takes all his strength to crank his borer through the perdurable wood.

In his first NSF summer, to his amazement and delight, Schulman "hits the pith" of three pines older than 4,000 years, all of them convenient to the military road. He names them, in order of age, Alpha, Beta, Gamma. To build an error-free cross-dated composite chronology for the locale—for in the idealized world of tree-ring science, there are absolute dates or no dates—Schulman decides he needs a complete slab. He chooses to sacrifice Alpha. He returns to LA, buys a two-man cross-cut saw, and brings along his teenage nephew and Frits Went to be the muscle. They drive in the dark to avoid overheating the Studebaker. The nephew sits in back while Ed and Frits talk science all night. In the glaring light of day, Ed shoots Kodaks of the cutting. Alpha's demise fails to make an impression on the sixteen-year-old, whose uncle conceals the weight of their deed.

In late September, once Schulman has counted and recounted every ring on the polished slab, the press office at the university announces, "UA Finds Oldest Living Thing." They say nothing about the thing being dead. Dr. Schulman has also, they crow, debunked the antiquity of El Árbol del Tule in Oaxaca. Newspapers across the United States pick up the press release, which includes a blurb on the utility of tree rings for meteorological "backcasting." It helps that 1956 is a bad drought year. The Office of the President, delighted with the publicity, congratulates Schulman personally.

Not everyone is happy that dwarf pines have pulled rank on giant sequoias. The *San Francisco Chronicle* runs a snarky, casually racist opinion. Ancient bristlecones look "terrible—half-dead, with little if any foliage, warped and writhing roots, and twisted, stubby branches." Schulman's bristlecone stand compares to a redwood forest "as an Indian shell mound does to the Cathedral of Cologne." At least it's still in California.[16]

The formal announcement of the discovery—again without acknowledgment of arboricide—appears simultaneously in Schulman's *Dendroclimatic Changes in Semiarid America*. This magnum opus, long in the making, contains analyses of one-third of a million rings. The bristlecone data is so new that it appears as "Appendix C: Millennia-Old Pine Trees Sampled in 1954 and 1955." In the acknowledgments to this technical volume, full of standard deviation calculations, correlation coefficients, and hand-drawn graphs, Edmund says his "indebtedness to Alsie" is "indeed great." He thanks her for seventeen years of discussing "so many research problems."

Publicity requests come thick and fast that fall. *National Geographic* is suddenly very interested. They'll send a photographer as soon as he drafts a story. In the meantime, Schulman makes the two-day drive back to the Whites to meet a cameraman from *Life*, the other leading pictorial magazine. He poses for a series of black-and-white shots. In one, he stands heroically, wearing a straw fedora, while screwing an extra-long increment borer into a hulk of a pine. In another, hat off, Schulman crouches at the remains of Pine Alpha, touching the stump with both hands, with the High Sierra in the background. Uncharacteristically, he looks happy. *Life* doesn't include either portrait in its brief article, "The Oldest Thing Alive," which reassures readers that cores can be taken without harming the tree. Only a "pinprick," Schulman tells Tucson's evening paper.[17]

Soon after his photo shoot, on the date 8 Cheshvan 5717, the Jewish newspaper in town runs a profile ("Meet Your Landsman") on Schulman and his find. This is the one publication that gets a partial admission from the professor—with a hint of remorse about the violation of something sacred. "There is something emotional to cutting down a tree you realize has lived more than 4,000 years," he says. "I remember feeling a pang of emotion when the saw bit through living core of the old bristlecone." The reporter notes this pine was old when Moses led the Israelites out of Egypt. Schulman, displaying his Americanist education, chooses a settler-colonial analogy: this pine looked over the Sierra during the Gold Rush and was there when the earliest Indian tribes roamed the valleys.[18]

In January 1957, *National Geographic* tries to remotivate Schulman with a $1,000 offer for a "personal-experience narrative" or "scientific detective story" at five thousand to seven thousand words. The next month, they press him for a target date. Schulman wants to delay publication—and postpone additional publicity—until the USFS has laid the groundwork for expanding the natural area. Not knowing the rationale, the editors think they're dealing with a procrastinator. Maybe they should send someone to Tucson to help him write something "humanized," as the "technical padlock" lies heavy on Dr. Schulman's publications.[19]

At semester's end, he responds to more pestering with a rough draft, "The World's Oldest Trees," and a request for an extension. He needs the upcoming field season to verify his cross-dating. His core samples have "phenomenal numbers of missing rings"—years in which trees didn't grow or grew partial rings on sections of trunks that have since eroded to nothingness. Fed up with the delay, one editor wants to relegate bristlecones to a black-and-white notice. The main editor keeps faith and gives Schulman an advance.

He can use the cash. His merit increase for documenting the world's oldest trees and landing on the front page of the *New York Times* is merely $400. Schulman doesn't know that Douglass gives Smiley a $1,000 raise, making him the highest-paid member of the lab. "Though we're poorer than a rattlesnake, we're more beautiful," Alsie tells her sister-in-law in Los Angeles, describing a recent splurge: a Frigidaire and matching GE washer in turquoise. "This end of the Schulman family is slow these days—and getting slower. Edmund undertakes too much all the time, works harder than he should, is continually under some pressure or another—all of which is not good for him."

His extra work that spring falls behind the scenes: he wants to ensure that all ancient bristlecone habitat in the Whites is federally owned and protected. He compiles a list of inholdings—old mining and timber claims—most of which have shifted to county ownership due to tax delinquency. Anyone can purchase "state tax lands," as demonstrated by two Anaheim businessmen motivated by Schulman's discovery. Within months of gaining title to a bristlecone parcel, they're making souvenir plaques, bookends, and lamp bases from the "oldest living thing" and collecting eroded pieces of "driftwood" for sale as modern home decor. Schulman anticipates he'll need such cross-datable pieces for research. Alarmed by data destruction, he reaches out to everyone he can think of—LA Water and Power, the Board of Supervisors of Inyo County, the county assessor, a state senator, all levels of the USFS. By the end of April, he has the ear of the chief in DC.[20]

His definitive survey begins midsummer, as soon as the road reopens. Schulman finds his ideal population on the southeastern side of the mountains, on steep slopes above a dry canyon. He names this locale the "Methuselah Walk," using an old-fashioned English term for a tract of forest under keep. Every mature tree is extraordinarily old. Within weeks, he finds a hundred alphas past 4,000 years; he hopes to cross the 5,000 barrier. For legal as well as scientific reasons, he needs to delimit the boundaries of this ancient population. The supervisor of Inyo National Forest has given him a topographic map and the freedom to mark upon it the boundaries of a future reserve that the chief has fast-tracked.

Schulman has more help than ever, thanks to the UC graduate students who pass through the bunkhouses of Crooked Creek in 1957. It's a thrilling, disquieting time to be atop the Great Basin, a military sacrifice zone. Every several days, in the brightest part of day, a brighter flash comes from the east—the Nevada Test Site—followed by the shaking of the earth. At night, around the campfire, the researchers talk nuclear physics, tree rings, and metaphysics. One graduate student will remember that Schulman shares his starry-eyed hope—out of character for a scientist known for exactitude—that the chemical basis of bristlecone longevity will be identified and extracted, and adapted for use as a life-extending "elixir."[21]

Mortal thoughts come too easily. Before driving up the mountain, Edmund confides in his brother, and his nephew overhears: "I'm walking on eggshells."[22] Dr. Cohen in Tucson has diagnosed him with arteriosclerosis and hypertension. He has difficulty breathing at high altitude. His ideal site is not ideally located—too far from the road. But he's tantalizingly close to a lifetime's worth of data. For some trees, the longest, cleanest cores aren't enough. He sacrifices Pine Beta, too, for the cause of understanding the slow-dying biology of the "pickaback," a morphological type he sees all over Methuselah Walk and nowhere else. Such trees have been shutting down for centuries, one root section, one bark strip at a time, all the way down to a single strip, only to restart in an undead sector where growth has been suppressed for a thousand years or more, producing a young strip of bark on the back of an ancient one. In a five-hour ordeal, Schulman and an assistant carry the cross-sectioned torso of Beta on a stretcher to the car.

In August, a *National Geographic* staff photographer arrives at the White Mountain Research Station. He takes a series of Kodachromes, including a posed picnic at the Patriarch, with Schulman in a flannel shirt and blue jeans playing the role of grandpa, smiling at a young couple and their son on a blanket. It's a wholesome Americana scene, complete with canned food in a Burgermeister box. The photographer also takes a portrait of Dr. Schulman beside his latest oldest tree. It's the face of a man who looks ten years older than his birth age.

Saying goodbye to the photographer, Schulman indicates he won't be able to write the article until late fall. He's lingering at Crooked Creek, then visiting Caltech. He's "right in the midst of some 'hot' stuff." In September, he returns the marked-up map to the forest supervisor and praises his field site's perfection. Not only will it benefit climatology—with thousands of multimillennial trees with sensitive rings—it will open new research into the genetics and ecology of longevity. So far, he's extracted 750,000 tree rings from Methuselah Walk—enough data for his next decade.

In November, *National Geographic* gives Schulman his first hard deadline, with repeated instructions: inject yourself into the story; include dialogue and first-person language; help your readers feel your astonishment. Right before Thanksgiving, Schulman submits a complete draft. In a clumsy attempt at readability, he opens with a personal anecdote—a tasteless joke about female beauty. Then he provides a historical overview, beginning with Humboldt, of scientists searching for the world's oldest tree; then he summarizes tree-ring science; then he describes bristlecones. The editors feel disappointment. The piece needs trimming, tightening, reorganizing. The "long, rambling dissertation" on dendrochronology must go. Schulman must "organize the plot." The climax should be the scientist entering the woods with a borer in his hand.

Over December, despite getting sick with the current bug, Schulman labors on literary revisions, undoubtedly with his wife's help. Over long-distance calls, the editors walk him through the process. The first typeset proofs arrive on December 20. Running up against a hard deadline, he cancels a planned vacation with Alsie during UA's Christmas recess.

On January 7, 1958, having received the second proofs, Schulman does a final phone consultation. The professor has proven to be educable. Playing up the drama, he describes his first trip to the Whites as serendipity—a "detour" to check on a "rumor." Ten years of searching for overage conifers becomes two summers. Various field assistants are reduced to one with a memorable name: Spade Cooley. The genre demands a eureka moment, and Schulman has invented one, complete with dialogue. The anecdote concerns Pine Gamma, the first tree "certified" past 4,000 years. To avoid confusion, Edmund has changed the tree's name to Pine Alpha. He doesn't mention the original Alpha.

Unlike all his previous work, he plays down climatology, suggesting that "underprivileged" and "worthless" bristlecones will be far more important for longevity studies. There is, Edmund writes, "something a little fantastic in the persistent ability of a 4,000-year-old tree to shut up shop almost everywhere throughout its stem in a very dry year, and faithfully reawaken to add many new cells in a favorable year." The senior trees have been "dying for two millenniums or more." How is this possible? "Maybe we cannot hope to find bristlecone pine trees very much older than those we have found already, for the days of the oldest studied are obviously numbered. But when research has been carried far enough in these Methuselah pines, perhaps their misshapen and battered stems will give us answers of great beauty."[23]

The next day, January 8, "Bristlecone Pine, Oldest Known Living Thing," gets released to the printer for the March issue. President and editor-in-chief Melville Bell Grosvenor reads the printed version and greenlights publication. That afternoon, *National Geographic* sends the first copy, with a check, by airmail to Tucson. Synchronously, at the University of Arizona, the author goes to the comptroller's office. The topic of discussion, as always, is money. During the meeting, he collapses into unconsciousness. An ambulance rushes him to the medical center, where his physician administers emergency care. Two hours later, all the days of Edmund Schulman are 49 years, and he dies.

Mom and Pop fly out from Los Angeles in time for the next-day funeral at Temple Emanu-El. The firstborn son of Sophie and Leon Schulman is laid to rest in Evergreen Cemetery, among the pioneer Jews of Tucson. The parents sit shiva long enough to see telegrams and letters arrive from admirers around the world. "The death of Dr. Schulman is a severe blow to the University of Arizona," says UA's president.[24] In town, the Jewish community, including the owner of the department store, raises funds for a memorial plaque in Edmund's name at the Hebrew University, and later a memorial scholarship for a US or Israeli student.

Schulman's tragic stroke leaves a leadership void at the laboratory. A distraught Doctor D., age 92, immediately asks to retire. In a memo to the president, he says Smiley could be director in the mechanical sense, though he lacks originality. Whatever has he done for tree-ring science? The old man has a better candidate in waiting, Bryant Bannister, who just needs to finish his PhD.

A few weeks go by, and Alsie writes to Edmund's surviving brother, Harry, and his wife. I can't eat or sleep, she reports—and yet she's teaching this semester. When her heart rate isn't dangerously low, she's having anxiety attacks. Please write to me, please sustain me, she says. I have so little family of my own. In the lockbox, she finds legal documents related to the land in Palestine, now Israel. They show that Edmund, after the war, had donated the family property to the Jewish National Fund. After probate, the paperwork goes to Harry, as does the first volume of Edmund's diary. Subsequent volumes will never be passed down.

Come spring, the USFS expands the natural area, renamed the Ancient Bristlecone Pine Forest Botanical Area, to encompass the new Edmund Schulman Memorial Grove. The agency is charged with protecting this resource for scientific study and public enjoyment. The federal government simultaneously announces a plan to buy out 1,360 acres of county-owned inholdings. At the signing ceremony in DC, the chief forester presents Melville Bell Grosvenor with a commemorative 4,000-year-old slab from one of three bristlecones that Schulman sacrificed for science. The LTRR owns a similar piece of Pine Beta, one that Schulman had been polishing to be a surprise gift for Alsie.

From Bishop, California, the forest supervisor writes to the tree-ring lab, inquiring about a dedication ceremony for the memorial grove. Smiley, in his temporary role as acting director, says there's no need: "All of us here feel that even Schulman would not have wanted it." Separately, Alsie writes to Bishop on "Mrs. Edmund Schulman" stationery. "Will there be a ceremony?" she asks. "I want to be there."

As per Smiley's request, the occasion is shelved.

SACRALIZATION AND CALIBRATION

The article "The Oldest Known Living Thing" is coy about the nature of the thing. The oldest is twofold—a species and an individual. *National Geographic* does not disclose the latter's location. Photographs have been doctored at Dr. Schulman's request. His pronouncement of the oldest known "right now" is definitive yet provisional. Were he alive in summer 1958, he'd be looking for something older than 4,600 years. But with so much ancient wood on the ground, he wouldn't need more data from living trees. Besides, for dendrochronologists, the category "oldest known" is less biological than epistemological—a subset of the oldest *knowable*: trees with uneroded piths that can be absolutely dated all the way to their first year of growth.

Conventionally, a formal name—a mark of personhood—would be conferred upon the latest oldest tree, but here the article stops short. The only named tree is "Pine Alpha," and it's not even the right wrong tree. The actual world record holder gets a silly nickname, a descriptive placeholder: "Great-Granddad Pickaback." The wording is the work of the copyeditor, who objected to Schulman's "Great-Gramps" in the draft. There's a telling dissonance between this informal Americanism and Schulman's biblical-cum-British place-name, "Methuselah Walk."

The foresters who redenominate the "walk" as "Schulman Memorial Grove" borrow a convention from Sequoia National Park, where tourists and rangers refer to populations of monumental trees as "groves." This nonscientific word is overgrown with cultural allusions. "Sacred grove" is the phrase used by English translators for a variety of consecrated places with trees in the ancient Mediterranean: the Greek *alsos*, the Roman *lucus* and *nemus*. Educated US Americans still know William Cullen Bryant's famous "Forest Hymn" (1824), which begins: "The groves were God's first temples."

Some journalists dutifully report that "Great-Granddad" is the patriarch of this grove. Other reporters prefer to describe a tree with no name. It's easy for casual readers to assume—and many do—that "Pine Alpha" is the Methuselah of Methuselah Walk. On their own, tourists turn this tree (the former Pine Gamma) into the initial destination. It's adjacent to the parking lot, it's named, and it's photogenic. The Forest Service lays down an easy one-mile loop, "The Discovery Trail."

Although not classically beautiful, ancient bristlecones are spaced widely enough to be composed in viewfinders as "lone trees"—a type of ecotonal landmark that entered the European imagination with Marco Polo's "l'arbre seul." In the nineteenth-century American West, settlers turned dozens of lone trees into trail markers, then tourist attractions. In the photographic era, the "1,000-Mile Tree" of Utah, the "Lone Pine" on Sentinel Dome above Yosemite Valley, and finally the "Lone Cypress" near Carmel-by-the-Sea created pictorial templates for lonely-looking bristlecones in the White Mountains.

Wes Ferguson, having taken over Schulman's NSF grant, wins a follow-up grant, and witnesses the transformation of a science field site into a scientific pilgrimage site. He's partly responsible. He starts to personify "Great-Gramps" as "Methuselah" after tourists keep asking him for directions to the eldest. The thing is becoming a being. In 1960, Ferguson shows rangers where to build a second, more strenuous loop trail so that visitors can walk right past Methuselah. The next year, the USFS employs Ferguson as seasonal custodian. In a series of memos, he documents tourists behaving badly: cairn building, trail cutting, cone collecting, bark carving. He objects that non-UA scientists without permits have cored and tagged "Schulman's trees." There are forty-two holes in Pine Alpha, and the tree-ring lab is responsible for only half of them. Ferguson recommends rerouting the trail because photographers have trampled the soil around this quadrimillennial. Soon a limb goes missing.

Tourists arrive in much larger numbers after 1961, when Inyo County crews eliminate the steepest grades and take the pavement nearly to the pullout. Automobile magazines publicize this road that leads drivers to the limits of biological time. To handle the seasonal crush, the USFS constructs an information center in 1964. The number one question asked by that summer's sixteen thousand visitors—including a high percentage of teachers and professors—is the expected one: Where's Methuselah?

In 1968, as annual visitation approaches fifty thousand, Ferguson writes an essay for *Science* subtitled "Science and Esthetics." In just one decade, "visits to bristlecone-pine localities took on the nature of pilgrimages," he writes. "And scientists were in the vanguard." He feels conflicted. The Schulman Grove is a world-class repository of radiocarbon material. A tourist's memento may hold the solution to a scientific problem. To safeguard the resource, the public must gain an education in tree-ring science.[25]

To complement his article, Ferguson writes a lengthy, know-it-all letter to the regional office of the USFS. Schulman Grove must have a full-time ranger to stop locals from carting off wood in their pickups, he urges. Secondary access roads should be gated and closed. Methuselah should never be marked; indeed, the trail should be rerouted away from it. The best practice would be guided walks only. This grove should be managed as a strict scientific reserve, not another multiple-use area.

Dr. Ferguson badly misjudges his audience. "We intellectuals who know all the answers will tell the masses what to do and when to do it," writes a forester in the margins. "Seems like I've heard that before, like every little dictator since time began." When the USFS issues a master plan for the Ancient Bristlecone Pine Forest, it ignores Ferguson's advice, except for hiring a seasonal ranger. Rangers place an identifying sign on Methuselah. The supervisor wants to give visitors the satisfaction of knowing they have seen the oldest known.

Summer of 1969 brings a "Bristlecone Pine Conference" at Schulman Grove, with attendance by the agency chief, a congressman, the president of UC Berkeley, and the director of the LTRR, Bryant Bannister. They debate suggestions by environmentalists that Congress or the Executive should designate a Bristlecone Pine National Monument to match California's latest federal reserve, Redwood National Park. The assembled experts agree that current management with a "low-key public sell" is preferable.

In 1972, the year *Reader's Digest* publicizes Schulman Grove, a researcher with the Institute of Forest Genetics notices a single cone growing on Methuselah. He obtains permission to harvest it to see if the organism remains viable after 4,700 years. He extracts 96 seeds and plants 36 of them. One hundred percent germinate—a "totally unexpected outcome."[26] The vitality of Methuselah makes TV news. After this round of attention, foresters wake up to the problem of their making. They take down the identifying sign only five years after putting it up.

After the record-breaking tree is honored with a personal name, its biogeographic population gets a species name to match. In the 1960s, a geophysicist in Boulder, Colorado—a Rhodes scholar who earned his bachelor's degree in astronomy from UA, who knew Douglass personally, and who bought a second home in Tucson—takes up tree-ring science as his hobby. He comes to believe that bristlecones in the arid Great Basin are distinct from bristlecones in the semiarid Rocky Mountains. He documents enough taxonomic evidence—mainly differences in resin ducts—to argue for a new classification, which he presents in 1970. Logically or sentimentally, or both, botanists quickly embrace his evocative binomial: *Pinus longaeva*.

Tourists of the 1960s who collectively sacralize Ancient Ones with their presence often encounter tree-ring scientists in the field. These men with increment borers are not botanists. The LTRR cannot advance Schulman's proposed investigation into longevity with the enigmatic collection of stumps, shoots, twigs, and needles left behind in his office. The collection goes to Ferguson, who focuses single-mindedly on core samples, with the goal of elongating and perfecting Schulman's cross-dated chronology—a master time series of cambial patterns for Methuselah Walk.

The exacting work suits Ferguson's obsessive nature. In extreme cases, his samples contain 750 rings per three-quarters inch. Counting them under a microscope isn't enough. A core sample along a given radius can be missing 5 percent of the tree's annual rings, meaning that a simple ring count can produce a 200-year error for a 4,000-year-old tree. There's the additional problem of "unknown rings"—dry years when sampled trees did not produce detectable rings. For example, 809 CE is absent from 95 percent of Ferguson's samples. In 1964, he brings to light yet another unknown: a new ring, the *real* 1498 CE, that requires renumbering the chronology. "Self inspection, based upon this incident, brings forth mixed emotions," he confides in his notes. "Are there more rings?"

Ferguson inherits material of great value to geochemists and archaeologists. The academic year of Schulman's death is the International Geophysical Year, when David Keeling of Suess's lab at Scripps starts making measurements atop Mauna Loa—the beginnings of the "Keeling Curve," a day-by-day, year-by-year record of C-12, C-13, and C-14 in the atmosphere. By 1965, there's enough data for the White House to release a preliminary report on the "vast geophysical experiment."[27] While Keeling works forward in time, Ferguson works backward. He begins distributing ten-gram, ten-year segments of bristlecone wood for use in radiocarbon labs at Scripps and Penn.

Within months of Schulman's death, the scientific community learns that the discrepancies between radiocarbon dates for Egyptian artifacts and their chronicled dates—a perplexing fact first noted by Willard Libby—are systematic anomalies present in wood around the world. Variously called kinks, warps, variations, wriggles, and finally wiggles, these little ups and downs of C-14 rumple the otherwise constant line of radioactive decay. This discovery necessitates a second round of calibration—a search for fixed points to add to the "Curve of Knowns," a phrase Libby will make famous in his Nobel Lecture in 1960.

By the end of the decade, Wes Ferguson has cross-dated *and* wigglematched Methuselah Walk past 7,000 years. The next step is site verification. Two newer members of the lab, Val LaMarche and Tom Harlan, select a second site in the Whites, higher up. From core samples taken in 1970, they develop a 5,000-year-long chronology that lines up with the Schulman-Ferguson chronology. Thus verified, the ancient pines of Methuselah Walk replace the regnal lists of Nile Valley pharaohs as the gold standard for radiocarbon calibration.

Much like Suess is Ferguson's key collaborator in geochemistry, Henry Michael, a scholar affiliated with the Penn Museum, is his primary link to anthropology. At Michael's request—accompanied by payments from Penn—Ferguson provides cross sections (cookies) of dead and previously felled bristlecones (Schulman's three, plus others cut by road and electrical crews) and various tiny chunks covering specific decades, samples that get consumed in the spectrometer. Michael and his supervisor, Elizabeth Ralph, are unconcerned with the origins of the "cosmic schwung" in C-14. They leave that to astrophysicists. Rather, they want to create a user-friendly calibration curve for reliably dating human remains and constructions.

By 1973, thanks to the painstaking work of Ferguson, Suess, Michael, and others, it becomes clear that radiocarbon dates initially adopted by anthropologists—with fanfare and commotion—were wrong by decades, even centuries. Ralph and Michael publish their improved curve in a museum newsletter, which quickly becomes the most photocopied manual in archaeological science.[28] The astonishing importance of bristlecones is suggested by Colin Renfrew's 1973 book, *Before Civilization: The Radiocarbon Revolution and Prehistoric Europe*. The first plate shows a skeletal pine: "Earth's oldest inhabitant." Renfrew reports that his colleagues in Britain have been known to mutter: "Why should I concern myself about this obscure California shrub?"[29]

Henry Michael wants the longest curve. Starting in the mid-1970s, he walks every square foot of Methuselah Walk, looking for wooden fragments that might be older than 8,000 years. His walking meditation for science is a welcome break from muggy, smoggy Philadelphia. He holds out hope for 10,000 years, a myriad, a semimystical metric. Michael looks high and low, including logjams in canyons and gravel quarries at canyon mouths. He arranges for radar equipment and tractors to dig for wood in alluvial fans. In 1983, his on-the-ground tenacity yields a lucky find: a splinter of sun-bleached pinewood with some 150 rings, radiocarbon-dated to 7070 BCE. This "floater" has no connection—not yet—to the chronology. By serendipity, Michael finds two more ultra-old pieces that summer.[30]

The quest for 10,000 becomes quixotic after 1984, when European tree-ring scientists catch up to their US counterparts. An international team announces a European oak chronology nearly as long as the Great Basin bristlecone chronology, with every expectation to surpass it. While a handful of Americans search for rare quantities of rarest wood in arid mountains, European teams sift through unlimited quantities of good-enough wood: oaks buried in anaerobic conditions in Irish bogs and the upper valleys of the Rhine and Danube.

Although radiocarbon gets the most funding, bristlecone research branches in multiple directions. In 1962, the year Doctor D. dies, Hal Fritts—Schulman's truest successor in the lab—begins using IBM punch cards to validate that tree-ring widths do indeed correlate to climate, albeit in some months and some years more than others. Unlike Schulman, Fritts has expertise in plant physiology, and he devises experiments to test his predecessor's assumptions that arid-zone conifers can be treated as weather instruments. He installs battery-powered dendrometers to measure growth in real time. He erects polyethylene tents around conifers and pumps in CO_2 to test the effect of atmospheric fertilization on ring widths. Fritts diagrams all the interconnected factors that can produce a narrow ring, including low soil moisture, rapid evaporation, increased respiration, and reduced photosynthesis. Even in the lab-like conditions of the White Mountains, individual pines respond to hyperlocal factors during the forty-five-day growing season, including slope aspect, stand density, and soil substrate. This fine-grained knowledge allows Fritts to better smooth bristlecone data for statistical analysis, growth modeling, chronology building, and climate reconstruction. He improves Schulman's methods of site selection and spatial networking of sites. In essence, though, he spends the mid-1960s confirming what Schulman knew from experience and intuition: tree rings from Methuselah Walk are exceptionally well suited for dendroclimatology.[31]

In 1974, while working on what will become the classic textbook in his field, Fritts arranges an international tree-ring conference—a first—bringing Americans and Europeans together in Tucson to set up a system for sharing data. First come punch cards, then magnetic tapes. The "Tucson format" for converting cambial indices into computerized spreadsheets becomes standard. With modest funding and crowdsourced volunteerism, Fritts oversees the International Tree-Ring Data Bank. In the 1980s, a lab member develops a free software suite in Fortran for analyzing sets of data in Tucson format.

Fritts does his best to avoid his testy, hard-drinking colleague, Val LaMarche. They barely speak, despite having overlapping research. La-Marche creatively fuses geology, climatology, and botany. He calculates erosional rates of mountain slopes based on the formation of bristlecone buttress roots. By mapping and dating relict wood on mountain peaks, he visualizes movements of pine populations, up and down, before and after the warm period in the mid-Holocene. Building on Schulman, La-Marche demonstrates that bristlecones at the upper forest border predominantly record a temperature signal, whereas those at the lower tree line record a precipitation signal, especially summer rainfall.

Most importantly, LaMarche interprets distorted tree rings from timberline specimens as "frost rings"—evidence of stress from abnormal cold. He speculates that some of these frost rings may represent planet-level climate discontinuities after supervolcanic eruptions. In 1984, he coauthors a *Nature* review of his vulcanological research. He uses bristlecone rings to give provisional dates to events of interest to geologists and anthropologists alike, including the Minoan eruption at the Aegean island Santorini/Thera.[32]

Separately, LaMarche argues in *Science* that the increase in bristlecone growth rates since the mid-nineteenth century—something noted by Schulman—results from rising CO_2 levels.[33] He speaks of "proxy information," "proxy records," and "proxy indicators," noting that scientists need more of them to understand how man is altering the climate.

Great Basin bristlecone pine, a wild-growing lab organism, contains an astonishing array of proxy data, not just ring widths and C-14 levels but also chemical compounds in the wood and in the variable-length needles, which persist on branches for decades as records of photosynthate gains and losses—records that can be autocorrelated with tree rings. LaMarche's insights, though brilliant, are upstaged by the scientific news about the oldest *non*living things datable to annual resolution: ice layers. As proxies for climate, polar ice cores take climatologists deep into the Pleistocene, beyond the temporal realm of bristlecone wood.

For the mid- to late Holocene, however, ancient pines remain unrivaled. With C-14 calibration largely achieved, and with "global warming" entering public consciousness, bristlecone studies go back to climatology. In 1986, an ecologist from Britain, Malcolm Hughes, succeeds Bryant Bannister as director of the LTRR, and reconstitutes the lab. Wes Ferguson, age 63, has died from a series of strokes related to a brain tumor. He leaves an unsorted collection of timewood—mixed with Schulman's samples—in the concrete warren under the bleachers. Within two years, LaMarche, the other great authority on *Pinus longaeva*, succumbs to heart failure. He dies without having shared his temperature reconstruction method.

After Ferguson and LaMarche pass away, their colleague Don Graybill takes over the Methuselah Walk project using a Department of Energy grant for carbon dioxide studies. After a few years, he becomes too frail to work at 10,000 feet; Tom Harlan goes in his place. Back at the lab, Graybill rechecks, rebuilds, and restandardizes Ferguson's chronology—originally made for the needs of radiocarbon scientists—to bring out climatic signals. Then, in 1993, the LTRR's streak of misfortune continues: cancer takes Graybill's life at age 51. The same year, Michael, age 81, ends his quest for ancient wood, after thirty summers in the field. His heart can't take the altitude anymore.

PROXY DATA

The passing of top bristlecone scholars is a crisis and an opportunity. Hoping to maintain the tree-ring lab's unique bristlecone legacy—and its near-exclusive relationship with Inyo National Forest—Malcolm Hughes hires Lisa Graumlich, an expert on foxtail pine, a closely related Sierran species. Graumlich is the lab's first woman faculty member. In graduate school, at UCLA, she had chosen foxtail because of its roadless habitat, meaning that she, as a backpacker, had the field to herself. UA scientists, middle-aged guys with trucks, had been territorial about the Whites. "The bristlecone pine world was about men owning things," she will recall later. "Research sites and samples were passed from mentor to mentee. There was something ritual about it. He begat him, and he begat him."[34] Upon arrival at UA, she inherits LaMarche's increment borers.

Hughes wins a big NSF grant, big enough to hire a full-time researcher, Matt Salzer. He's done temporary part-time underpaid work for the lab and impressed everyone with his technical skills. Salzer has experience with Rocky Mountain bristlecone pine. Hughes and Salzer go to the White Mountains with the goal of understanding accelerated growth. They set up a transect. It will take years to collect and analyze the data, but they will prove that temperature, not CO_2, is the controlling factor, and that this recent growth spurt—confined to the upper tree line—has no precedent in nearly four thousand years.

Meanwhile, Hughes and Fritts transfer operation of the International Tree-Ring Data Bank to the National Oceanic and Atmospheric Administration. By the 1990s, there's a century's worth of instrumental weather readings from the North American West—enough to verify the utility of western tree rings for climate reconstructions and to calibrate chronologies with observed inputs. NOAA serves as a central depository for climate data derived from lakebeds, lake terraces, and ice sheets. All this data production and archiving allows for a new subdiscipline: multiproxy analysis. Hughes, a tree-ring expert, begins collaborating with authorities on ice and corals, plus a young postdoc with mathematical training who studies atmospheric science. His name is Michael Mann.

In 1998, this research group, with Mann as lead author, publishes a multiproxy analysis of planet-scale temperature patterns and climate forcing over the past six centuries. The group concludes that greenhouse gases emerged as the dominant forcing agent in the twentieth century. This technical paper, published by *Nature* on Earth Day after a warm winter in the Global North, attracts considerable media attention. In the long run, the group hopes to extend their analyses beyond one thousand years. At the moment, though, this is barely possible, and only at the hemispheric level—North America, where bristlecones grow. In 1999, to little initial notice, they publish a letter (not a paper) full of conditionals, including the word *uncertainties* in the title. The letter includes a temperature reconstruction graph derived largely from bristlecone data, indicating a steep increase in temperatures after an extended period of relative stability.[35]

Two years later, 2001, the Intergovernmental Panel on Climate Change (IPCC) lifts Mann's temperature graph out of context into an executive summary for policymakers in its Third Assessment Report. The IPCC thus elevates the data visualization—soon known as the "hockey stick graph"—to iconic status, attracting the attention of climate science skeptics and climate change denialists. In July 2003, on the floor of the US Senate, the chair of the Committee on Environment and Public Works, Jim Inhofe (R-OK), says: "With all the hysteria, all the fear, all the phony science, could it be that manmade global warming is the greatest hoax ever perpetrated on the American people? I believe it is."[36] Inhofe likens the IPCC process to a Soviet purge and ridicules evidence of global warming coming from one set of tree rings from one mountain range in one hemisphere.

The manufactured controversy about the "hockey stick" peaks in 2005. An economist and a mining consultant coauthor a lengthy rebuttal to the graph, including a section on Great Basin bristlecone pine. Their paper is a combination of exaggeration, spuriousness, and legitimate minor critiques. The authors point out correctly that climate scientists don't fully understand the biology of this peculiar species. The oldest specimens are not "mystical antennae," they rightly say. They go off a cliff when they contend that bristlecone serves as the "dominant arbiter of world climate history," and that pine data must be expunged until basic problems in plant physiology are systematically worked out.[37]

Inhofe's ally in the House, a Republican from Texas who chairs the Energy and Commerce Committee, sends letters to Mann and Hughes and others, demanding they turn over emails and raw data. Denialists allege that contradictory tree-ring evidence has been "censored." Hoping to stop the political circus, the Democratic chairman of the House Committee on Science commissions the National Academy of Sciences to examine the data underlying the hockey stick. A panel of academy members upholds Mann's findings, with the caveat that uncertainties in the climatic record for the period before 1600 CE were not communicated as clearly as they could have been. Bristlecone pine data—in the absence of other multimillennial tree-ring data sets—slightly distorts the graph, but the synoptic picture is correct.[38] However, the politics of falsity continue, and Hughes and UA counsel fight a costly and time-consuming battle to prevent the disclosure of university emails. The lawyering demoralizes the lab.

In the Whites, far away from the field of politics, Tom Harlan continues Henry Michael's quest. After early retirement in 1999, Harlan receives an anonymous grant that allows him to revive the LTRR's chronology work. To save him from walking too much—he has arthritis and a bum knee—he takes walkie-talkie directions from volunteers who scan the slopes with binoculars. Harlan is looking for a single piece of wood to bridge the gap to form an 8,702-year-old chronology. "Schulman worked on that slope when there was no trail," Harlan says. "I'm not surprised that he had a heart attack a few months later."[39] Schulman, unlike his successors, never tagged his trees, so Harlan consults his predecessor's field notes. Harlan's team affixes aluminum tags with new alphanumeric identifiers, takes digital photos, and records GPS coordinates.

Back in Tucson, another volunteer attempts to bring order to all the bristlecone samples collected over decades. Due to multiple filing systems and poor curatorial practices, the situation is a mess. Of some nine thousand bristlecone cores, barely one-third have been mounted and inventoried, even fewer cross-dated. Certain important specimens exist only as plots on slips of graph paper, the originals having been misplaced. Unlike an ancient tree, each core sample looks virtually the same.

In 2007, on the seventieth anniversary of the LTRR, the University of Arizona receives a transformative gift from a donor, Agnese Nelms Haury—enough money to erect a stand-alone, state-of-the-art building for tree-ring scientists. Dedicated in 2013, it is everything the "temporary" lab was not: large, airy, light filled, climate controlled. In preparation for the move, the lab hires a curator. "The best part about working under the stadium was that six times a year, during home games, you could get nachos delivered to your office," he recalls dryly.[40] He's in charge of clearing out the old storage room—stacked so high and deep with wood that he must turn sideways to walk through it. Exposed overhead pipes from a men's bathroom run above the samples that calibrated the C-14 dating method. The mingled stench of stale sweat and fresh shit overwhelms the cologne of ancient conifers.

To adorn the foyer of the new building, UA installs a cross section of a giant sequoia, formerly housed at the Arizona State Museum, where A. E. Douglass once gave presentations on tree-ring science and Southwestern archaeology. The building isn't named for the founder of dendrochronology but for his golden boy, Bryant "Bear" Bannister, who oversaw the growth of the lab and who rectified its abysmal salaries. Everyone loves Bannister, who in turn remembers Doctor D. as "just the kindliest, most gentlemanly man you can imagine."[41]

The last person who might have spoken differently dies in 2008. She is Gladys Phillips, a close friend of Alsie Schulman, who preceded her in death by almost three decades. In the late 1940s, Phillips had joined the LTRR as office secretary, a position for which she was overqualified, given her degrees. She resigned after a falling out with Douglass in 1953, and became a writing and literature teacher at a Tucson high school. In 1958, she moved in with Alsie during her grievous transition to widowhood. Late in life, Phillips and her sister endow an undergraduate fund they call the Alsie French and Edmund Schulman Memorial Scholarship, eligible to promising students in either English or tree-ring science.

In the Bannister Building, researchers can now devise bristlecone research projects without ever having cored one or visited the White Mountains. Thanks to advances in microscopy and spectrometry, they can do much more than analyze tree-ring widths as proxies for precipitation and temperature. Additional climate signals may be read in cellular density, chemical composition, and isotopic signatures. Variations in the ratio of oxygen isotopes may relate to shifts in Pacific storm tracks. Analyses of calcium, manganese, and zinc may reveal new proxies.

The most exciting development—one that would gratify Douglass— relates to astrophysics. In 2012, a Japanese scientist documents in the growth rings of sugis an extra-solar cosmic ray event that caused a major intensification of C-14 in 774–775 CE. Once colleagues verify this C-14 spike in European oak and bristlecone pine chronologies, they go looking for more "excursions." "Spike-matching" joins "wiggle-matching" and "cross-dating" as techniques for linking trees temporally across space. In 2017, a research team documents an extraordinary excursion in global radiocarbon that lasted ten years—the effect of a cosmogenic event that reached Earth in 5480 BCE.[42]

Excited by Holocene "horizon points"—key years, temporal anchors— an investment banker turned Aegean historian funds an initiative at UA: the Center for Mediterranean Archaeology and the Environment. The immediate goal is a new-and-improved calibration of C-14 dating at the annual scale, compared to the decadal scale of the prior calibration. In 2020, the center provides a new estimated date for the supervolcanic eruption at Thera: 1560 BCE. The Curve of Knowns is getting more and more populated.

From an institutional perspective, the current LTRR is a consilience of the tripartite program—astronomical, archaeological, climatological— that Douglass charted in 1937, when hiring Schulman as his first employee. UA's goal to be the world leader in tree-ring science has been maintained, despite the vicissitudes of funding and a series of untimely deaths.

The current keeper of Schulman's research legacy is Matt Salzer. He has extended Ferguson's Methuselah Walk chronology to the brink of ten millennia using wiggle-matching. He has updated and extended LaMarche's research on tree line dynamics and frost rings, documenting volcanic forcing of climate in the era before fossil fuel. And he has refined Fritts's analyses of hyperlocal variations in bristlecone signals. Because of slope gradient (how steep), slope aspect (how shady), and patterns of air drainage, a single population of *Pinus longaeva* may contain a mix of signals—some individuals limited in growth by temperature, others by precipitation. By segregating according to microtopography, Salzer and his collaborators improve the signal-to-noise ratio and produce cleaner climate data. The relevance of their work cannot be denied: the twenty-first century brings a megadrought so severe that policymakers speak openly about Hoover Dam's future inoperability.

Like Dr. Schulman of the early 1940s, Dr. Salzer has an untenured, unsalaried position, meaning he relies on soft money from NSF grants. Bristlecones are data containers, he says, useful record keepers.[43] Like most tree-ring scientists, Salzer avoids metaphysical language, though he's artfully arranged on his shelves cross sections from some of the oldest conifers ever sectioned. It smells heavenly in his office. There he also looks after large-format original prints from *Life* showing Schulman and the quadrimillennial stump. Salzer himself would never cut a tree, not even a 200-year-old. Inevitably, people ask him about the "curse" of the bristlecone, a morbid joke among dendrochronologists. "Always thought it was stupid," Salzer tells the *New Yorker*. "Had second thoughts when I had to get that stent put in."[44]

Salzer serves as UA's contact person for media inquiries about world-record longevity. How old is Methuselah? Is there an older one out there? He has answers, though not exactly satisfying ones. He relates a story about his predecessor, Harlan, who died in 2013. It goes like this: Schulman runs out of time before dating all his cores. Decades later, Harlan examines one of these forgotten samples and uncovers proof of a tree older than Methuselah, older by centuries. Using Schulman's oblique but topographically precise field notes, Harlan eventually locates the pine—a perfectly healthy specimen. He vows never to tell another soul, given the damage inflicted upon Methuselah. Harlan's protectiveness competes with his desire to be known as the co-discoverer of the oldest known. On national television, he says: "Anonymity is its absolute best defense."[45] He takes his secret to his grave.

Salzer inherits the mystery timewood, whose innermost ring dates to 2798 BCE, making the pine 4,811 years old when Harlan passes away. No one in the lab thinks he fibbed, even though longevity records—both for people and for plants honored as persons—generate fallacies, hoaxes, and self-deceptions. After Harlan's death, the secret tree with no name exists between the known and the unknown—until Salzer finds it himself in 2019 using photocopies of Schulman's field notes.

In a deeper sense, the identity of the true oldest living bristlecone is simply unknowable. That's not just because no one has the time—or the funding, or the imperative—to do an exhaustive search throughout the roadless areas of the Great Basin. The effort would be futile. On most ancient bristlecones, the oldest wood has long ago been ablated, speck by speck, by desert winds.

In effect, the tree that Schulman dubbed "Great-Gramps" remains the popular champion, though its exact age is a matter of confusion, with different numbers listed in *National Geographic*, on the Forest Service's website, and by other sources presented as authoritative. The innermost rings on Schulman's core samples are extremely suppressed and partly eroded, making cross-dating difficult. The most accurate statement about Methuselah's age is equivocal: whereas Harlan entered 2490 BCE for the oldest ring, Salzer thinks 2555 BCE might be the actual year.

This cryptic ring is irrelevant to science. If premodern pilgrims once thought of ancient tree age in relation to an external object (*as old as the temple*), and if modern tourists think of it numerically (*exactly this many years above or below 1,000*), tree-ring scientists approach time differently. They think of age in relation to an internal object (*at least as old as the oldest sampled ring*) and also abstractly (*as old or as young as needed to bridge a gap in a chronology*). Dendrochronologists don't need to know the exact age of a tree to derive chronometric data from it. They only need to assign calendrical dates to whatever usable rings they extract. Through cross-dating or wiggle-matching, even a short sample can be elongated.

The instrumental advantage of *Pinus longaeva* is twofold: primarily the extraordinary temporal length of coreable samples, and secondarily the sensitivity of the rings to climatic signals. It almost seems miraculous that the longest-living individual plants on Earth have turned out to be perfect for Earth system science. Then again, Edmund Schulman, the historical figure who, because of his academic precarity, searched the hardest for arboreal longevity, selected for plants with the qualities of "Methuselah pines." He needed nonclonal organisms that record annual-resolution climate data and preserve that linear data in stable storage devices within cold, dry, archival environments. Bristlecone growth rings are natural media in a readable format.

Datafied under the code "PILO," *Pinus longaeva* now live in many places at once. They are mounted, sanded, and accessioned core samples in envelopes on shelves; slices in spectrometers; hand-drawn plots on graph paper; published tables in bound serials and online articles; numbers in computerized spreadsheets; smoothed data sets in the cloud; and, all the while, photosynthesizing organisms in complex relationships with fungal symbionts. In its current synchronous states of being, this lab-ready tree has become something new again, something postmodern. Ironically, plants that exist as hyperlocals and ultraterrestrials—rooted in rocks for four or five millennia—have become proxies for general circulation models at the planetary scale.

For nonscientists, the "real" Methuselah retains totemic power. Visitors to Schulman Memorial Grove want to see the oldest living thing known to science, four and one-half millennia old, knowing ahead of time they won't know when they've seen it. Their walk past the tree is an act of agnostic faith. No one seems dissatisfied when they return to the parking lot, having crossed the grove—what the Forest Service calls a "scientific resource"—off their bucket list.

Besides remoteness, and a single seasonal ranger who doesn't live on-site, there's nothing to protect this "grove"—nothing except the values that visitors bring with them, including stories of sacred groves. More than a few people cause soil and root damage in pursuit of lone-tree photo opportunities. The majority come in reverence, and leave in humility, believing they have experienced the numen of Methuselah, the aura of scientific discovery—and, increasingly, a foreboding. Whereas Edmund Schulman drove up the mountain without local prohibitions or premodern fears about felling elderflora, visitors in his tracks bring global anxieties. What would it signal about the planet if its oldest trees died? The postwar history of Schulman's over-overage conifers is the history of ancient trees in modern times, concentrated and accelerated.

VII.
LATEST OLDEST

LIFE-EXTENDING SITES — ARBORICULTURAL BEINGS — LIVING FOSSILS — CLONAL BEINGS — ARBORESCENT ETHICS

Since the documentation of bristlecone age, no one has found a species with older datable specimens. The Golden State remains for now the hotspot of longevity. However, scientists keep finding extremely old things around the globe. Elderflora grow in uninhabited spaces—swamps, escarpments, lava fields—and also woodlands and forests with extended histories of human habitation. In the Amazon, a region once imagined as timeless, long-lived trees are living artifacts of ancient gardeners. Newfound appreciation for the widespread occurrence of longevous plant life applies to genomes, too. In the 1990s, canyoneers in Australia chanced upon relict populations of *Wollemia*, an unknown "dinosaur" conifer. This was the greatest botanical disclosure since the 1940s, when Chinese scientists located living members of *Metasequoia*, a genus previously known only in fossil form. Both "living fossils" have been widely propagated. To lesser fame, people have identified other plant forms that propagate in perpetuity through clonal regeneration. A handful of colonial plants, thousands of years old, have become environmental symbols as quasi-individual quasi-trees. There is a boundary below which most people withhold attention and moral consideration, and for the time being that boundary is shrubby.

LIFE-EXTENDING SITES

Trees keep on living in place—until something happens, as something will, sooner or later, or later still. Some external force will end a woody plant's indefinite experience of placetime. In a biosphere dominated by *Homo sapiens*, a fire-starting and a tree-felling species, elderflora achieve longest life by being as remote as possible from the depredations of people, or as close as possible to their care. For a plant to become a perdurable, it must be pre-adapted to long living, and also fortunate in time and place.

To be precise, there are five main *placeways*—in addition to genetic pathways—to extreme longevity in lignophytes. One such place, as described in the opening chapters, is a shrine, churchyard, or temple compound. Cultivated plants live for centuries at such consecrated sites, within or adjacent to population centers, as seen throughout South Asia, Southeast Asia, East Asia, Ethiopia, the British Isles, and Mexico. Consecrated trees may achieve enormous sizes and irregular shapes, supported by props maintained by caretakers. Or they can be miniature. Bonsai trees are, in a sense, mobile sanctuaries that can be passed down in caretaking families for centuries.

Inversely, stunted trees live old lives in submarginal habitats: cold and dry, or hot and dry, or steep and exposed, or high altitude, or nutrient poor—or, in the case of Great Basin bristlecone pine, all the above. The White Mountains are rare for having whole populations of super-elders, but Edmund Schulman's maxim "longevity under adversity" applies to many lands. In China, Vietnam, Pakistan, and Italy, tree-ring scientists have documented millennial junipers, cypresses, and pines hanging on in sites too harsh for competition, including human competition in the form of agriculture or commercial development.

A striking example of this second placeway occurs in Canada with northern white cedar, a familiar Great Lakes plant. Historically, this species was the source material for cedar-strip canoes of First Nations and split-rail fences of British settlers. White cedar grows a hundred feet tall in Ontario—or used to, before colonizers cleared the forest. They had no reason to cut the cedars that grew atop the Niagara Escarpment, the craggy ridgeline that runs from Lake Huron's Bruce Peninsula to the world-famous falls that separate Lake Erie from Lake Ontario. Never coveted by farmers, the escarpment attracted aggregate mining companies that served Toronto's postwar growth and weekend recreationists escaping that growth.

In the 1980s, Doug Larson of the University of Guelph started a cliff ecology lab. His initial project related to recreation management on the escarpment, comparing forest structure in private quarries against that in public parks. The latter had no understory: hikers and climbers had trampled the plants beneath the northern white cedars. To understand "wood productivity value"—biomass divided by years—Larson and his collaborators started taking core samples. They were stunned by the number of rings, higher than expected by an order of magnitude: five hundred instead of fifty. "We all assumed our European ancestors had nuked the forest vegetation," Larson recalled. "We just assumed, like everyone else, that starting around 1850, the place was a blank slate."[1]

Years into his research, Larson's curiosity turned from the moderate-sized conifers on the forested plateau to the small, deformed ones growing in isolation from the cliff itself. Rooted in cracks or caves, suspended from tiny ledges, these trees appeared aerial, and sometimes held the aeries of vultures. Their progenitor seeds had lodged in place long ago due to swirling winds or caching rodents. To core such a liminal tree required rock-climbing equipment. From some of their perches, scientists could see the CN Tower on the horizon.

By the late 1990s, Larson's lab had processed samples from hundreds of cliff specimens and documented ten white cedars beyond 1,000 years. This quorum received names, including "The Hunchback" and "The Ancient One" (date of birth: 688 CE). Canadian media gave extensive coverage to the news, which surprised Torontonians as well as big-tree old-growth activists in British Columbia. Larson's specimens almost immediately went from science to heritage. People worried about the impact of rock climbers on Canada's current senior trees and the threat of illegal bonsai collectors on up-and-coming elders. Once again, curiosity, care, negligence, and despoiling commingled.

A variation of submarginality occurs in ecotones—transitional habitats—where conditions shift from wetter to drier, with feast-or-famine variability. Some plants specialize in living on this edge. For example, blue oak can endure up to five centuries in the woodlands where California's Central Valley meets the lower flanks of the Sierra Nevada and Coast Ranges. These drought-resistant old-growth oaks contain informational rings that register California's off-and-on "atmospheric rivers"—trans-Pacific bands of dense water vapor. In the contemporary megadrought, though, blue oak seedling recruitment has suffered. One severely dry year, 2014, did a number on mature specimens, causing diebacks and deaths. "Plants we thought were probably more resilient to the drought weren't," said a UC Berkeley researcher. "We saw them pushed to their limits."[2]

Similarly, on lower tree lines in the Basin and Range geological province, piñon pines have lived long, and junipers longer.[3] Because the "PJ woodland" is neither scrubland nor high-canopy forest, it resists conventional aesthetic categories. When piñons and junipers expire, their deaths don't attract the same media attention as do lone trees or tall trees. In New Mexico, piñons massively died back in the early twenty-first century as a result of drought and bark beetles. In Utah and other western US states, ranchers and land managers continue to "chain" junipers—kill them by dragging a heavy anchor chain between two bulldozers—thus encouraging the growth of annual grasses grazed by beef cattle.

Farther east, in the Cross Timbers, the transition zone between the treeless Southern Plains and the forested Ozarks, post oaks and red cedars grow, barely, year after year, twisting and turning into handsome knobby oldsters. On bluffs and rocky outcrops, trees of three hundred to five hundred years are still common. These plants survived the onslaught of the United States of America. After visiting in 1832, Washington Irving famously wrote of the "mortal toil, and the vexations of flesh and spirit" of traveling through "forests of cast iron."[4] In the same decade, the US government forced Cherokee, Choctaw, Chickasaw, Creek, and Seminole populations to abandon their verdant homelands for reservations in the Cross Timbers region of Oklahoma. Compounding the trauma of the Trail of Tears, the government later broke up and sold off reservation lands. Forced privatization led to habitat fragmentation. Today, in Oklahoma's east, less than 10 percent of old growth abides, though that amounts to tens of thousands of acres—patches of oak trees shunned by historic sawlog producers as overly short and stout. In these relict patches, the soil community—grasses and lichen—stayed intact, having been spared the impacts of grazing.

Since around 2000, dendrochronologist Dave Stahle has tried his best to bring attention to the oldest and least-disturbed old growth in the nation's midland. He's organized the Ancient Cross Timbers Consortium with support from his employer, the University of Arkansas. He's done personal outreach. "I've never met a landowner who wasn't excited to learn they had a tree older than American democracy," Stahle told me.[5] Thanks to Stahle and comrades, a few preserves have come into being.

The larger picture, with fracking and pulp logging, looks grim: "Owners are degrading the land. It's being assaulted by private development." Bit by bit, old growth gets eaten away, without any activists tree-sitting or chaining themselves to trunks. Fayetteville is not Berkeley. Besides, half-millennial half-size post oaks lack the romance of California's elder-flora, though 500 years is extremely old by the global standard of oaks.

The most spectacular longevity report of late—the eldest North American life-forms east of the Mississippi—represents the third placeway: productive environments where big old growers dominate. Most such conifer zones were logged long ago, but stands persist in rugged coastal highlands (like British Columbia) and remote coastal lowlands like the bogs, bayous, and blackwater rivers of the South. This is prime habitat for bald cypress, a species closely related to Mexico's ahuehuete. Counter-intuitively, given its swampy habitat, bald cypress has scientific utility as a gauge of precipitation. Its tree rings are extremely sensitive to that one climatic signal.

In the mid-1980s, Stahle began sampling a large population of bald cypress located in the backchannels of the Black River in Sampson County, North Carolina. A local botanist, Julie Moore, guided him there by canoe. Many of these stocky trees had what Stahle called the "gnarl factor." In local parlance, they "got shook." Their deformed tops resulted from regrowth after decapitations from hurricanes. Individuals have stood their ground with the help of knobby support roots that look like miniature aboveground trunks, with moss instead of leaves. These non-feeding roots function as cooperative infrastructure between individuals. The forest holds itself up together.

Each canoe trip, Stahle cored older trees, and eventually realized the Black River hosted one of the planet's greatest old-growth stands, with thousands of millennials. It rivaled Schulman Grove. Like Edmund Schulman, Stahle reached out to local politicians and land managers, and gave public talks. The state legislature considered purchasing the backwoods—near-pristine habitat in a region known for hog factories and timber plantations—to form a state park. At the end of the day, the Nature Conservancy made a deal. The next year, 2019, Stahle announced his documentation of a living cypress (tagged as BLK-227) more than 2,600 years old—centuries older than the previous record holder, which locals had started calling "Methuselah."

The fourth placeway to oldness is less photogenic: productive environments where dominant young growth gives cover to subdominant old growth. A kind of "sweet adversity" (to paraphrase Shakespeare) happens here. A tree may grow quickly in its early years, taking advantage of a windfall that opens a window to the sun. Then, faster- or taller-growing trees may overtake it, casting shade, and slowing down growth. A time-tested alternative to fast living is slow dying. Although the subdominant "suffers" in terms of energy, it "thrives" in terms of longevity.

A perfect example is black gum, a broad-leaved angiosperm of eastern North America. Although it occurs widely, it can be hard to spot because of its low population density. Slow growing by nature, black gum tolerates shade, fire, drought, and flooding. The oldest occur near wetlands. Loggers shunned the species for its combination of hardness and hollowness. Lumbered specimens preserved poorly. Taken on its own terms, black gum's capacity to live nearly seven hundred years in squelchy soils of New England is phenomenal. One thousand seems unfairly and arbitrarily high as a standard for timefulness in an environment that gets nor'easters and hurricanes. In such a high-energy climate, trees in the midstory last longer precisely because they're not on top. They avoid the worst of the wind, ice, and snow that regularly shreds and topples the overstory.

Unlike millennial isolates (e.g., stunted white cedars on Niagara cliffs) or millennial dominants (e.g., redwoods on the North Coast of California), there's no globalized aesthetic for not-quite-millennial subdominant medium-sized growers. They're hard to see, hard to photograph. They don't appear in calendars and coffee-table books alongside pictures of lone pines on mountaintops. Their connection to the *longue durée* is subtle, and their rings are challenging to interpret. The discipline of tree-ring science, as established at the University of Arizona, historically gave preference to easy-to-read rings from overage conifers on semiarid mountain gradients. To this day, angiosperms, particularly tropical angiosperms, remain understudied. Not coincidentally, amateur naturalists located many of the oldest trees in the Northeast.[6]

Old Ones can be found everywhere, if people take time to look.

ARBORICULTURAL BEINGS

The fifth and final placeway comes back to human care. In woodlands, trees can be planted, or selected by humans, and tended for centuries. Sometimes these semi-domesticated ancients end up outlasting the cultural groups that nurtured them into being, as shown by Mediterranean and Amazonian examples.

Sweet chestnut is the only native chestnut in Europe. The spiky burr (cupule) contains a smooth nut (calybium) with high nutrition value. That being so, European chestnut arboriculture developed surprisingly late, millennia after olives. Late-antique Romans ate chestnuts, though not typically as a staple. Aristocrats shunned the fruit as lowly food for rustic shepherds. They preferred cereals—and had a labor system to provide both bread and circuses. Romans valued chestnut trees for timber more than food. When coppiced, or cut to a stool, a chestnut will typically resprout from the root with dozens of stems of high-quality wood in convenient size. The region of Insubria (today's Lombardy) was the Roman center of chestnut pole production, with rivers from the Alps allowing water transport to the Po Plain.

The sweet chestnut didn't become the "bread tree" until the early medieval (Carolingian) period. It was the perfect plant for changing times. While the western Roman Empire had existed, rural peoples could produce grapes and grains for export to urban centers. After it fell apart, there were fewer laborers as well as consumers. Economies had to become localized and self-sufficient. Groves of chestnuts required little labor compared to vineyards and wheat fields, and they thrived in hilly topographies unsuited for cereal crops. Thus, the "decline and fall" of the empire looked vibrant from sweet chestnut's point of view. With human help, the species expanded all over the Italian peninsula, and throughout the western Mediterranean, from the ninth century onward. The same post-Roman depopulation that created conditions for rewilded forests in Europe—the deep dark woods of folklore—produced humanized woodlands, too, with chestnuts as the dominant species. The fruits of this biocultural landscape helped to sustain regions such as Campania and Lombardy until the revival of coastal trade around the first millennium CE, at which point smoked chestnuts themselves were a tradable commodity.

The next eight centuries comprised the golden age of Italian chestnuts, as people perfected techniques of breeding, grafting, pruning, coppicing, and pollarding. Tasty varietals could be grafted onto wild rootstock, then carefully pruned each year in a kind of slow-motion botanical sculpting. Other chestnuts were coppiced on a long-term rotational basis for splints, stakes, staves—and charcoal. Other fruit trees were pollarded, or selectively delimbed, to encourage both wood and fruit production. Stewards encouraged cycles: feces from goats and sheep became fertilizer for the trees; leaves from trees became litter for stables; and discarded cupules became additional fertilizer for cereals intercropped between trees. Even dying chestnuts could be useful as sources of tannic acid for leather making. A well-managed chestnut woodland was sustainable centuries before Europeans invented the idea of sustainability and practices of forestry.

Year after year, century after century, chestnuts reliably provided for humans and their animals.

The one hectic time in the chestnut calendar came in fall, when women harvested, boiled, roasted, and milled the nuts. With a bountiful crop, peasants would use the surplus to fatten their porkers before slaughtering them and salting or smoking them, producing hams renowned for flavor. Renaissance epicureans and modern connoisseurs inherited from classical authors a classist disdain for raw chestnuts, yet chestnut-finished pork became a delicacy for the rich—the kind of thing you would find in restaurants, once restaurants became fixtures in the new urban centers of the eighteenth and nineteenth centuries.

The menu of a nouveau Parisian restaurant, with ingredients from around the world, represented a new kind of global capitalism. Changes in food production affected rural peasants and urban proletarians alike. Agricultural regions of Italy became major exporters of people in the nineteenth century. Many emigrants left behind chestnut groves when they sailed for manufacturing cities such as Boston, New York, and Philadelphia. In these unfamiliar settings, Italians could not make chestnut polenta and pasta, so they purchased cornmeal and factory-made wheat noodles.

The rise and fall of chestnuts happened forcefully in Corsica, an island conquered and reconquered over the centuries. During the early period of Genoese control in the late sixteenth and early seventeenth centuries, Italians tried to "civilize" Corsicans by instituting compulsory chestnut cultivation. After a period of resistance, Corsicans adapted to silvopastoralism and made it their own, particularly in the northeast region, La Castagniccia (Chestnutland). When French subjugators entered the incipient Corsican nation in the late eighteenth century, nationalist resisters drew sustenance and symbolism from their chestnut economy. Their biocultural woodland became politicized. The House of Bonaparte fought to modernize Corsica (birthplace of Napoleon) by breaking down the communalism and self-sufficiency that chestnuts provided mountain people. The French looked down upon the silvopastoral economy as premodern, backward, and lazy.

French modernizers won in the end. In the late nineteenth century, once Corsica was integrated into the coal-powered industrial economy with steamers and railroads, labor dislocations hit the island. Unable to compete with products from the mainland and Algeria, Corsican agriculturists and woodworkers left the island en masse—an outmigration similar in scale to the Irish diaspora. Whole villages and their chestnut groves were abandoned. Many absentee owners sold their derelict trees to tannic acid producers. Over the first half of the twentieth century, one-time extraction replaced a prior system of sustainable production.

The decline of chestnut culture in Corsica and other western European regions had causes beyond industrial capitalism and associated rural depopulation. Microbiota acted as intensifiers. The same kind of global exchange that diversified and enriched Mediterranean cuisine with tomatoes and potatoes brought unwelcome things. "Ink disease"—the consequence of a species of water mold—arrived in Europe in the eighteenth or early nineteenth century, causing root and collar rot in chestnuts. A century later came chestnut blight, a pathogen that had annihilated the mighty chestnuts of eastern North America. After arriving in Genoa in 1938, the blight spread throughout Italy, then France and Spain. People assumed the worst outcome before something unexpected happened— the appearance of hypovirulence, or a virus that attacked the fungal pathogen. The hypovirulence transmitted quicky and widely enough to prevent complete devastation, an example of all-natural biological control. More recently, when the Asian chestnut gall wasp appeared in Italy, authorities responded with classic biological control. They introduced the parasitoid wasp from Asia that had evolved to use chestnut gall wasp as the host for its larvae.

Cankered but not killed by blight, old chestnuts have survived in great enough numbers to permit a partial revival of foodways. Corsicans no longer fight for independence, but they can assert difference from the French mainland through chestnuts, while attracting mainland tourists with chestnut festivals. In Italy, several varietals of chestnuts are now covered by EU protections of denomination of origin or geographical indication, much like "Prosciutto di Parma" is a legally protected chestnut-finished ham. The transhumance that defined silvopastoralism has largely vanished, though. People don't need sheep and goats to top-dress the soil when they buy fertilizer by the bag.

Overall, far fewer chestnuts exist today in the western Mediterranean than existed in the late nineteenth century. Land abandonment has favored oaks and pines; former farms have been rewilded through reforestation. Sweet chestnut woodlands require the active presence of caretakers, so their continuing oldness is a function of the arts of rejuvenation. Neither wild nor fully domesticated, the European chestnut beautifully embodies the ugly phrase "anthropogenic impact." The challenge for the next Europeans will be to create, with plants as their partners, new landscapes and economies that might last as long as chestnut culture did after the fall of Rome. If not the end of an empire, or the world, staggering changes lie ahead.

An equally remarkable example of long-term mutualism can be found in a forest often imagined as wilderness: the Amazon.

Before the arrival of humans, the forest here rarely burned. As Paleo-Indians transitioned from foraging to mixed farming, they directed flames to make clearings in the forest. They used ashes, combined with ceramic, compost, and feces, to prepare the soil for their annual crops. To complement their clearings, they planted and tended long-lived beings, choosing from the incredible assortment of native trees the species that benefited humans most. Although the tropical rain forest provided an abundance of edible plants and animals, the environment was not conducive to food storage, making arboreal food patches vital for security. Numerous Amazonian tribes still practice this kind of agrosilviculture; it is at least 8,000 years old.

In the colonial period, the Europeans who claimed Amazonia used narrow definitions of agriculture and domestication, and expected to find neither. Their limited vision confirmed their expectations. They misperceived Amerindians as hunter-gatherers—people without history, adrift in a sea of trees. This racist interpretation was facilitated by mass mortality. The Amazon appeared wild to missionaries, settlers, and ethnographers partly because Old World diseases had depopulated large areas of the forest before the newcomers arrived.

Within Western academia, the recognition of Amazonia as a region rich with human influence came shockingly late, long after its rain forest had became a symbol of Eden and global endangerment. In the 1990s and early 2000s, geographers, anthropologists, and historians engaged in a debate about the so-called pristine myth, part of a larger discussion on the "trouble with wilderness." At the end of it all, the intellectual consensus came full circle: experts described the precolonial Amazon as a "manufactured landscape," an "anthropogenic forest," and an independent center of domestication, complete with "garden cities."

Forest demographers have tested this hypothesis by looking at the composition and age structure of the current forest, using ground and aerial surveys to map the overstory. They have inferred historic manipulation by assessing the distribution and abundance of useful tree species in dominant positions. The picture that has emerged is hybrid: a "natural" forest thoroughly interspersed with patches of "anthropogenic" woodland in which specific tree species achieve "hyperdominance."

One of these species is Brazil nut—called *castanha* in Brazil, after the Portuguese word for sweet chestnut. It can reach heights up to 200 feet and ages up to 500 years. A specimen takes decades to reach reproductive age, then starts dropping large, heavy capsules. The fatty, protein-rich "nuts" encased inside are technically seeds. Very few creatures can open the woody capsules—large rodents, monkeys, humans. Some scientists speculate that the tree is anachronistic, because the gomphotheres (elephantine megafauna) that presumably dispersed the seeds went extinct some 10,000 years ago.

Today, Brazil nuts generally appear in well-spaced population clusters. This pattern does not fit models of random distribution. Of the estimated sixteen thousand tree species in the Amazon, castanha is one of a handful that is wildly overrepresented. Others include maripa palm, rubber tree, and cocoa tree—all similarly useful to humans. Whether or not Paleo-Indians and their descendants managed the entire Amazonian forest, they shaped sectors of it by creating and managing stands of trees that provided food perennially, supplementing annual crops such as squash and cassava. All the evidence about Brazil nut—including the near-uniform genetic composition of many stands—suggests that humans have for millennia been its primary dispersal agent. It is a "gap dependent" species, meaning it requires disturbances for establishment, and sunlight for peak seed production—and humans are nothing if not adept at creating gaps and disturbances. The species also favors "dark earth"— the ancient soil legacy of slash-and-char techniques. On protected bluffs above rivers, where castanha trees rise from rich black soil, archaeologists have located funerary urns, rock mounds, geoglyphs (trench-like geometric earthworks), and other remains of fire-tending forest stewards.

Tree-ring evidence indicates that most of the big old Brazil nut trees in Amazonia date to the sixteenth and seventeenth centuries. The age structure is top heavy, with a shortage of specimens in early maturity. The missing stands of trees represent missing cohorts of people. The period Europeans call the Enlightenment was a devastating time of depopulation for Amerindians. Those who planted the current oldest trees managed to think ahead in the long term, even as the world they knew came to an end.

With opening of the Trans-Amazonian Highway in 1972, a new period of destabilization arrived. The Brazilian state has vacillated between establishing Indigenous reserves and encouraging the incursions of loggers and miners who would murder local Indians, burn their forests, and claim the spoils as commercial property. In the context of deforestation and climate change, the future of the Brazil nut as a cultural and ecological cornerstone looks uncertain. In Peru and Bolivia, governments have experimented with contracts and concessions to incentivize nut collection and to deincentivize illegal logging. From December to March, collectors go into the forest with machetes and sacks, walking ancient trails, hunting for husky seedpods half-buried in the mud. Because Brazil nuts are a valuable export commodity—far more valuable today than tapped rubber—Indigenous and non-Indigenous peoples can engage in "conservation-through-use" by harvesting this "non-timber forest product" (to use the clinical terms of the UN). Various cooperatives attempt to capture new wealth through an agroforestry system developed by Natives millennia ago.

According to some climate models, the habitat suitable for Brazil nuts may expand, provided that enough Amazonians see value in anthropogenic forests. Right now, many prioritize clearing, ranching, and soy growing—all for fattening beef. In terms of sustainability, the food trees of the rain forest put cattle to shame. They are storers of greenhouse gas, not emitters; supporters of biodiversity, not destroyers.

LIVING FOSSILS

The evolutionary life of a tree species begins and ends in hyperlocal places—an initial recruitment ground and a final refugium. Because of competition and the vicissitudes of glacial periods, species have been "on the run" for tens of millions of years, living through multiple planets in the process. In refugia, the duration of finality can span deep time. A single island or an island mountain range can be good enough for long-term persistence if the habitat is sufficiently large, with enough topographic relief, to allow for local microclimates within global climatic fluctuations. New Caledonia and the "island of California" are two centers of conifer endemism—the quality of being restricted in place. Humans can hasten the extinction of endemic species, or they can provide them evolutionary life extension by releasing them from their refugia or domesticating them. Monterey pine, a previous isolate now grown commercially throughout the temperate zone, exemplifies one kind of human-assisted nonfinality.

A few species avoided extinction on their own for so long they become "living fossils." This phrase comes from Darwin and describes "remnants of a once preponderant order" that persist in confined areas and exhibit "anomalous" morphology that suggests an evolutionary dead end. It is a literary descriptor rather than a taxonomic classification—a way to imagine the contingencies of natural selection over deep time. Among flora, the iconic example is *Ginkgo biloba*, a plant with incomparable foliage that goes back almost 300 million years. It lives out of time, with no surviving kin in its genus, family, order, class, or division. From mountain refugia in central China, modern people spread ginkgoes throughout East Asia and the wider world, as edibles and later as ornamentals.

In the twentieth century, in central China and then southeast Australia, two more rare, endemic, anachronistic conifers—genera with one species each—burst into world view.

The first was *Metasequoia*. Its Linnaean name came into being in 1941 thanks to a Japanese paleobotanist who realized that a common type of fossil had been misassigned to the genus *Sequoia*. As luck would have it, a couple of years later, in China, a government forester took samples from an unidentified deciduous conifer in Moudao, a remote village in the Three Gorges region.[7] Scientific knowledge about the interior provinces increased rapidly during the Second Sino-Japanese War, when Chiang Kai-shek made Chongqing the provisional capital. As part of the evacuation from Nanjing, the government relocated China's first modern university, which included an arboretum and herbarium. Most of the plants and plant samples didn't make it. In the new herbarium, the samples from Moudao stood out and prompted a dendrologist to dispatch a graduate student to obtain additional cones and needles. Finally, in 1946, a US-trained botanist in Beijing matched the second samples to the recently renamed fossil genus *Metasequoia*, and announced the discovery in a Chinese journal.

The author, Hsen-Hsu Hu, immediately forwarded the news to his friend Ralph Chaney, a paleobotanist at UC Berkeley, and to his mentor, Elmer Merrill, director of Harvard's Arnold Arboretum. Hu had been the first Chinese botanist to receive a doctorate from Harvard, and Merrill relied on him and other Asian alumni to expand the arboretum's East Asia collection. When Merrill learned about a "forest giant just on the edge of extinction!" he boasted of his ability to "bring home the bacon" without mounting an Asiatic expedition.[8] Thanks to a donor, Merrill had a special fund for Chinese collecting. He cut a $250 check for Hu, supplemented by $25 from Chaney (for a total gift equivalent to 10 *million* yuan in war-torn China). Hu used the money to hire a student, who gathered roughly one kilogram of seeds from three trees in Moudao and about a hundred more in neighboring valleys. Through diplomatic channels, precious seed packets arrived in Boston in the first week of 1948. Arnold Arboretum redistributed seeds to the National Arboretum in DC, the Royal Botanic Gardens, in Kew, and scores of research institutions in the Atlantic alliance, thus spreading out the risk of propagation failure.

Merrill, the world's greatest living plant taxonomist, issued research notes on *Metasequoia*, which he called a living relict and a living fossil. At the annual meeting of the AAAS in early 1948, people hailed him as the "Johnny Appleseed of Metasequoia." Newspapers in the Northeast printed stories about Harvard's plans to cultivate "Trees from Earth's Early Ages" from "Dinosaur-Age Tree Seeds."[9]

On the opposite coast, papers ran sensational features on Ralph Chaney, who made a whirlwind journey to East Asia under the sponsorship of the Save-the-Redwoods League, which had an interest in all things (distantly) related to *Sequoia*. In a letter to Merrill, Chaney had explained that he was "extremely anxious to visit the occurrence of these trees before China disintegrates politically." His botanizing trip began with a series of plane refueling stops at airfields made famous in the recent Pacific War: Honolulu, Midway Island, Wake Island, Guam, Tokyo, Shanghai, Nanjing, and finally Chongqing. Chaney traveled in the company of Milton Silverman of the *San Francisco Chronicle*. The US consulate provided an interpreter; an American reverend arranged for porters; and Hsen-Hsu Hu recommended a guide. The party swelled to fifteen lifters, eleven porters, and twenty-two bodyguards, one of whom killed a threatening "bandit." In the mountainous portion of the trip, Chaney traveled in a shaded sedan chair mounted on bamboo poles lifted by relays of laborers who worked for twenty cents per day. The Americans ate Army surplus K-rations, and they "bombed" the interiors of their village accommodations with DDT to kill bedbugs and mosquitoes.

In Moudao, Chaney exulted at the sight of three living *Metasequoia*, though the deciduous conifers looked dead: It wasn't quite spring. The way Chaney understood the situation, locals used the largest of the "water firs" for divination and made offerings at a tiny shrine at its base. This plant became the taxonomic "type tree." Standing about one hundred feet tall, it occupied a geomantic position on the bank of a rice paddy. "The big tree is revered by the villagers as the abode of a god," wrote Silverman, simplifying the matter. He would later remember that Chaney deposited, at its base, a handful of needles from the coast red-wood that grew in his backyard in Berkeley. After this act of personal reverence, the scientist positioned his increment borer and, without asking permission, cranked out a core sample.[10]

Chaney's guide led him to a neighboring valley where isolated stands of *Metasequoia* grew naturally. They were the only woods of their kind anywhere. According to Silverman's news dispatches—missives sent by runner to the Yangtze and rushed by boat and plane to San Francisco in less than two weeks—Americans had found a "lost world" in the "Valley of the Dawn-Redwood" (now called Xiaohe Valley). An accompanying drawing showed a triceratops and a pair of tyrannosaurs in a forest of redwoods, palms, and tree ferns. "We have come a hundred miles by trail and a hundred million years in history," Chaney dictated. "Now, for the first time, we can see with our own eyes how the world really looked in Cretaceous times, in the Age of the Reptiles."[11]

The UC professor wasted no time: he shot photographs of the "prehis-toric redwoods," uprooted four seedlings for transport, and cut down one mature tree in a desperate attempt to find cones with stray seeds inside. It was simply the wrong time for collecting. A few months after Chaney's visit, an expedition led by a China-based US entomologist counted over 1,200 *Metasequoia* trees and recorded the fallout from Chaney's felling of the seed tree near Moudao. The mayor's wife and one child had un-expectedly died, and locals attributed the ill luck to the Californian's vandalism.[12]

At the end of his trip—a hybrid of imperial botany, publicity junket, and doom tourism—Chaney encountered a customs inspector in Ha-wai'i, who informed him that the seedlings must be confiscated by the rules of the US Bureau of Entomology and Plant Quarantine. According to Silverman's recollections, Chaney made a scene, "shouting defiantly and citing millions of years, tens of millions of years, a hundred million years." The exasperated inspector at last found a provision in his rulebook that allowed a way out. "Please lay off all that crap about a million years," he snapped. "Tell me just one thing: *Are they more than a hundred and fifty years old?*" After Chaney replied, "Hell, yes!" the agent redefined Chaney's contraband plants as antique objects.[13]

Even before Chaney returned to California in early April, Silverman's serialized reports had appeared in the *Chronicle* and aired on NBC's coast-to-coast network. Scores of smaller media outlets picked up the "absorbing adventure story" about Chaney's "discovery" of a "100,000,000-year-old race of redwoods." In a follow-up story, Silverman lauded the Berkeley fossil hunter for bringing the "dawn redwood" back to America after an absence of over 20 million years. "Finding a living dawn-redwood is at least as remarkable as finding a living dinosaur," said Chaney, on the record. As a scientist, he knew that *Metasequoia* was more like a cousin than a grandparent to *Sequoia*. But as an active figure in the Save-the-Redwoods League, he believed that the name "dawn redwood" generated publicity for the social and ecological good.

In subsequent scientific publications, Chaney took pains to acknowledge the international scientists who preceded him. A guilty conscience lay behind the grandstander. He realized that hyperbole in the *Chronicle*—repeated uncritically by other writers, year after year—did professional damage. Merrill's relationship with Chaney turned sour, then toxic. Harvard deserved all the credit for discovering and disseminating the species, grumbled the scientist in Boston. He preferred "Chinese redwood" as the common name: "'Dawn Redwood,' my eye!" Congenitally bilious, Merrill railed against the "utterly false Chaney propaganda." His one-sided vendetta became so intense that the *Harvard Crimson* saw fit to run a story on the "squabble over seeds." In 1954, with his health failing after repeated heart attacks, Merrill composed a diatribe against "Chaneyism" in the form of a six-page single-spaced letter—"the last long one I will write." It was a dismal coda to the life of "the American Linnaeus."[14]

For his part, Chaney wanted to help China establish redwood reserves like those in California. China wanted no assistance. The communist takeover in 1949 effectively closed *Metasequoia* to outside scrutiny. Western scientists did not return for three decades. In the meantime, the PRC nominally protected the redwood stands of Xiaohe Valley, but did not institute anything like an endangered species recovery plan. In 1984, when researchers belatedly performed a detailed survey of the microhabitat, they counted fewer than six thousand "mother trees" amid intensely modified agricultural landscapes. Subsequently, the total has decreased by hundreds. The IUCN placed *Metasequoia* on its Red List in 1997.

Compared to giant panda, the other iconic endangered species from central China, *Metasequoia* is less likely to go extinct, for plants can be bred asexually as well as sexually. However, the original germplasm from 1948 was less than ideal, and Chinese redwood in cultivation sometimes displays characteristics of inbreeding depression. Since the 1990s, the Dawes Arboretum in Newark, Ohio, has maintained a living *Metasequoia* collection as a gene bank. It's unclear whether the species can become naturally occurring again or, like *Ginkgo*, must depend on humans for survival. It has barely reproduced in the postwar period, despite being germinated by the thousands, propagated by the millions, and planted throughout the world.

Chinese redwood prefers a humid temperate climate but tolerates a variety of conditions, including urban ones. On Houston Street in Lower Manhattan, a beautiful specimen stands tall thanks to guerilla gardeners from the 1970s. Pizhou, a city in Jiangsu Province, China, has planted over five million *Metasequoia* to beautify its roads and highways and to experiment with timber uses. More recently, a Japanese town honored the scientist who named the genus by planting an "Ancient Forest" alongside life-sized models of dinosaurs.

Paleobotanists have debunked Ralph Chaney's wishful notion that the "Valley of the Dawn-Redwood" was a fastness for Cretaceous flora. Instead, it appears that redwoods came to mountainous central China as their last stop in a long, slow, incredible history of migration. The range of the species expanded in warmer, wetter periods and contracted in colder, drier periods. The original *Metasequoia* probably originated in eastern Siberia approximately 100 million years ago, and from there spread to Central Asia, North America, and eventually Europe, reaching an initial maximum expanse in the late Paleocene epoch (about 60 million years ago), a time of intense warming, when the Arctic was ice-free and wooded. A super-photosynthesizer, *Metasequoia* could make energy in low light. During the long darkness of winter, its deciduousness conferred advantage. When cooling prevailed in the late Eocene (about 40 million years ago), *Metasequoia* retreated from far northern latitudes, only to recolonize the boreal zone in the warmer Miocene epoch (roughly 5 to 25 million years ago). When climate oscillated again, the once-dominant genus retreated again, disappearing from most of its former habitat and possibly becoming restricted to the Japanese archipelago. Finally, in the late Pliocene or early Pleistocene, the one surviving species (of at least four preserved in fossils) reentered Asia and migrated south into its current natural habitat. For an inconceivable length of time, including a mass extinction event that killed whole orders of life, *Metasequoia* persisted. The warming climate of northern and western Europe may be conducive to a regional rebirth.

On the continuum of endangerment, no conifer is closer to extinction than the plant Australians call "Wollemi pine." It counts as a "living fossil," though less definitively than *Ginkgo* for it still has relatives at the family level. With its primitive branching system, *Wollemia* bears a spindly resemblance to the monkey puzzle tree, its Chilean kin. Like *Metasequoia*, it was first identified by scientists in fossil form and assumed to have gone extinct in the geologic past. A few hundred persevere in the wild—four stands total. The final refugium for the species is a single system of sandstone slot canyons in the Blue Mountains of New South Wales, less than a hundred miles from Sydney.

In 1994, a group of canyoneers made a helicopter reconnaissance of ravines in Wollemi National Park, looking for adventure routes. From the air, David Noble spotted a peculiar plant he didn't recognize. He promptly rappelled down for a closer look. Still stumped, Noble made return visits, looking for subpopulations and trying to obtain a cone. He finally got one by dangling from a copter with clippers. Dissection of the cone proved it was *Wollemia*. The announcement became huge news. Like the *San Francisco Chronicle* writer a half century before, Australian journalists followed a neocolonial script: white man discovers living dinosaur hiding in secret canyon within virgin wilderness. The largest specimen received the honorific "King Billy."

Right after thrill came anxiety, first about poachers. Then, when a fire broke out in the Blues, people sorrowed over the possible extinction of a species they didn't know existed weeks before. The biggest threat was curiosity seekers. Although New South Wales National Parks and Wildlife Service immediately declared the undisclosed location off-limits to visitors, trespassing bushwalkers made their way, taking only photographs, leaving footprints with spores of *Phytophthora*, a genus of pathogenic water mold. In its first twenty years as a managed species, *Wollemia* encountered two novel pathogens that cause root-rot disease. The government responded with fungicide treatments in the not-secret-enough canyon.

Endangered upon discovery, and jeopardized by discoverers, *Wollemia* became the subject of intense scientific and commercial activity. The Royal Botanic Garden, Sydney, supervised the transplantation and "uncaging" of one specimen, and the extraction of genetic material from others. In the ravine, researchers wore protective gear to prevent further spread of pathogens. Seed collecting proved impractical, so recovery efforts focused on vegetative propagation. Within months of extraction, tens of thousands of King Billy clones grew in greenhouses. Arboretums throughout Australia established ornamental Wollemi pine sections, and the Blue Mountains Botanic Garden introduced (or, technically, translocated) a population intended to be a genetic backup in native habitat. Meanwhile, a commercial nursery obtained exclusive license from the government to market Wollemi pine to the public.

By the time the potted "pinosaurs" were available for sale—at a premium price—the revenant species had achieved national fame on a par with koalas, wallabies, and kangaroos. To mark the twentieth anniversary of the unveiling, New South Wales designated *Wollemia* an "iconic species"—the only plant on its list.

Icons receive veneration and also protection from external threats. In the austral summer of 2019–2020, a megafire burned through the Blue Mountains—the most extensive fire in national history, which followed the hottest and driest calendar year on record (so far). The flames got right to the rim of the gorge in which *Wollemia* hangs on. In the immediate aftermath, the government announced it had "saved" the species in a "military-style operation": bombers had dropped water and retardant from above, and hotshots had deployed an irrigation system below.[15] The victory announcement may have been premature: subsequent field surveys revealed significant scorching of juvenile plants. To showcase their seriousness, the ruling Liberal Party announced a new bureaucratic overlay for this critically endangered iconic species in a World Heritage Site: "asset of intergenerational significance." The same politicians continued to give devotion and protection to Australia's coal industry, a stance climate activists called intergenerational thievery.

At the genomic level, the 90-odd-million-year-old *Wollemia* genus is moribund. Barely any diversity exists in the four remaining stands. That by itself does not spell imminent extinction. Many plants that people have tended for millennia—including bananas—have low genetic diversity. Through the ancient art of domestication, horticulturists can keep germplasm on life support indefinitely. With new technologies like genetic sequencing, genetic editing, and transgenic engineering, living information from this Gondwanan conifer could someday find new life in novel species. Such a beginning would signal the end of something, though not the final ending.

CLONAL BEINGS

Wollemia was one of two types of elderflora that Australians suddenly appreciated in the late twentieth century. The other example—a clonal example—came from southwest Tasmania, involving a species called Huon pine, which, like Wollemi pine, is not a pine. Its closest relative, a fellow podocarp—a family of southern conifers—occurs in New Zealand.

Despite being an extra-slow grower, adding to its girth one or two millimeters per year, a Huon pine can reach heights well over 100 feet in its typical lowland riparian rain forest habitat. Its resinous yellow wood is fine-grained, soft, and phenomenally rot-resistant—an ideal combination for boatbuilding. In the 1820s and 1830s, convict "piners" cut and floated Huon timbers for the Crown. The wood fetched such a price that industrial loggers later extracted almost all the big old trees from the riverbanks of Tasmania.

One anomalous, disjunct population exists below the summit of Mount Read, a volcanic peak on the island's soggy northwest coast. Despite activities of gold and copper miners, these medium-high-elevation Huon pines escaped notice until the 1980s, when the government commissioned a survey of the species. By this point, 90 percent of all stands had been logged. Ed Cook, a dendrochronologist from Columbia University, went to Mount Read in the 1990s and cored living specimens over 1,000 years old, containing tree rings that registered the El Niño-Southern Oscillation. His Australian colleagues observed that every Huon pine on Mount Read was male. After determining that this hectare-sized population represented a single genet—one clonal superorganism—they tried to measure its placetime. By radiocarbon dating on-site wood as well as pollen from an adjacent lakebed, they assembled strong evidence that the organism had been growing in place for at least 10,000 years.

In 1995, Australian newspapers hyped the discovery of this "tree" as the "world's oldest known living organism." The government soon banned logging in the area, voided a mineral lease, and established a strict reserve. "Part of the stand burned over in the 1960s," Cook told me, "so mostly what you see—if you gain access to this restricted area—are the bleached branches of dead trees. To see the living part, you got to get down on your hands and knees." Half-jokingly, Cook recalled a nomenclatural scuffle among tree-ring scientists. "People in Tucson started screaming bloody murder: That's not a tree; it's a distributive thing. We caught a lot of grief."[16]

Another clone with debatable "treeness" is a population of quaking aspen, a familiar feature of Canada, the Great Lakes, and the Mountain West—indeed, one of the most widely distributed woody plants on the current planet. Similar to Huon pine, Wollemi pine, and many other vegetal species, aspen can reproduce both sexually (by seed) and asexually (by runners). In the fall, it's easy to see outlines of clones on western mountainsides: one patch of foliage will change a different hue on a different schedule than a neighboring patch.

In the 1960s, researchers began hypothesizing that particular aspen genets might be as old as the Holocene. These populations had taken advantage of changing habitat conditions millennia ago and never surrendered their space, thanks to clonal reproduction. In 1966, ecologist Burton Barnes argued that clonality was aspen's typical growth habit, particularly in the semiarid Mountain West. Here, the species enjoyed fewer germination opportunities, but lesser competition. Through fieldwork, Barnes mapped genets in Colorado and Utah that were much larger (and presumably older) than those in Minnesota. In 1976, he and a colleague described a supersized male clone on the Fish Lake Plateau in central Utah.

Not many made note of this finding until the early 1990s, when a series of slightly ridiculous events occurred. First, scientists bragged in *Nature* that they had unearthed the world's largest organism—a 38-acre fungus in Michigan. Land managers in Washington State countered with a 1,500-acre mushroom colony. In a letter to *Nature*, a Colorado professor weighed in on "big science" and "extremism."[17] Citing Burton Barnes, he nominated aspen, a common North American species, as the champion. The professor followed up with an article for *Discover* that publicized the 106-acre Fish Lake clone. He and his colleagues gave the stand a name: "Pando" (from a Latin verb *pandō*, "to spread out"). As he would recall later: "It was simple. It was easy to say. It had nice phonemes."[18] It also slant-rhymed with "panda," the poster species for endangerment. Thanks to a catchy name and a publicity push, Pando got the attention of *Good Morning America* and the *New York Times*. Soon the mega-tree became a standard entry in compendia of amazing facts: its forty-seven thousand stems and interconnected root system weighed roughly thirteen million pounds—the equivalent of thirty-five blue whales.

Pando stands on Forest Service land zoned for "multiple use." In the late twentieth century, this grove served as an overflow camping area for Fish Lake recreationists, who left behind fire rings and hundreds of now-black carvings on off-white aspen trunks. Defacers favored names and dates but managed to knife in a cross section of Americana: hearts, happy faces, peace signs, swastikas, and an endless supply of penises. Even after Pando earned fame, new "arboglyphs" appeared. The site has not been sacralized to nearly the extent of Schulman Grove.

By 2006, when Pando appeared on a US postage stamp as one of "40 Wonders of America," land managers had larger worries than tree graffiti. Aspen stands throughout the southern Rockies were declining. Over the subsequent droughty decade, a spate of scientific papers and journalistic reports described "sudden aspen decline" (SAD). Many commenters presented aspen as the "canary tree" for climate change in the Mountain West.

The truth is more complicated. SAD does not result from a monopathogenic invasive disease like sudden oak death. Rather, the ongoing decline of western aspens results from aggregate factors: primarily drought and water stress; secondarily pests, pathogens, and land-use history. The current contraction follows a previous period of expansion caused by a different cluster of contingent factors, including industrial logging of late-successional conifers and fire suppression. As an early successional species, aspen benefited in the short run from US conquest.

In the specific case of Pando, however, nothing about the national period has been kind. Aerial photographs clearly show attrition over the decades. Anthropogenic climate change is simply the latest in a series of new stressors, starting with the introduction of livestock by Mormons and the extermination of apex predators by settlers and US government agents. In the absence of wolves and cougars and grizzlies, populations of aspen-nibbling deer and elk exploded. The hooves and teeth of sheep and cattle caused additional damage to tender aspen runners. Over time, only the mature survived, leaving little vertical diversity and limited chronodiversity in the aboveground part of the colony. Its age structure is top heavy. Any given aspen lasts on the order of one century, meaning most of Pando's stems are approaching their life expectancies. "To put it in human terms, there's a missing generation," said Paul Rogers, director of the Western Aspen Alliance, based at Utah State University.[19]

Rogers and his colleagues installed forest-monitoring plots arrayed according to three possible management regimes: no fencing, fencing with passive management, and fencing with active management. At each plot, scientists regularly measured demography, looking for dead stems and new suckers. To obtain proxies of ungulate presence, they measured the height of the understory and the number of pellets on the ground. Eco-activists added motion-triggered cameras.

Given the inordinate attention conferred upon this "forest of one," journalists inevitably inquired how old it might be. Aspen researchers came up with a "super rough estimate" of 80,000 years based on known rates of clonal expansion, and reconstructions of past climates.[20] (The Fish Lake Plateau escaped glaciation during the last ice age.) This eye-popping number was nothing more than a guess at the maximum theoretical age of an aspen genet. Pando's status as the "oldest" plant, much like its status as the "largest," is largely a matter of faith—and convenience. Like the bristlecones of the White Mountains, Pando is accessible by car. A highway runs through it.

Rogers has expressed skepticism regarding eighty millennia, but that doesn't change his sense of purpose. He wants to figure out how best to nurture Pando—and, by extension, other aspen groves in the Mountain West—toward a new state of resilience after the megadisturbance known as US settlement. His research suggests that people must provide continuous protection from ungulate herbivores, mainly cattle, if they want aspen runners to survive into maturity. To save this grove, people must now serve the ecosystem function of the "wasters and destroyers" they previously exterminated. "If the colony dies on our watch," said Rogers, "we're doing something majorly wrong."[21]

After the sudden, surprising fame of Pando, it was only a matter of time before people claimed to find bigger, or older, or more arborescent clones. In 2008, a Swedish press release went global: a professor of physical geography at Umeå University and one of his students had discovered the "oldest living tree in the world" in Fulufjället National Park, some 250 miles northwest of Stockholm. The professor had named this Norway spruce "Old Tjikko" after his dear departed dog. The tree appeared to be an individual—a classic mountaintop "lone tree" with a spindly trunk photogenically rising above a tangle of ground-hugging branches. From the site, the researchers had recovered a piece of spruce wood that radiocarbon-dated to approximately 9,550 years BP. On that insubstantial evidence, they conjectured that the same living plant had stayed in place since the retreat of the ice sheet. With spruces, the sole pathway to extreme longevity is clonal regeneration, also called vegetative reproduction, or layering.

Like a pet or an icon, Old Tjikko has personage. Or, to be more precise, Old Tjikko personifies planetary change and the global mobility that drives so much of that change. Rachel Sussman, a Brooklyn-based photographer, trekked to Sweden to shoot a portrait, which became the cover image for her book, *The Oldest Living Things in the World* (2014). Taking the Umeå professor at his word, Sussman called her art "a portrait of climate change" because it seemingly showed how the spruce had, in response to global warming, changed its morphology from multitrunked krummholz to single-trunked arborescence. German forester and international bestselling author Peter Wohlleben made his own pilgrimage, then returned with a film crew, despite expressing concerns that excess visitation threatened the fragile tundra site. Tree tourists now hire guides and share GPS coordinates. Instagrammers post images of their limber bodies in tree pose next to the "world's oldest tree."

Like so many claims, going back centuries, of ostensibly ancient things, this one found acceptance before verification or falsification. Only after Old Tjikko had been named and publicized and shared on social media did a follow-up investigator cast doubt. There was no evidence of genetic continuity between dead wood and living tree; no evidence of clonality; and ample evidence of sexual reproduction. It's possible this spruce switched from clonal to sexual reproduction (trees are known to do this), and it remains plausible the organism is old, even extremely old. But its current high ranking on Google searches for the oldest living thing is less about botany than the never-ending search for symbols. The *latest oldest* is a mobile site where the modern fetish for the new and the novel meets the modern fetish for the ancient and the original.

And yet, the same temporized site always contains a nonhuman time-being beyond human temporality.

ARBORESCENT ETHICS

This is no Ur-tree ancestor common to all plants that people honor as trees, by which I mean largish single-trunked plants that live a longish time. Arborescence (treeness) has happened—and *un*happened—many times in evolutionary history. Plants are nothing if not plastic. Some herbaceous plants like strawberries have woody ancestors, while some woody plants like mulberries have herbaceous ancestors. In other words, arborescence exemplifies convergent evolution. Different evolutionary pathways can produce similar treelike outcomes in gene expression. Taking the form of a tree brings specific disadvantages (slow growth, immobility) along with distinct advantages (stability, longevity). Even grasslike angiosperms (monocots) can achieve treelike form if they produce enough lignin to rigidify and thicken their outer tissue. Palms are monocots that can grow taller than most lignophytes. Other monumental monocots include dragon trees and Joshua trees, both in the asparagus family. Arborescence is dynamic: not an evolutionary state but a strategy shared by various genera. Its expression varies within the same species. Depending on habitat, the same plant might present as a tall, single-trunked individual—a classic "tree"—or a lowly, multitrunked shrub.

Cycads in particular defy categorization. They have palmlike fronds, but they are gymnosperms, unrelated to palms. They produce some of the most amazing cones in nature without being conifers. They contain wood yet lack growth rings. In terms of evolutionary age, they rank among the oldest plants that people call trees, though people rarely do. In the main, cycads are small and easily potted—and, by extension, easily stolen, or uprooted and smuggled. On the black market, rare specimens fetch a small fortune.

In 1911, the leading authority on cycad biology, Charles J. Chamberlain of the University of Chicago, took a solo "round-the-world botanical excursion" to see "oriental cycads" in Australia and South Africa. On the way, he stopped in New Zealand as a doom tourist. He wanted to see kauri—the big conifer of the North Island—before the timber monopoly liquidated the oldest stock, which he misestimated at 5,000 years in age. Moving on to Sydney and Brisbane, the professor found imperial arboreta with world-class collections of conifers. In the wild, Australia was largely bereft of cone-bearing plants, with the notable exception of cycads. In Brisbane, the director of the botanical garden greeted Chamberlain and guided him to Tamborine Mountain, a forested plateau north of the city. Here they saw cycads with enormous cones—up to two feet long, weighing fifty to eighty pounds. They also saw an exceptionally tall cycad, a pineapple zamia, roughly twenty feet tall. This plant possessed treeness. Locals called it Great Grandfather Peter, though it was female.[22]

Amateurs queried the expert: How old could this patriarch be? Given that cycads grow upward only, leaf upon leaf, Chamberlain postulated that the number of leaf scars could, with a multiplier, stand for age, because the leaves are so durable that years go by between each unfurling. After counting Peter's scars, he conjectured a life span beyond Methuselah's 969 years. He rounded up to one millennium for effect. This guesstimate became the basis for puffery: Someone in Brisbane added a zero, aging the plant by a factor of ten. Irresistibly, locals pushed the age up to 10,000, then 12,000, finally 15,000 years.

Broadcasting fake news brought unintended results. In March 1936, Peter was chopped down by unknown person(s)—or, in the words of a local newspaper, wantonly "murdered" by "some thoughtless boy."[23] Upon impact, the hoary crown came to pieces. Police looked for a culprit and gave temporary police powers to volunteer rangers. Even small-town papers in the United States ran stories about this vandalism.

Arborists attempted a rescue. They treated a section of the trunk with antiseptic solution, then partially buried it, like a fence post, hoping to stimulate roots and leaves in either direction. A crowd of Brisbanites came out to see a Queensland University botanist supervise the planting. The professor cautioned that the event might in fact be a burial. "A gambler's chance of survival," he said. "Like winning the first prize in the Golden Casket." A local headline writer took a chance by announcing, "Famous Giant Tree Resurrected."[24] Months later, a different local newspaper called the transplantation a failure, and months after that, a syndicated story expressed hope that the plant's second coming might yet occur.

Chamberlain himself did not lose hope. For his 1938 Christmas mailings, he sent a photograph of a woman perched in the once-mighty cycad. On the accompanying card, he expressed blind faith that many years could be added to the tree's great age, which was already "greater than that of Methuselah."[25] At least the genome survived. Chamberlain's Australian contacts had, years before, mailed him a cone from the tree, from which he had distributed seeds to botanical gardens and college campuses.

Because of the Chicago connection, California's state forester accepted the claim from Brisbane about the oldest living thing and downgraded the age rank of the Golden State's state tree. Californians responded with indignation and disbelief, which prompted the forester to verify the claim. He shared with the press a letter he received from the man who had replanted Great Grandfather Peter. "Somewhere out in those mountains there must be a champion liar," wrote the Queensland professor.[26] A syndicated correction issued by the AP ran in the *Press Democrat* of Santa Rosa, the hometown of Robert Ripley, creator of the phenomenally popular "Ripley's Believe It or Not." Ripley, relocated to Chicago, didn't get the news, or didn't care, and he released a cartoon on the "Oldest Living Things on Earth" that showcased the cycads of Queensland as the "Methuselah of Trees" at 12,000 years.[27] Whereas initial reports had referred to Great Grandfather Peter as a giant, Ripley highlighted the diminutiveness of cycads. Plants "only 20 feet high" made sequoias "look like mere infants," he marveled. Into the 1950s, before bristlecone pine became renowned, listicles continued to rank cycads as the world's number one elders. Despite their ambiguous treeness, they retained the standing of trees because of their purported oldness.

If a cycad could (briefly, mistakenly) be the oldest treelike thing, why not something smaller? This issue came up in 1958, when *National Geographic* received letters in response to Edmund Schulman's article on bristlecone pine. We expected readers to challenge Schulman's claims, the editors relayed to the Laboratory of Tree-Ring Research; we never expected them to defend a bush in Pennsylvania.

Like all plants, this bushy Appalachian species goes by many names: juniper berry in West Virginia, ground huckleberry in Kentucky, bear huckleberry in Tennessee. It resembles blueberry, a close relative, without tasting as good. The commonest common name, box huckleberry, suggests the likeness of its glossy foliage to boxwood shrub. As a rarity, it grows in the shade of mixed pine-oak forests. Repeated glaciations caused local extirpations and habitat fragmentations. The relict populations are too widely spaced to cross-pollinate. The fruits produce sterile seeds. In the absence of sex, the plant survives by self-propagation, sending out underground runners.

A government botanist, Frederick Coville, discovered the clonality of box huckleberry in 1918 and reported it to *Science* the next year. This beautiful, useful plant is threatened with extinction, he warned. He could locate merely two populations: one in Sussex County, Delaware; another near New Bloomfield, Perry County, Pennsylvania. The latter population had been found by nineteenth-century botanists, before being lost, and found again.[28] It covered eight acres. On the theory that this was one genet joined at the root, and on the assumption that it had grown outward from a center point at the rate of six inches per year, Coville estimated an age of 1,200 years—older than the "thousand-year rose" of Hildesheim, Germany.

Coville wanted to both protect the habitat and domesticate the species. He gave directions to a nursery firm, which extracted truckloads of rare plants from Delaware and Pennsylvania for cross-pollination in USDA greenhouses in Washington. Science would rejuvenate oldness through horticulture, Coville anticipated. Box huckleberry was a "charming little thousand-year-old lady of the forest," he concluded. Let's keep her "living forever."[29]

After the publicity, amateur botanists went looking for additional patches and found several, including an extensive one near Duncannon, Perry County, opposite a Pennsylvania Railroad flag station called Losh Run. This one-hundred-acre population aligned with the theory that each patch consisted of one genet. Samples went to Harvard and the New York Botanical Garden. Experts from the USDA, the Carnegie Museum, and Penn each made "pilgrimages." In their excitement, naturalists extrapolated: If this clonal patch is ten times larger than the other Pennsylvania patch, it must be ten times older. One of the rarest plants in North America became the "oldest bush in the world," possibly even the "oldest plant in the world" at 12,000 years—as old as the Holocene.

The year before Schulman cored his first bristlecone pine, a Pennsylvania representative submitted to the *Congressional Record* an article from a state forestry magazine, "The Oldest Living Thing in the World." This "most unique shrub" was 11,000 years old when Christ taught the disciples at Galilee, it read.[30] Schulman himself responded to a letter from a Pennsylvania partisan before he died. The esteemed tree-ring expert dismissed box huckleberry longevity as "another enthusiastic overestimate."[31]

In the period between bristlecone celebrity and the Endangered Species Act, box huckleberry faced new threats. It lacked enough defenders, and lacked enough treeness, to avoid institutional vandalism. Road crews decimated the Losh Run mega-shrub while upgrading US-22/US-322 along the Juniata River. As a result, the species is now listed as threatened in the Keystone State. If Pennsylvanians had valued old plants as much as they valued divided highways, wide and straight, they might have preserved this population long enough for gene sequencing. Now it's too late to know whether a single genet existed. The theory was never disproven.

This dynamic—a tension between short time and long time—recurred in the California desert. During the last glacial period, when humans arrived in western North America, piñons and junipers dominated the Mojave. As the climate got warmer, then hot and dry, gymnosperms retreated to the mountains, and plants better adapted to new conditions moved north from the Sonoran Desert. The best adapted was creosote bush, one of the toughest, hardiest plants on the planet. Its resin-covered leaves, small and thick, can desiccate without dying. The whole plant is a volatile organic compound factory, repelling floral competitors, insect predators, and faunal grazers. (The Late Pleistocene extinction of the American camel, the only large mammal known to browse on its leaves, benefited the species.) Over a century-long life, each oil-stained, yellow-flowered bush can reproduce both sexually and asexually. In its clonal form, it expands outward from the original, which eventually dies and disappears. As the root mass widens and hollows out, an elliptical ring of genetically identical bushes forms aboveground.

In the 1970s, Frank Vasek, professor of botany and founder of the herbarium at UC Riverside, worked with a colleague to conduct an environmental assessment of Mojave Desert lands where Southern California Edison proposed to build gas and electric lines. As part of their research, they examined US Air Force aerial photography from the 1950s. From the high-altitude perpendicular angle (the "plan view"), Vasek could easily spot creosote rings that were hard to see at ground level. He began inventorying and measuring them, and found hundreds over thirty feet in diameter. To the largest he gave a name: King Clone. He found a motorcycle track bisecting it. In 1979, he snapped a picture of twenty-three UC students encircling the waist-high plant like a coven.

The next year, Vasek published an article in which he extrapolated the age of the clone through a combination of modern growth rates of living wood and radiocarbon dating of dead wood. He came up with an estimate of 11,700 years, assuming constant growth over time. Compared to the named cycad in Queensland and the unnamed huckleberry in Pennsylvania, the science behind King Clone had more credibility.

In vain, Vasek waited for *National Geographic* to ring his office phone. In 1983, he managed to get the attention of the *Los Angeles Times*, which turned King Clone into a half-mocking meta-story: "What would you do if you discovered the oldest living thing on Earth and nobody cared?" Vasek tried to compare the outer ring of a creosote to the outer layer of a redwood. "I'm just sick about this," he said. "It could be as important as the pyramids of Egypt. But it could also vanish with the snap of a finger." If subdividers don't destroy it, off-road riders will, he warned. He had failed to convince the feds, the state, and the county to buy the private land. He tried to raise $20,000 himself, with the same sad result. He couldn't believe it: How could people be unmoved by plants twice as old as bristlecones? A local real estate agent weighed in: "Maybe the professor is right. Maybe those bushes are the oldest living thing on Earth, but when you have millions of creosote growing all over the desert and they all look alike, it's hard to get too excited about it."[32]

Such talk concealed thoughtless, clueless racism, for creosote bush (aka greasewood) is the preeminent medicinal plant for Indigenous peoples of the Mojave, Colorado, and Sonoran Deserts of the United States and Mexico. It figures in creation stories as well as ceremonies. Organic compounds from this cure-all can be absorbed topically or orally, or inhaled in steam baths. Historically, Natives used *la gobernadora* (its Spanish name) to treat Old World diseases such as smallpox, influenza, chickenpox, and TB. As sacred as it was common, creosote demanded respect. Before gathering therapeutic material, a Western Shoshone healer would address the creosote as "you," a fellow healer, telling the plant-person which medicine was needed for what ailment and, most importantly, saying: "Thank you."[33]

In contrast to traditional ecological knowledge, Vasek applied scientific, global, and materialistic perspectives to creosote. In his attempt to force treeness upon a bush, he even accepted an invitation from the *Ripley's Believe It or Not!* television show, hosted by Jack Palance. Finally, he found allies among members of the California Garden Clubs, whose financial donations facilitated a Nature Conservancy land transfer in 1985. At that point, the *New York Times* deigned to cover the "world's oldest known plant." The University of California initially agreed to manage the micro-reserve, but it was never clear what kind of management was appropriate. King Clone sits a few miles away from near-identical creosote habitat that the Bureau of Land Management has zoned for off-road vehicle "free play." Does the rickety fence around the plant offer protection, or invite footfall and vehicular vandalism? This specimen and its species have yet to be sacralized by science like Methuselah and Great Basin bristlecone pine. Compared to tree rings, creosote rings are barren of data. Without more precise age verification, Vasek's bush cannot be transfigured into full arborescence. King Clone thus represents failures of Western thinking in Vasek's time and place. Even as the botanist worked to break down tree trunk bias, he replicated parallel biases for named individuals, gendered champions, and manly discoveries.

In practice, only megaflora and elderflora have automatic dignity in modern Western thought. Big old trees have the highest standing, then big trees, then little old trees, then olden quasi-trees such as Great Grandfather Peter. Among clones, the lofty collective of Pando outranks the short but striking form of King Clone, which outranks formless, nameless huckleberry patches. Like a name designation, the scientific assignation of age enhances the treeness of trees and, in select cases, grants treeness to life-forms not normally personified as such. As ethical shortcuts, exact cambial age is superior to estimated radiocarbon age. Modernity privileges exactness.

Resistance to extending the circle of plant dignity became apparent after Swiss voters amended their constitution by referendum in 1992 to offer vague protections to "living beings," including plants and animals—a reaction against genetically modified organisms. A follow-up federal report, "The Dignity of Living Beings with Regard to Plants: Moral Consideration of Plants for Their Own Sake" (2008), was widely lampooned in the media and castigated by lab scientists. What next—would the Swiss ban agriculture? The editorial board of *Nature* characterized the vegetal outgrowth of Kantian philosophy as silly and absurd.[34] Tellingly, federal ethicists admitted they had failed to reach consensus about plants, unlike animals. The committee majority assumed that the object of moral consideration was the "individual plant," whereas a minority focused on "plant networks."

The traditional concept of treeness is both an intellectual impediment and a moral precedent. Plants misunderstood as individuals have had cultural standing for millennia, and certain specimens have enjoyed legal standing in city codes and state laws for decades or centuries. Whether the trajectory of animal protection movements—from fringe to mainstream, from large mammals to a wide variety of species—will provide a model for arboreal ethics is unclear. Humans have a general bias for animals and, regarding plants, a special bias for trunks. It's conceivable that enough people in a society could mature in their plant thinking to permit the dignification of vascular plants as whole organisms—leaves, cambium, roots, mycorrhizae. People could learn to relate to these modular, social, communicative beings on their own multitudinous terms, including sexual, unisexual, and asexual reproduction, in all kinds of forms, big and small, trunked and shrubby. Biological treehood, and the fullness of tree time, could at last supplant anthropomorphic treeness.

If it happens, it will take time. In 1831, in the first long-form scientific treatise on longevity in plants, Augustin Pyramus de Candolle speculated that certain small shrubs might be much older than people were accustomed to believing. No one had investigated the question. Descending to "vegetables still more humble," Candolle spoke of lichens that didn't change in size over decades. Some of these colorful specks may be as old as the rocks themselves, he wrote. Yet Candolle himself set aside small things "too obscure perhaps to attract general attention" and confined himself to trees, which had "universal interest."[35]

Doug Larson, the scientist who documented Canada's oldest trees, started out as a lichenologist, working on *Rhizocarpon* on boulders in Iceland and other extreme environments. "You need a microscope to appreciate the beauty of lichens," Larson told me. "It's a form of existence that defies nomenclature. Lichens aren't really alive. They're not species. They're microscopic ecosystems."

Midway through his career, when applying for a new round of funding, he received a reality check from a high-level administer: "Doug, you're gonna have to shift over to higher-level plants. No one gives a shit about lichens. Doesn't matter if they're 4,000 years old. Doesn't matter. People don't care about rock scum." This dictate prompted a discussion in Larson's lab: What besides lichen grows on rock? That's how they arrived at white cedar. Science followed funding, which followed aesthetics and ethics.

I queried Larson: Are you moved by lichens? "Oh, yeah," he replied immediately. "They've put up with far more. You can have a nuclear war and they'll do fine. Lichens and cockroaches—a perfect food chain! When environmentalists say humans are going to destroy the Earth, that's arrogant. There's nothing that's going to destroy the Earth. We might destroy ourselves; we might influence the food chain." He imagined a thought experiment: What if humans discovered another planet that had nothing but rats, lice, and cockroaches? "That would be a big deal. A planet with nothing but bacteria? We would think that's fantastic!" Yet here on Earth we struggle to appreciate perfectly adapted, wonderfully timeful microbiota.

"What is it about trees?" said Larson, before answering his own question. "Trees have bodies."

VIII.

TIME TO MOURN

DYING FOREST — BRISTLECONE FUTURES — DISCOVERY AND LOSS — KAURI PASTS — UNDEAD WOOD

In an age of novelty, millennial trees are both culturally precious and ecologically precarious. In a fast-changing climate, forests will reorganize or die. Reorganization will be the norm—the Earth won't soon lack for woody plants—but even in resilient ecosystems, premature death will visit the Holocene's eldest. Slowly, or all at once, it will happen. Living connections from one age to another will succumb to the temporal change that is climate change. Forest dieback does not always spell forest doomsday, yet the demise of olden trees represents an ecological loss, a cultural impoverishment, and a social problem. How can artists and storytellers commemorate lost elderflora, and how should people nurture the future oldest? Would novel humans—genetically enhanced, technologically augmented—still care about arboreal antecedence? When the planetary becomes this personal, and emotional, it's appropriate to return to perdurable conifers: bristlecone pine, bald cypress, metasequoia, and especially kauri, a species more threatened and timeful than previously known.

DYING FOREST

When I was a child, living in Provo, Utah, I used my allowance to purchase made-in-Korea plush toy monkeys, and pretended they were the final four breeding pairs of their species—ecological refugees from the Black Forest. Somehow, I'd heard the news about the doom of Europe's conifers. Along with nuclear winter and the ozone hole, the dying forest was my Gen-X introduction to planetary thinking.

Concern about "acid rain"—an effect of industrial coal burning—began in Scandinavia in the 1960s, when scientists noted acidification in freshwater lake ecosystems. After Nordic countries called for international controls on sulfur dioxide emissions, British and US coal industries cast doubt on the science, then cast doubt on cost-benefit analyses of the proposed controls. In a rhetorical arms race, purveyors of doom opposed these merchants of doubt. Prominent scientists in West Germany went directly to the public with their prognosis of forest collapse. The media sensationalized their message. In 1981, *Der Spiegel* ran a three-part cover series on acid rain. When the respected magazine announced, "The Forest Is Dying," it used *sterben*, the verb for human mortality. *Waldsterben* was a multispecies disaster: *First the trees die, then the people!* Discolored and defoliated firs and spruces in Lower Saxony, Baden-Württemberg, and Bavaria become icons of "global eco-death." In West Germany, the Green Party capitalized on this apocalyptic fear, which echoed wood famine phobias of previous centuries. A national ritual—a death watch in the form of annual government reports—continued well after the predicted calamity failed to materialize.[1]

Today, from a risk-averse perspective, the survival of the German forest is a vindication of stricter air pollution controls enacted in the European Union (and the United States) in the 1980s and 1990s, and a demonstration of the "precautionary principle." This concept entered German law in the 1970s and international law in the 1990s. At the Earth Summit, signatories agreed: "Where there are threats of serious or irreversible damage, lack of full scientific certainty shall not be used as a reason for postponing cost-effective measures to prevent environmental degradation."[2]

From a risk-tolerant perspective, Greens couldn't see the forest for the trees. Monocultural conifer populations, a legacy of plantings by royal and imperial foresters, were declining in health anyway. Sulfur dioxide hurt some species more than others, while nitrogen dioxide brought selective benefits. No one knows what might have happened if unscrubbed coal burning had continued indefinitely. It's possible that central European forests would have shown the resilience of a dynamic ecological system. They might have changed without dying.

Waldsterben previewed the challenge of evidence-based long-term thinking about anthropogenic climate forcing. Long-termism is hard enough by itself. It becomes harder when short-term incentives, financial and political, prop up business as usual. It becomes purposefully impossible when industry-funded denialists and industry-supported politicians sow confusion about certainty and risk. In response to doubtmongering, environmentalists and journalistic allies have often resorted to fearmongering, which further adds to the misimpression that climate science itself is an ideological field—or a belief system. Although environmentalist prophecies about the end of the Earth have proven no more reliable than Christian prophecies about the End Times, scientific projections about global warming, habitat loss, and the Sixth Extinction have in fact been remarkably accurate. If anything, scientists have erred on the conservative side.

In 1992, there was more than enough knowledge about the risk of unmitigated greenhouse gas emissions to justify precautionary action. Tragically, climate scientists got caught in the political trap of "further research is needed." They competed for grants to compile additional data to run better models to make higher-resolution scenarios and likelihood scales to fill thicker reports. The potential for Earth system modeling is endless. In the three decades following the first IPCC report in 1990, many uncertainties about the long-term effects of unmitigated climate change were resolved. In the meantime, timely opportunities for decarbonization went by the wayside. Even if engineers eventually "come to the rescue" with carbon capture and storage (or geoengineering or nuclear fusion), a risky quantity of heat has already been baked into the future. Because of incautious inaction, the world will soon find out the true resilience of ecosystems and even what lies beyond their resilience.

For Europe, a key reference point is the last "primeval" forest, Białowieża, located on the Polish-Belarusian border. The ice sheet did not reach this mixed forest during the last stage of the Pleistocene, and humans did not clear it for agriculture during the Holocene, so it represents an ancient—if constantly changing—ecological community. Slavic barrows dot the forest, as do modern relics of Polish honey-gatherers, but the single most conspicuous modification is the grid system of dirt roads, a nineteenth-century legacy of Russian managers who fattened big-game animals, including bison, for the benefit of tsarist hunting parties. Prior to the Russian partition of eastern Poland, Białowieża had been reserved as the hunting ground of Polish-Lithuanian kings of the Jagiellonian dynasty.

After independence in 1918, the democratic Polish state turned the heart of the royal forest—the part undamaged by German troops in World War I—into a strict scientific reserve. When Germans returned in World War II, they herded locals through the national park's wooden gate and massacred them, while making plans to manage the bison as living monuments of German nature. The reserve itself survived the war, as did the tsar's hunting palace—only to be torn down in the 1960s, its bricks sent to Warsaw for the reconstruction of Old Town. During the communist era, the inner forest became a national park again and a World Heritage Site.

Soon after Poland shook free from Soviet influence, the larger-than-life biologist Simona Kossak helped spread the fear that the primeval forest was dying as a result of mismanagement.[3] Partly in response to this sensational idea, the government enlarged Białowieża National Park and created buffer zones around it. After Poland joined the EU in 2004, the greater forest complex received designation in "Natura 2000," a network of protected areas. In the early years of the capitalist era, the government interpreted Białowieża as the inheritance of a half-millennium-long conservation program of Jagiellonian stewards, violently interrupted by Russians and Germans. The new caretakers, state ecologists, believed it was their patriotic duty to protect the forest as European eco-heritage and to save it from commercialism.

Then came Law and Justice, the populist party that swept into power with the national election of 2015. As luck would have it, an outbreak of bark beetles accompanied this political upheaval. The new environment minister, Jan Szyszko, a former forester, co-opted the rhetoric of the "dying forest": Poland must rescue Białowieża from insects as well as pesty experts who don't understand forests like locals. Ethnonationalists claimed that Poles had properly used and managed Białowieża from the time of Christ through the partitions. Foresters had a God-given duty to restore and protect the forest as Polish heritage. Old growth was killing the ancient forest, supposedly. Szyszko wanted to replace passive neglect by "pseudo-ecologists" with active conservation management by (male) foresters. For the ruling party, environmentalism represented one more external infestation—akin to Nazism, communism, atheism, and the catchall threat dubbed "gender." To accomplish his goals, Szyszko opened Natura 2000 lands to commercial loggers, over the objections of Brussels. Urban activists from western Poland, pronouncedly women, came to stand with trees marked for cutting—"Solidarity with Białowieża!"

Despite degradation of its buffer zone, Białowieża National Park remains a dynamic ecosystem for now. At the community level, it is ancient precisely because it's changeable and changeful. In the 1920s, ecologists projected a future forest without linden; today, linden thrives. Hornbeams march forward, with maples close behind. On the other side of the ledger, ash falls back from fungal disease, and spruce declines from beetles. In general, current climate stressors select against tall conifers and massive oaks.

If Białowieża must contain all these iconic plant forms to count as "primeval," the forest is indeed dying. But if the essence of its antiquity is processual, the forest can endure even as change accelerates. Here and throughout Europe, mortality rates for mature trees increase as global warming amplifies natural disturbances: stronger storms, drier droughts, hotter heat waves. As my mother-in-law in Poland told me: "There is no winter anymore." As a result, the overstory shifts in favor of younger, shorter, deciduous trees. Endowed with more than fifty woody species, Białowieża should have enough biodiversity to keep up the olden process of changing with the times. Its resilience benefits far more than megafauna. The greatest ecological importance of Białowieża—as Europe's hotspot of mushrooms, mosses, and lichens—may be small-scale.

Not every forest is equally equipped to adapt to climate forcing. For different perspectives on ecosystem reorganization, I spoke to two US scientists, one based in the mainly deciduous Northeast, another in the mainly coniferous Southwest.

Neil Pederson works at Harvard Forest, a research station sixty miles west of Harvard, one of countless forested tracts in New England that used to be farmland. Anglo-Americans chopped down the Northeast's former old-growth forests while dispossessing Native peoples. Over the twentieth century, economic and demographic changes—and a sustained wet period, what Pederson calls an "epic pluvial"—have produced forest recovery. The new growth swallows abandoned stone walls. This community-level resilience inspires Pederson. "I go to Ecological Society of America meetings, and it's really frustrating," he told me. "Everything is doom and gloom. I want to raise my hand and say: 'Evolution—it's still working!'"[4]

At the organismal level, too, Pederson marvels at the adaptability of plants through thick and thin. He has tree-ring-dated one chestnut oak that "hung out" in the understory for four centuries before suddenly having a growth spurt—an opportunistic reaction to a neighboring eastern hemlock succumbing to aphid infestation. "If your grandma could slam dunk—that's what this oak's doing," he quipped. "We can't conceive that, but that's what trees and other plants do all the time." In other locations, Pederson has cored red maples that barely grew for years at a time. On one radial sample, covering ten years, he counted a single tree ring. This "biological impossibility" turns out to be relatively common. A revenant maple might live one-tenth of its 200-year life as if it were dead.

Pederson prefers process-based definitions of old growth over arbitrary ages. In his neck of the woods, 500 years is the maximum effective age for any one plant, and post-settlement second growth is getting up there.

"It's happening right now," he said. "Red oaks are falling over, creating pit-mound topography. Old growth is coming if we leave it." This forest-in-the-making won't replicate the Indigenous-managed forest that existed at the time of European contact. Pederson doesn't mind: "We can't be so pure about it anymore. That's not fair to us, not fair to the people who came before." The most important oldness goes beyond any one organism or species: "The processes are still working." In the eastern deciduous forest, the large number of species—what ecologists call "species richness"—should buffer the system from future disturbances, known and unknown. "Life is adaptable. That's what I take solace in. That's why I care about old trees."

In Santa Fe, I heard a contrasting message from forest ecologist Craig Allen. He acknowledged that temperate and boreal trees are, in the aggregate, thriving temporarily, especially in hydric (very wet) environments like the North American East. Thanks to CO_2 enrichment and the elongation of the growing season, humid temperate forests are growing faster than ever. Unfortunately, fast-growing forest trees are more vulnerable to dieback as climatic systems approach tipping points. Good times can end suddenly. Allen knows this from experience. When he moved to the Southwest in the 1980s, he encountered a forest that embodied the then-popular phrase "Don't worry, be happy." Three decades later, as he prepared to retire, he was witnessing large-scale, nonlinear transitions to shrubland, even grassland. He had become a coroner of trees.

Northern New Mexico fits a subcontinental pattern. Throughout the Mountain West, the early twenty-first century has been a time of "mega-disturbance" and massive forest reorganization. Compared to the Northeast, the overstories of the wooded western mountains are species poor and coniferous. Conifers have adaptations for persistence in marginal environments, but they also have limits. From the spruces of the northern Rockies, to the whitebark pines of the central Rockies, to the piñon pines of the southern Rockies, many landscapes have, in one human generation, turned from healthy green to sickly yellow to sooty black. After recent fires, ponderosa pine and Douglas fir, once dominant in low-elevation zones, have struggled or failed to recruit new cohorts.

Historical and environmental feedback loops explain this coniferous crisis. First, rising temperatures and CO_2 levels allowed for faster tree growth, which created more biomass—more potential fuel. Fire suppression policies in the early twentieth century further boosted fuel loads, which contributed to hotter fires by the late twentieth century. Exurban developments added opportunities for human-caused sparks. Finally, drier droughts and longer, hotter summers produced stressed-out trees susceptible to native bark beetles and nonnative fungal diseases, with tree mortalities adding excess combustible material. Year after year of drought in the early twenty-first century added up to megadrought, and each terrible fire season became the latest worst. Snowpack melted sooner and faster, and soil moisture declined, as did summer humidity. In this regime of accelerated change, superhot infernos scorched instead of rejuvenated ecosystems.

Allen believes the Southwest is a harbinger. His research suggests that arboreal vulnerability has been underestimated worldwide. Dynamic global vegetation models that climate scientists run on supercomputers are crude compared to general circulation models; they don't yet account for dieback. The prospect of dying trees extends far beyond semiarid zones: "Forest systems are more fragile than you think."[5]

He explained why. Trees operate close to their hydraulic safety margin for their given place and time. Unusual heat becomes a growth-limiting factor and, close behind, a mortality factor. Beyond biomechanical thresholds, overheated trees undergo "vapor pressure deficit," which leads to "hydraulic failure"—embolism of the xylem. Growth spurts become liabilities in sudden hot droughts, when trees struggle to transport water to their lofty crowns. In a hydraulic crisis, certain species can close the stomata in their leaves, saving water, but potentially inducing "carbon starvation" in the process. A short period of intense arid heat can do the same damage as chronic drought. Hundreds of crowns can cavitate and collapse at once. Trees that manage to survive such embolism may be unable to return to former physiological states, making them less likely to survive whatever disturbance comes next.

Even the North American East is vulnerable to dieback, Allen warned. A major drought hasn't hit the region since the 1960s. The next one will be hotter. He recognized that western Massachusetts, like eastern Poland, has advantages. In a biodiverse temperate deciduous forest in a nutrient-rich hydric system, there are bound to be arborescent species that thrive in any given condition, even counting for invasive predators and pathogens. Some species will migrate northward, others will decline or disappear, yet global warming may not fundamentally alter the system for centuries.

By contrast, greater climate velocity exists in northern New Mexico, an environment with a baseline of fewer species and harsher conditions. While the global atmosphere gets more humid, the regional atmosphere of the Southwest gets more arid. Here, the system could switch from mesic (moderately dry) to xeric (very dry), and from forest to shrub, on a compressed timescale. This is called vegetation type conversion, and whether it represents new resilience or the end of resilience may be in the eye of the beholder. With additional heat in the pipeline already, the near-future US West from the Coast Ranges of California to the Front Range of Colorado looks to be less wooded and less coniferous. In the Golden State, the chapparal that defines coastal flora may one day characterize Sierran foothills, too. There's ecological precedent for diminution, Allen told me. In Australia's xeric center, ankle-high woody plants are the norm.

Overall, Allen and other authorities expect the forest cover of the planet to get younger and shorter, with fewer total species and lower carbon storage. Signs of reduced resilience and forest dieback in the Amazon—textbook tipping elements—have already been documented.[6] As one of Allen's collaborators said: "When old trees die, they decompose and stop sucking up CO_2 and release more of it to the atmosphere. It's like a thermostat gone bad. Warming begets tree loss, then tree loss begets more warming."[7]

From an evolutionary point of view, every absence is an opening. In the long term, many plants, including new species, should thrive under the atmospheric chemistry that disfavors mid-latitude conifers. The blue planet will remain a green planet.

In the medium term, existing tree species may survive in obsolescent habitats with individuals that deviate from "classic" trees. When times get tough, woody plants put energy into inner stability rather than outward growth—dying slowly rather than living fully. Forest ecologists speculate that the emerging climate of high variability and high intensity will select against large, fast-growing, full-canopied plants and select for slow growers that hold up to chronic stress and regular disturbance. Plasticity—the ability to adapt in place to changes over lived time—is a hallmark of lignophytes. In an altered climate defined by instability, "deformed" or "low-grade" trees that old-time foresters would have deemed "unhealthy" could in fact embody resilience. This scenario has been dubbed the "Age of the Mediocre Forest."[8]

Big old straight tall trees—the lucky ones that escaped logging—are living anachronisms of the Holocene. They are winners of lost climates. In the North American West, most old-growth stands are half a millennium old—outcomes of the Little Ice Age. They were surviving out of time, outside their original climate space, even before global warming, in a demonstration of magnificent resilience. If their mortality now doesn't signal the death of the forest, it feels like something worth grieving.

BRISTLECONE FUTURES

The ancient bristlecones of the White Mountains have been "mediocre" for millennia. In a sense, they have the least to lose from climate change, for their habitat can scarcely get more adverse. They tolerate low-nutrient carbonate soils. They withstand droughts and resist beetles. Given the wide spacing of the trees, lightning fire isn't a serious threat at the population level. The famous population known as Schulman Grove stayed in place for the entirety of the Holocene, more than ten millennia, as entire forests in temperate zones vanished or transformed during the cold centuries comprising the Little Ice Age. Prior to that, this population persisted through the Medieval Climatic Anomaly, when the Great Basin was warmer and drier than anything Californians have experienced—so far.

Even with these Methuselahs, people envision the end of placetime.

In 2017, a group of botanists prepublished a peer-reviewed article on "divergent responses of tree species and life stages to climatic warming in

Great Basin subalpine forests." To summarize: In the White Mountains, tree lines are ascending. Bristlecone pine has so far been less successful at colonizing the new growth zone than associated limber pine. The scientists described limbers "leapfrogging" bristlecones "in slow motion." Over time, *Pinus longaeva* could face "overall range contraction and possibly local extirpations," they suggested.[9]

What once might have been slow news, or no news, entered the fast news cycle. The lead author, an entrepreneurial PhD student, worked with a UC media officer to prepare a press kit. They changed the metaphor "leap frog in slow motion" to "race to the top" and implied that the USFS should try to fix the game so that bristlecone wins. A local reporter recycled the press release, and then AP issued it for syndication with a headline optimized for the algorithms of Facebook and Twitter. In a matter of days, a technical study completed its conversion into clickbait: "Future of Oldest Tree Species on Earth in Peril."[10]

This headline, like so many, was bogus. Ecologists don't worry about the extinction of Great Basin bristlecone pine in the manageable future. For the moment, at least, the bristlecone-limber zone expands up *and* down—up because of higher temperatures, down because of *lower* temperatures. The latter is the counterintuitive effect of localized cold air drainage on a gradient.

Ecological mechanisms contributing to "leapfrogging" remain enigmatic. The preferences of nutcrackers, who collect and cache the larger, unwinged seeds of limbers, probably matter. (The smaller, winged seeds of bristlecone pine depend on wind dispersal.) Another possibility is that bristlecone roots are less adept than limber roots at initiating symbiotic associations with fungi. Rather than being novel, limber "leapfrogging" may be part of an ancient dynamic that prepares the soil for bristlecones, which appear later and last longer. Or perhaps bristlecones simply abide in extremity, while limbers march up and down the mountain in response to climatic variations.

All things being equally adverse, bristlecones live twice as long. Unlike limbers, they can compartmentalize growth—and damage and death—with their sectorial root system. They have superior "constitutive defenses": greater wood density, higher incidence of resin ducts, higher concentrations of volatile organic compounds. They also have superior "induced defenses," meaning they will, when attacked, apply energy to terpene production, when they would seem to have no energy to spare. A bristlecone's thick needles can live up to one century, approximately ten times longer than limber needles. In lean times, *P. longaeva* takes energy from this emergency supply.

Ecologists make educated guesses about how Great Basin bristlecone communities—which occur with and without limber pine—might adapt over long time. That's different from soothsaying. Whether climate change is "good" or "bad" for *P. longaeva* depends on the time frame. Tree-ring scientists have demonstrated that subalpine bristlecones grew faster in the second half of the twentieth century than in any fifty-year period in the previous 3,700 years. Is faster better for a slow plant?

For Connie Millar, the proper time frame covers millions of years. "To me, the Pleistocene is recent," she told me over Starbucks coffee by a BART station. "I think of that as just last year." Millar has expertise in paleoclimates and proto-pines, having earned a PhD in evolutionary genetics from UC Berkeley in 1985.[11]

The foxtail group of pines, including Great Basin bristlecone, ranks among the oldest *Pinus* groups, going back to the middle Mesozoic era, over 100 million years ago. Proto-pines evolved in the middle latitudes of North America. Intense warming in the early Cenozoic forced pines into refugia, northward and southward—and upward. The land that would become Nevada was then Great Basin Altiplano (aka Nevadaplano), a vast plateau with a temperate climate. A variety of conifers, including sequoia, thrived there.

A series of tectonic changes ended the reign of conifers. The plateau began pulling apart, causing parallel horst-and-graben (up-and-down) block faulting. Through millions of earthquakes separated by spans of stillness, mountains rose and valleys fell, even as the High Sierra uplifted to the west. Accompanying this regional rain shadow was global cooling, meaning drier air. Only a handful of conifer species survived what the Great Basin slowly became in the late Cenozoic: a cold desert.

Bristlecone and associated plants made it not by migrating but by staying put on land that became snow-catching island mountains.

With snow, as with everything, there can be too much of a good thing. Millar described glacial ages within the Pliocene and Pleistocene epochs as "sweepstakes events" in the Great Basin. Some conifer populations lucked out; others ran out of luck. Of the 319 mountain ranges in Nevada, bristlecone now occur in less than 10 percent. Over the Holocene epoch, other pine species have outperformed bristlecone in terms of establishing new habitat space. "This is the era of [single-leaf] piñon," Millar said. "Piñon is so happy. It's going up, it's going down, it's filling in."

When genetics became genomics, Millar made a career change. "I didn't want to chase grants, and work in a lab, and be exposed to carcinogens," she recalled. She chose dendrochronology: "Tree-ring work is a lot cheaper, a lot more fun, and it smells nice." Working for the USFS, she did her best to help the agency transition from an extraction mindset to a management approach, and from a backward-facing philosophy to a forward-facing one. She described the old paradigm: "'People fucked up everything and we just need to go back to presettlement conditions.' I would try to tell my colleagues, 'Don't go back, because that's the Little Ice Age!'"

Millar's pet peeve is *unprecedented*. The word is meaningless without a timescale, she says. Current climate forcing is novel in its anthropogenic origin but hardly novel in scale or rapidity. "We're talking about the Earth—4.5 *billion* years." A high rate of change does not require an asteroid or a supervolcanic eruption. Regular variations in the Earth's movement—known collectively as Milankovitch cycles—can throw the planet into disequilibrium. When the Pleistocene ended, the climate of the Great Basin and its two great lakes changed almost overnight.

"It drives me crazy when people talk as if climate change started in 1975," continued Millar. "But if you start talking about natural climate change and historic climate variability, you get tied up with climate deniers, which also drives me crazy." Seeing humans as a product of evolution, Millar tries to think beyond "anthropo-blamism." "Of course," she added, "we can't continue like we are. I'm 1,000 percent for doing whatever we can politically to mitigate climate." However, being a geohistorian with expertise in ecology and evolution, she can't categorically declare that planetary change is always and forever "bad." Same for extinction. Over 99 percent of species that ever lived no longer exist, she reminded me. From death comes diversification. This *long*-long-term perspective—faith in the yin and yang of evolution—gives her equanimity. In contrast to young people in her circle, she's managed to keep pessimism at bay.

Shortly before Covid-19, Millar retired, sold her house in the Bay Area, and moved permanently to the western Great Basin. Now she's "living the dream," doing full-time field research, tracking down rumored stands of bristlecone in remote Nevada ranges. She and colleagues have documented recent *P. longaeva* mortalities from the combination of megadrought and bark beetles. There's no precedent for this in the (short) historical record. She predicts extirpations at the local level in the coming decades and centuries. But she takes comfort knowing that bristlecone is uniquely adapted for persistence. If native bark beetles appear high up in the White Mountains, she won't necessarily freak out about that "unprecedented" event. It's hard to imagine they've not been there before.

Millar disapproves of scientists who use social media to push their data in service of apocalyptic narratives—and their own career advancement. She has a suggestion for environmental science journals, which already require authors to submit keywords. She thinks researchers should also submit a "key message" for peer review instead of relying on nonexpert journalists and headline writers to tell the story.

Millar's friend Scotty Strachan at the University of Nevada is even more critical of for-profit scientific publishing: "We're not telling you anything new, but we're going to make an alarmist story about it." A lifelong resident of the Silver State, Strachan came to science by his bootstraps. After community college, he picked up dendrochronology on the spot as a lab technician, worked his way through a PhD program in geography, then won a national grant to create the Nevada Climate-ecohydrological Assessment Network (NevCAN).

The network consists of a pair of valley-to-mountain transects with monitoring stations. The resulting real-time data is meant to help planners in Las Vegas—the largest water user in the driest US state—prepare for the future. Strachan figured out where and how to build solar-powered infrastructure that constantly measures soil and atmospheric conditions, and also tree growth. It required "sweat equity." After the building phase came the troubleshooting phase; now he performs long-term maintenance. Before the initial grant ended, he purchased a shed's worth of spare parts. He visits his remote sites at least once a year, maintaining the network on a shoestring budget.[12]

The northern of the two transects runs to the top of the Snake Range. One summer day, I went along for the ride. The switchbacking two-track access road repelled my rental SUV. Strachan drove his truck fast, in full control. Near the limestone cliffs of Mount Washington, he pulled over to show me instrumented bristlecones, which, with wires and wraps and probes, looked more like cyberinfrastructure than sacred trees. From these pines, Strachan gets readings on sap flow and trunk diameter. Nearby, a phenocam measures the greenness of the foliage; a conventional camera captures the setting to show the presence or absence of snow.

All this data, having little immediate utility, gets stored on servers. Strachan espouses process-driven "slow science" and rejects citation-driven "fast science." His methodical research program gained traction with University of Nevada administrators once they understood the gravity of early-melt snowpack on regional water budgets. Strachan described how his data mining will, in the long run, allow refinements of western tree-ring chronologies—the foundation for regional climate modeling. "Ring width doesn't tell you much about hydrology. What is the tree responding to? What is the controlling factor? Temperature? Precipitation? Carbon dioxide? Solar availability? Soil column charge or soil water balance?" By demonstrating that Great Basin conifers respond to microtopographic signals—slope, aspect, shade, and so on—Strachan and his peers have uncovered statistical biases within older networks of proxy data. He says modelers don't necessarily want to hear this: "Just give me the numbers," they say.

Strachan predicts the big payoff will come fifty to eighty years down the road, long after he retires, when multimillennial chronologies have been smoothed to the utmost. "We go back to old tree-ring records all the time because there's little or no reward system now for creating them," he said. As a faculty member at a "low-status" research university, Strachan has a knowing perspective on the Laboratory of Tree-Ring Research in its formative decades. He suspects the Tucson lab benefited, on a purely scientific level, from being peripheral and nondepartmental before academia became corporatized. He notes that many chronology builders were "socially dysfunctional" and "obsessive" men who did slow science on remote trees, publishing "low-impact" articles in relative obscurity. In today's publish-or-perish system, such researchers would not be promoted.

"To understand the problem with fast science, you need to understand behavioral economics," Strachan told me. "Scientists are economic agents acting rationally in an established reward system." At the Anglo-American university—now copied in Asia—scientific "success" demands scientific "product." One result: an excess of sloppy science or plain bad science with un-reproduceable results. Perversely, given the bureaucratic work demanded by grant writing and grant reporting, scientists have less time to do research. Invoking Stewart Brand, Strachan compares the "pace layer" of contemporary science to the trends of fashion. He believes the system rewards short-term thinking.

Brand is a legendary figure in California's technophilic counterculture, the founder of the *Whole Earth Catalog* and later the Long Now Foundation. Based at Fort Mason in San Francisco, Long Now operates as an "incubator" for thinking about duration. Its evocative name comes from electronic composer Brian Eno. Its signature project, decades in the making, with no announced completion date, is the "Clock of the Long Now"—a monumental timepiece with pendulum and gears designed to keep moving in place for a myriad. To respond to modern civilization's "pathologically short attention span," Brand called for a slow version of the key instrument of Western modernity and capitalist time.

"Such a clock, if sufficiently impressive and well-engineered, would embody deep time for people," he wrote in his manifesto on long-term thinking. "It would be charismatic to visit, interesting to think about, and famous enough to become iconic in the public discourse. Ideally, it would do for thinking about time what the photographs of Earth from space have done for thinking about the environment. Such icons reframe the way people think."[13]

Trees do that, too—in a nonmechanical, all-organic way. And they provide the service freely.

Brand liked the idea of placing the clock adjacent to living timekeepers. Scarcely any bristlecone stands are privately owned, and the Long Now Foundation bought the best one: Mount Washington, Nevada. This inholding within Great Basin National Park used to be the highest-altitude mineral operation in Nevada. After the silver played out, the 12,000-foot-high mountain was worth a paper fortune because of the presence of beryllium, a strategic material. When Long Now made its initial inquiry with the property owner, a "very cantankerous one-man mining company," he held out for a millionaire's ransom. Later, after experiencing a health scare, he dropped the price, contacted the foundation's realtor, and gave the San Franciscans three days to cough up the money. Long Now called up its big donors, Silicon Valley founders, and closed the deal. The year was 1999. A group of Y2K preppers had set up camp in the adjacent valley, armed and ready for techno-apocalypse.[14]

A few years into the new millennium, Stewart Brand invited friends, including Jeff Bezos, to inspect Mount Washington as the box for the Clock. The foundation ended up choosing a different mountain, located in west Texas, within the vast ranch that Bezos purchased for his spaceport. The property has better weather, less seismic activity, better access—and, Texas being Texas, no permitting at all. With financial backing from the multibillionaire bringer of one-day delivery, the Clock of the Long Now is happening. The clockmakers compare their work to the pyramids. Bezos sees the mega-chronometer as the ultimate expression of his business philosophy. In every annual letter submitted to Amazon shareholders, he declared: "It's All About the Long Term." With no apparent sense of contradiction, the founder developed another catch-phrase: "It is *always* Day 1."

For Mount Washington, the Long Now Foundation has less pharaonic goals, namely an art installation that complements the NevCAN stations. It gave permission to a conceptual artist to build a kinetic calendar in the bristlecone zone. The artist plans to arrange a double spiral of ten limestone pillars around five pines. The pillars will mark expected tree girths in five hundred years, one thousand years, and beyond, as extrapolated from recent average radial growth. The numerals of a future calendar year will be incised into each stone. Notionally, when a tree reaches a growth boundary, it will topple the corresponding stone, thus marking the approximate date. However, the artist expects that climate change will alter bristlecone growth, meaning that the arboreal clock will fall out of sync with the Gregorian calendar. The installation will become a monument to temporal instability and predictive inability.[15]

Thought experiments take time, and when it comes to climate, some people are richer with time than others. West Coast techno-futurists, typically white men, can afford to feel greater equanimity about the geologic now, and hope about a geoengineered "Good Anthropocene," than the hundreds of millions of poor inhabitants of low-lying megacities in Asia and Africa. According to risk assessments, the Global South, which historically burned far less fossil fuel, will experience the worst effects of rising temperatures, advancing seas, and intensifying storms. In a worst-case climate scenario—global environmental collapse—the tech elite may steal more time by fleeing to island mountains, private islands, and backup countries. But not even Bezos has access to Planet B.

DISCOVERY AND LOSS

Given all the human suffering a worsening climate brings, it may seem inhumane to dwell on the future of old trees. Certainly, no one should waste their worry on the general death of woody plants. This form of vegetal growth independently evolved more than once, and has persisted through deep freezes and iceless periods. If lignophytes endured the end-Cretaceous extinction event, they will survive whatever humans throw at the planet. Nonetheless, the pace of biodiversity loss and chronodiversity decline is alarming. Although the current planet's "tree cover"—three *trillion* large plants covering roughly 30 percent of all land—has expanded of late, the new canopy mostly consists of shelter-belts, temperate-zone timber crops, and tropical plantations that supply eucalyptus pulpwood and palm oil. It's young stuff. A dwindling percentage of tree cover consists of species-rich old-growth communities.

What would humans and nonhumans stand to lose if these persisters all died prematurely? A world of things. The argument for helping elder-flora live out all their remaining years is multitudinous.

Olden boughs sustain forest communities. They produce copious seeds and life-sustaining litter; they host epiphytes and roosting animals. In ecologist Meg Lowman's formulation, there's an "eighth continent" in the canopy. The ecosystem underground might as well be a ninth. Trees share excess nutrients through mycorrhizae, the symbiotic association between fungi and plants at the root level. Fungal threads and tree roots come together in two-way connective tissues. Preliminary research on these mycorrhizal networks (the "Wood Wide Web") demonstrates that big old trees have outsize importance. In network terms, these trees are "central nodes" or "hubs" connected to hundreds of other trees. Megafloral hubs redistribute nitrogen and carbon—first to their own kind, secondarily to out-of-kin plants, even competitor plants. During a seedling's precarious recruitment phase, the cooperative assistance of a big old tree may mean the difference between death and a long, long life. Suzanne Simard, the leading ecologist in this field, refers to well-connected givers as "Mother Trees." The clearance of old-growth forests—a form of ecocide that continues in British Columbia as well as in the tropics—destroys not just standing trees but also the underworldly links among them.

Each millennial itself (themselves?) is a precious genetic repository. According to simulation models, one-quarter of the trees in an old-growth forest will be triple or quadruple the median age, and one-one-hundredth will be ten or twenty times the median age.[16] This demographic 1 percent has disproportionate significance, for chronodiversity aids biodiversity. Each Old One represents a specific moment in the past—a matrix of favorable conditions that existed upon establishment, and may not recur for centuries. As bridges between past conditions and possible futures, ancients contribute genetic resilience to the population. Moreover, any tree that has experienced placetime that long has persisted through hundreds of stochastic disturbances within hundreds of annual cycles, meaning its genome is adapted to withstand a range of conditions.

As genomic resources, perdurables are irreplaceable for science, too. Thousand-year growers generally come from ancient taxa of conifers. Their code—the product of hundreds of millions of years of evolution—contains information scientists have barely begun to analyze. As the technology of genetic sequencing advances, people may find new biochemical applications for the DNA of *Pinus longaeva* and other elderfloral species.

A subset of elderflora—conifers with climate-sensitive tree rings—has distinct scientific utility. Their unbroken dendrochronological record is a unique living data source for reconstructing past climates and modeling future ones. Sensitive conifers register cycles and disturbances at local, regional, planetary, and cosmogenic levels. As a class, their significance can be compared to monitoring stations like Mauna Loa Observatory.

Populations of Old Ones—particularly climax stands of megaflora—provide planetary services to humankind as carbon sinks. The slower that big old trees grow, the higher their potential for "negative emissions"; the longer they live, the longer they keep greenhouse gases locked inside their wood. Extremely old trees have already reached their saturation points, so technically they don't do anything to ameliorate excess CO_2. However, as parts of multispecies, multiage communities, they continuously delay carbon recharge at the ecosystem level. By comparison, large-scale tree-planting programs—monocultural and monochronic—are inferior in longevity and carbon storage. Because afforestation often fails, and because reforestation takes time, "proforestation"—the management of intact forests for maximum sequestration—makes global sense when the local fire risk is acceptable.[17]

Finally, elderflora make people more human by encouraging them to be themselves: *Homo sapiens*, the "wise people," the earthly beings who tell cosmic stories and who contemplate their future as well as their past. Stories of sanctified plants are among the oldest living stories. These narratives make reality. Around the world, at shrines and temples and churchyards, local adherents give care to storied trees planted centuries ago—or, just recently, the latest in a long, unbroken sequence of consecrated plantings. Meanwhile, agents of modern states guard secular sacred groves: natural monuments and national parks devoted to naturally occurring megaflora and elderflora. By keeping up or starting long-term relationships with long-lived plants, caregivers maintain (or recover, or reinvent) links to premodern pasts. Simply by hoping to prolong interspecies attachments through the great diminution, people exercise intergenerational sapience. Their hope is a rejection of The End, an affirmation that there will be—must be—tomorrow.

All that being said, more and more Old Ones *will* die.

We will see it on our screens if not in person. People visualize climate crisis with disturbing images taken with telephoto lenses, underwater cameras, and drones: desperate refugees displaced by severe floods or intense droughts or unlivable heat waves; haggard polar bears stranded without ice; bleached coral reefs devoid of colorful fish; monumental sequoias with scorched crowns; and dead standing trees near encroaching coastlines.

Science journalists refer to low-elevation mortalities as "rampikes" composing "ghost forests"—a consequence of saltwater intrusion. In the long run, ghosting will hit the Sundarbans of India and Bangladesh, the greatest mangrove forest on the planet. With dense human populations claiming the next-level land, the mangroves and their tigers will have no upward retreat, as they did in past periods of rising sea levels. On the US Atlantic coast, blackwater forests and coastal woodlands are not mangroves, and thus less tolerant of increased salinity. The 3,000-year-old bald cypress population along North Carolina's Black River stands just two meters above current sea level. In New Jersey, ghosting imperils old conifers already.

To call attention to lowland casualties, the celebrated memorial architect Maya Lin installed the bare bodies of Atlantic white cedars from New Jersey's Pine Barrens in Madison Square Park in Manhattan, coinciding with New York City's partial reopening from Covid-19. "That does become part of the piece," said the artist. "There is a sense of mourning."[18] Her temporary memorial, *Ghost Forest*, included a soundtrack of birdsong from endangered and extinct species. Disconcertingly, visitors had to download the avian voices to their phones.

The large-scale demise of megaflora and elderflora may be compared to the Late Pleistocene extinctions and the lost world of megafauna that haunts our mammalian imagination. We wonder: What was it like for our remote ancestors to live alongside mastodons and other giants? Our not-so-distant descendants may wonder: How did it feel to visit mammoth and millennial trees? Some of the greatest plants that ever lived, organic landmarks of the Holocene, will be relegated to libraries, electronic databases, and other technologies of memory. Modern people barely had time to cherish age-old plants like giant sequoias before beginning their goodbyes.

The local or general disappearance of largeness *with* oldness—another "extinction of experience," like undarkened night skies—is different from species extinction. As long as genomes survive, the potential exists for a future time when humans again coexist widely with multimillennial megaflora. But near-future people may be deprived for an interregnum, centuries long, when vascular plants adapt to a new regime of variability. The Earth will not instantly start cooling whenever the global economy at long last reaches global net zero. In the coming time of overshoot, elderflora will continue to time out.

As living monuments fall, the services of artists and designers like Maya Lin may be necessary to maintain collective memory. Text-based remembrances of slow plants are quite ephemeral if recent history serves as a guide. In the long nineteenth century, Europeans and their colonists—also the Japanese state—inventoried big old trees, both for harvesting and for protecting. This cross-purpose activity led to unprecedented public awareness of the lives and deaths of outstanding specimens. Few of these once-celebrated organisms are remembered today. There may be nothing so lovely yet forgettable as a tree.

In the American East, multiple species, including some of the tallest growers in North America, have declined or virtually disappeared because of nonnative invasive pests and pathogens. While it happened, people bemoaned the dieback of American chestnut, American elm, white walnut, and American beech. Today, most easterners have no idea what was lost. Dendrophiles, meanwhile, feel the ongoing bereavements of eastern hemlock, Carolina hemlock, white ash, and green ash—losses that future Americans may again not register.

The concept of "plant blindness" dates to the 1990s, when researchers noted that US students, particularly boys, lacked interest in botany. Two biology professors coined a slogan, "Prevent Plant Blindness."[19] Ignoring the influence of gender, they made an evolutionary argument: Bias against plants was less zoological chauvinism than human adaptation. Our optical system evolved to detect movements, patterns, and subtle facial cues. We remember faces. Big-eyed moving animals command greater attentional resources in the human brain—faster processing, better recall—than do stationary plants.

Given that trees are the most visible of flora, my own limited ability to remember them pains me. When I lived in Brooklyn, a tornado passed right by my walk-up apartment. Stupefied by the eerie green light, I stood by the window: I couldn't fathom what was happening. A half block away, the twister sundered two large trees I had learned to love on my daily walks. I sorrowed when I touched their shattered trunks later that day. Weeks later, looking at their chainsawed stumps, I realized I couldn't visualize the leaves or canopy structures, or identify the species.

I've met one person with perfect recall for trees. Her name is Valerie Cohen. After working as the first woman law enforcement officer at Yosemite National Park, she developed a second career as a watercolorist and pen-and-ink artist. She hikes off trail in the White Mountains, looking for bristlecones that speak to her. "No one would ever find my fucking trees," she told me. She refuses to take photographs. She stares and sketches for hours until she creates an indelible memory.

The cover of Cohen's *Tree Lines* shows a twin-boughed pine that resembles a cantilevered sculpture. The year following publication of her artist's book, she returned and found the tree collapsed, its mechanical tension released. "It looked like a bomb had gone off. It looked like the ruins of a war-torn city." She was flabbergasted, awestruck, spooked: "What are the chances? That one insignificant person, one tiny little artist, would get to see and sketch this tree, this perfectly balanced thing that had been growing for fucking four thousand years—and then, two years later, it's gone!"[20]

I asked her what it felt like. It felt like grief, she said.

Mourning is at root an act of care. Organizing and mourning can be complementary when losses motivate future-minded actions. If the Holocene's perdurables survive in memory, it will be thanks to painters, photographers, videographers, and installation artists. Storytellers, too. The soon-to-be untimely dead will be more memorable if they fit into a narrative. However, the primal tale of the tree desecrator—an individual who profanes a local sacred tree—doesn't capture the dynamic of industrial emissions contributing indirectly to global and regional patterns of local diebacks of big old trees. Who's the wrongful agent of this foreshortening: Nobody? Everybody? The One Percent? Fossil fuel firms? The parties of Thatcher and Reagan? The Global North?

A conventional perspective foregrounds Westerners as heroes as well as villains. Indeed, the dominant Western narrative about nature in the modern period can be summed up in a phrase: discovery and loss. Explorers and scientists, imperial white men, "discover" new species, unknown peoples, unnavigable rivers, unclimbed peaks, latest oldest things. They navigate, they summit, they survey, they measure, they name, they core, they claim. They create knowledge that permits forests to be cleared, roads to be blasted, rivers to be dammed, minerals to be extracted. They feel pride about progress, mixed with regret. They have "imperial nostalgia" for old landscapes, old languages, old ways, and old trees they helped to eliminate.[21] So, they apply a portion of their developmental ethic and their accumulated capital toward saving pieces of depopulated Eden, and restoring damaged parcels to imagined presettlement conditions.

In restoration ecology, there's a textbook concept called "shifting baseline syndrome" that attempts to capture the relationship between collective memory and changing ecosystems. The concept originated in fisheries science because the earliest available population data—the historical "baseline" or "reference point" for potential restoration—is purely anecdotal. For generations, commercial fishers told the same story: back in the day, the boats came back with catches *this big*. The reference point for the "way things should be" was relative; each generation started with its own baseline of plenitude. Until the threshold of collapse, a fishery's rate of decline could be acceptably slow on the scale of a human lifetime. As a result of "generational amnesia," steady depletion became normalized, step by step, again and again.

The insuperable problem with restoration ecology is that time goes one direction, and no one can stop environments from changing—not ever, especially not now, with species introductions and what ought to be called global *heating*. Historical anomalies are the new normal. There's no trajectory to "climax," "steady state," "stasis," or "equilibrium"—jettisoned keywords of classical ecology. At most, given enough time between disturbances, an ecosystem reaches "dynamic stasis."

Ecologists draw a distinction between disturbance and catastrophe. The former is frequent and ubiquitous, the latter infrequent. In effect, every functioning ecosystem exists in recovery mode. How well it bounces back—its resilience—depends on the scale of the exogenous disturbance as well as endogenous factors like species richness and the adaptable traits contained within any given genome. A sequence of steps will become disturbingly familiar in coming decades: First, woody plants and their forest ecosystems persist under stress. Beyond a threshold of persistence, populations suffer mortalities yet recover. Beyond a threshold of recovery, communities reorganize by species assemblage and morphology. Finally, beyond a threshold of reorganization comes type conversion. A catastrophe basically fast-forwards this sequence to the end. After a supervolcanic eruption, for example, adaptation breaks down. Species and communities are not preadapted for rare events that exceed evolutionary tolerance. Instead of another waypoint, an endpoint arrives. A reset occurs.

Extinction becomes catastrophic in the ecological sense, in addition to the moral sense, when the scale of genomic loss forces resets of ecosystems. So far, global heating has been a disturbance—even a megadisturbance—for forest ecosystems, not yet a catastrophe. In abstract evolutionary terms, the climate-forced dieback of trees is simply strong selection. Ideally, deaths open niches for diversification and migration. On the current planet, though, biodiversity continues to plummet due to land use and resource extraction. Destroyed habitats and obstructed corridors are everyday catastrophes.

In a "best-case" scenario for global heating—1.5°C above the preindustrial average—most protected forests should be able to change into new assemblages of species and morphologies. But linear increases in temperature bring nonlinear transformations. The difference between 1.5°C and 2°C is extreme and fiery. And at 3°C, an unbearable scenario that seems increasingly likely, boreal and tropical forests may meet ruination. Expressed as an aphorism: if people want old coniferous trees growing on the ground, they must, as part of degrowth, keep older Carboniferous trees in the ground.

Maintaining life support for megaflora and elderflora will require global political action and a variety of local interventions: prescribed fire, mechanical treatment, assisted migration, genetic banking, and possibly genetic engineering. Without increased stewardship, natural forests on a rapidly warming planet will experience sudden diebacks and slow recoveries instead of gradual transitions. Given the multitude of forest crises—and the availability of funding—tree lovers will face upsetting choices about resource allocation. Ignoring for the moment our feelings for emblematic megaflora, saving any specimen or stand is less important than sustaining the conditions for continued biodiversity and chronodiversity.

A culture of sustaining includes a system of triage: stopping change or slowing it, ameliorating or accepting. The acceptance of all change is perilous, and heartless, for we are, in fact, emotional animals with place attachments. If there is no continuity between the baseline conditions of one generation and another, intergenerational alienation ensues. Discontinuity becomes deworlding. The term *solastalgia* attempts to describe this affective state—desolation at home, or homesickness in one's own place, when change occurs too rapidly for attachments to stabilize.[22] The youth climate movement is in part a reaction to chronic temporal instability, a condition in polar opposition to a tree's placetime. Many young people feel they have everything and nothing to lose, having lost hope in the future. They cannot even grow into wistfulness when baselines shift so fast, right before their eyes.

Climate change widens the gulf between neophobes and neophiliacs. Holocene pessimists fear that technological, cultural, and ecological novelty could in combination produce an evolutionary crisis for *Homo sapiens*—an omega point beyond which the stress of the new could debilitate the species itself. Anthropocene optimists look forward to directed evolution ("*Homo deus*") and the achievement of life extension through genetic enhancement and technological augmentation. Both scenarios— the inhuman and the transhumanist—assume the virtual world as primary. In the metaverse, it's hard to imagine people caring, or remembering to care, about ancient trees and chronodiversity. In science fiction terms, an Empire of Day 1 could deliver a monochronic world where everything is new, and nothing lasts; a time of permanent juvenescence, devoid of wisdom, where the calendar reverts annually to Year Zero.

I hope against that story, which is one more variation on the end of the world. I'll take waypoints over any endpoint. I try to keep faith in the garden of life beyond fossil fuel capitalism, without knowing what that garden will be. A great transition will happen one way or another—either catastrophically, with a fast-changing planet undoing its changers, or deliberately, with enough people working together to revitalize the cycle of the young becoming the old who sustain new youth. In oldness is the rejuvenation of the world.

Here, to my own surprise, I find selective value in Stewart Brand's thinking. In contrast to the *Bulletin of Atomic Scientists*, which uses the Doomsday Clock to perpetually count down the final minutes to Zero Hour, the Long Now Foundation places a zero before each date in the Gregorian calendar. The year 2020 thus becomes 02020. I would take the exercise further, and backward: to accompany the long now, the *long then*. To account for all the non-Western times before the Common Era, I would place two additional zeros—Y2K thus becomes 0200000. The extra terminal digits give a more accurate sense of the potential duration of hominin species, which is still a fraction of the evolutionary life span of gymnosperms.

As far back as knowledge goes, dignified olden trees have existed. They help people make sense of—and make amends for—their destructive capacities. Could humanity abide without the moral grounding provided by elderflora? Probably. Humans are adaptable.

Do we really want to find out?

KAURI PASTS

Of all the stories I've collected of people who feel for ancient trees, the one that resonates with me most comes from Aotearoa.

Francis Rei Paul Hamon, born in 1919, grew up on the North Island, near Poverty Bay, an appropriate place-name for his upbringing. One-quarter Polynesian (Ngāti Porou and Te-Aitangi-a-Māhaki), he grew up at a time when Māori internalized colonialism. He never became fluent in the language, despite growing up around it. His second mother was his Māori grandmother, who reportedly died with a full set of teeth at age 118. His Māori grandfather had converted to the Church of Jesus Christ of Latter-day Saints, based in faraway Utah. LDS missionaries published the first Māori-language edition of the Book of Mormon in 1889. Rei's father, Hixon Hamon, baptized him into Mormonism at age eight. The Hamon family suffered during the Depression, particularly after their small dairy farm flooded. Rei, the senior of fourteen children, left home to work as an early teen. He made his way to the forests of Te Urewera and learned the subsistence timber trade from Māori and Pākehā bushmen.

Rei later reported that he and his logging comrades saw three huias, an iconic bird of Aotearoa made endangered by feather hunters. It was an unforgettable, unrepeatable encounter. Today, the huia cannot be seen, though an echo can be heard. After the species went extinct, a Māori bird-trapper accompanied a Pākehā historian to a sound studio in Wellington to make a recording of a mimicked call, now available online. Listening to a dead man sing is the only way to experience the departed species.

Rei and his brothers ended up in Thames, New Zealand, on the Coromandel Peninsula, working as loggers. They felled relict stands of ancient kauri in rugged, remote locations that previous loggers couldn't reach. In a family film from 1955, the smiling Hamons use a bulldozer to remove a twenty-five-ton "monster" from the hills above Tapu, a place-name that literally means *sacred*. Dragged to the water, the giant floated like a harpooned whale. "And so we say 'farewell' to this king of the forest," says the narrator of the home movie.[23]

Rei constantly struggled to provide for his prodigious family. Between biological, adopted, and foster children, he and his second wife raised more than thirty descendants. He loved the parenting but felt compunction about the breadwinning. "She's starting to talk!" loggers would yell when a kauri groaned and crackled. Sometime in the late 1950s, Rei began to hear that sound as a protest. When a kauri uttered its last, he would heedlessly rush forward and embrace it right before it fell. Finally, in 1961, he killed his last "monarch." "It had been standing there for maybe a thousand years," he recalled later. "I went back later to where it had been standing, and there were birds fluttering around there, kaka and kereru, that had nested in that tree for generations. That was the finish. I handed in my resignation. I vowed never to fell another healthy tree."

He experienced a second life-changing fall into grace. Still in need of money, he worked for the Ministry of Roads, blasting roads in rugged timber country, where an accident took him off the edge of a cliff. At age forty-six, near bankruptcy, living on bread and milk, with screws in his spine, he felt overwhelmed by darkness. His wife, Maia Hamon (Ngāti Porou), led him in kneeling prayer. She asked Heavenly Father for assistance. After intoning, "Amen," Rei opened his eyes and noticed that his six-year-old daughter had forgotten to take her drawing pad to school. He felt inspired to pick up a pen. For the first time in his life, he sketched a picture of a tree. It became the first of hundreds.

Within months, he switched from ballpoint to mapping pen, from strokes to dots. With intuition and practice, he became a master of pointillism. There is purity in the dot, he would say. Each picture took months—hundreds of hours, millions of dots. After his children went to bed, he would shave, put on a clean shirt, complete with bolo tie, and work through the night, dipping and dotting and blotting. He began at the corners and worked toward the middle. He never sketched or traced in pencil before applying the India ink. He worked from images in his mind. His motto: "To see things just as God made them." In the still of the night, a veil was lifted.

Within years, his drawings sold for tens of thousands of dollars in Auckland galleries. New Zealand newspapers played up the "discovery" of his pure, authentic, true, naive talent. Rei Hamon was a "bush miracle": this "tough, tanned, and nuggety" timberman had never taken a lesson in his life, only graduated from the University of Nature. When Elizabeth II came to New Zealand on one of her royal tours, the government presented her with a Hamon original and a bound volume of his work. In return, the Queen bestowed upon him the title Commander of the Most Excellent Order of the British Empire.

Leveraging his stature as "World Famous Artist of the New Zealand Bush," Rei became an activist, calling for the government to halt kauri logging on Crown land in Coromandel. The species may go extinct as a result of greed, he warned. He admitted his entanglement: "I helped to destroy the bush. Until one day after felling a kauri I put down my axe and said, 'No more.'" His activism directly contributed to the establishment of Manaia Kauri Sanctuary, though not before twenty giants succumbed to chainsaws. To memorialize their loss, and re-create their lives, he made a pen-and-ink in large format. In the background, half-dotted outlines show kauri spirits. "The ghost trees represent our LOST HERITAGE," he wrote. Later, he made a composite portrait of New Zealand's preeminent kauri specimens, living and dead, titled *Ngā Ariki*. This was the term for hereditary high chiefs, also the term adopted by Latter-day Saints for senior officers in the priesthood. Rei himself served as a High Priest.

Arthritis set in, and Rei produced little art in the last quarter century of his long life. He persevered with his bum hand in two special cases. He dotted a love picture for Maia in her declining months. And he produced a large-scale work intended to hang in the Salt Lake Temple at Temple Square, the inner sanctum of Mormondom. For this task, he felt prompted to draw people for the first time. While he could effortlessly depict kauris, silver ferns, and extinct huias, the human face seemed out of reach. So, he fasted with Maia on a Sunday in preparation. Then he dotted in the Angel Moroni guiding Joseph Smith to the ancient text of the Book of Mormon. To represent the American setting for the prophet of the last dispensation, Hamon added the supreme American tree, giant sequoia, to the background. Aotearoa ferns and flowers adorned the foreground. "The picture is a testimony of my wife and I," said Hamon in 1993, when presenting it to an apostle on ecclesiastical tour in New Zealand. Upon arrival in Salt Lake City, it went into storage, where it still waits to be hung.[24]

I keep this allegorical art in my mind's eye as a surprising composite of themes I turned to after turning away from my homeland: globalization, modernity, purported antiquity, and sacred trees. In the context of Utah Mormon culture, which in my lifetime aligned with hypercapitalist anti-stewardship, Rei Hamon's syncretic life offers a reminder: The past contains rootstock for various futures. The fruit depends on what people choose to tend. I think of the apocryphal birthplace of the LDS religion—the backlot of a family farm, where, circa 1820, young Joseph Smith knelt to pray among cordwood and sugar maples, and received a heavenly visitation. One century later, Mormon leaders consecrated the second growth as the "Sacred Grove." By the bicentennial of the "First Vision," the Church referred to it as "one of the few tracts of primeval forest in western New York."[25] Among these outgrowths of history, I see things worth tending.

In my corner office in College Hall, Philadelphia, alongside my grandmother's souvenir plate of Temple Square, I recently hung a framed copy of Rei Hamon's ghostly memorial, *Lost Heritage*. This print was a gift from Rei's granddaughter, a musician. Her name is Huia. She's not LDS, but she serves a calling valued by Māori and Mormons alike—family historian. Huia Hamon showed me documents from Rei's original gallery show. The sale list included his first piece, the one created after his wife's prayerful intervention. Rei gave it the title *Inspiration (Dead Kauris on the Kiri Kiri Pinnacles)*. This clunky hatch drawing (no dots) depicted five defoliated giants on steeply eroded badlands in Coromandel. Killed by a bush-clearing settler fire, the trees remained standing on rhyolite slopes. It had been Rei's job to extract them.

Isn't it eerie, Huia said to me in a hushed voice, that his dead kauris look just like kauri dieback?

She was referencing an ongoing demographic crisis in the North Island's northern forests—the consequence of an invasive microorganism. The genus *Phytophthora* has a deadly modern record. One species caused the potato blight that led to the Great Famine in Ireland. Another produced sudden oak death in North America and Europe, and yet another threatens cocoa crops worldwide. The species named *Phytophthora agathidicida* (literally "kauri-killing plant-killer") infects trees at the root. By the time people notice oozing trunks and dying branches, the disease has reached its chronic, fatal phase. Because forest trees are interconnected at the roots, one tree may infect its surrounding kin—the malevolent version of a "mother tree." Probably of exogenous origin, the pathogen has taken advantage of kauris stressed by two hundred years of chronic disturbance, including fire, habitat destruction and fragmentation, the loss of coevolved birds, and the introduction of pigs, goats, and possums. Mammals, including hikers, spread the deadly spores by walking through the forest.

First identified on Great Barrier Island in 1972, the pathogen eluded notice on the North Island until 2006. Although New Zealand, like California, has strong biosecurity protocol, the state prioritizes vineyards and nonnative kiwi fruit over native trees. Māori pushed the government to do more. Te Kawerau ā Maki iwi placed a travel restriction (*rāhui*) on the forests of the Waitākere Ranges in 2017. Hikers ignored this customary rule until the Auckland Council followed with a formal rule, and started issuing fines. The ban may have come too late, given the lag between infection and death. Aerial surveys of Waitākere show ghostly white crowns in the upper canopy. The Department of Conservation now lists *Agathis australis* as "nationally vulnerable." Some activists and scientists speak openly of extinction.

Others pledge to fight back with science, including early detection with dieback-sniffing dogs or hyperspectral camera drones, followed by medicinal injections of phosphorous acid. The race is on to locate disease-resistant specimens that could become the basis for a selective breeding program. Lab engineering may be feasible, too, as evidenced by the new blight-resistant transgenic American chestnut. This kind of "restoration" divides Western environmentalists. Whether command-and-control science could be harmonized with Indigenous knowledge (*mātauranga Māori*) is another issue.

Tāne Mahuta, the kauri with honorary status as the eldest elder in Aotearoa, remains open to visitors for the time being. Heritage tourists pass through a scrubbing and disinfecting station, then walk on an elevated track. Now that dieback has been detected in soils of the Waipoua Forest, the visitor experience includes anxiety and the forefeeling of loss. The death of Tāne Mahuta would be devastating to the local Te Roroa iwi, and heartbreaking for the country, for kauri has become a bicultural treasure (*taonga*). Te Roroa "ambassadors" stationed on the track encounter tourists who weep at the sight of the tree, or pray, or apologize for colonialism. Meanwhile, surveillance cameras in Waipoua show that some visitors cannot stop themselves from jumping the bounds of the boardwalk to hug old trees.

Even as kauri has become more mortal, it has become more timeful. In the Northland, whole stands of prostrate trees—possible windthrow of cyclones—lie just below the surface of pasturelands. These preserved kauris fell into anoxic wetlands—or, as Pākehā call them, "swamps." In the absence of oxygen, the resinous wood barely rotted over tens of thousands of years, retaining the power of treeness. When excavated from paleo-bogs, the attached leaves remain momentarily green, before browning from oxidation. The technical value of these ultra-old trees greatly increased after scientists demonstrated in 2000 that kauri tree rings can be used as proxies for El Niño-Southern Oscillation activity.

On the global market, the amber-colored paleo-wood had greater exchange value. In the 2000s, by which point it was illegal to cut old-growth kauri, New Zealanders began digging for "swamp kauri" like gold. Drought in the Northland caused soil contraction, revealing the outlines of treasure trees easily unburied with earth-moving machines. On pastures that had been burned and cleared in the nineteenth century, subterranean deforestation commenced. Farmers accepted cash from digging operators with names like Swamp Cowboys. Millers cut gigantic boles into twenty-foot-long knot-free boards the likes of which no one had seen since the bad old days of redwood clearance in pioneer California. Furniture makers polished the "world's oldest wood" to an opalescent shimmer, making spectacular novelty dining tables for China's nouveau riche.

For different reasons, Māori, environmentalists, and scientists objected to exhumed taonga leaving the country. Lawsuits, journalistic exposés, government reports, and a Supreme Court ruling at last produced a regulated system in 2018. With get-rich-quick operators now gone, established extractors continue their business. They self-report to Auckland, and government scientists travel to digging sites to retrieve stratigraphic metadata and samples of wood—smaller "biscuits" or bigger "cookies." Time is of the essence for reliable radiocarbon dating. In return for fresh on-site samples, the owners receive government-issued certificates stating estimated ages, which increases sale values.

The point man for this scientific salvage work is Drew Lorrey. I spoke to him at the headquarters of NIWA, the Realm's equivalent of NOAA. Lorrey moved to New Zealand from the United States and adopted Aotearoa as home. After surveying glaciers in the Southern Alps, he switched to dendroclimatology, becoming the overseer of a paleo-archive consisting of multiple floating chronologies across 70,000 years. He dreams of bridging all the gaps to form one "ultra-long" kauri chronology, five times longer than the bristlecone chronology, covering the entire Holocene and deep enough into the Pleistocene to reach the previous interglacial, which ended circa 115,000 years BP. Comparable supplies of wood don't exist in the Northern Hemisphere: glaciations destroyed the evidence. Thinking in the long term like a slow scientist, Lorrey refers to his collection as a "time capsule" for future researchers who will have better tools. "I'm a 'steward,'" he said. "If that's all I do in my career, it will have been a service to New Zealand."[26]

For the moment, commercial swamp kauri is the only source for kauri tree-ring data. Māori never liked dendrochronologists drilling into kauri, and now dieback makes the practice unallowable. "Did I spread it myself?" wonders Lorrey. The arboreal crisis reminds him of Aotearoa's glaciers. It's emotional to witness their attrition, he said, akin to seeing a family member waste away from cancer. "I may be here when the glaciers die. Will I also see the end of kauri? I'm here for the end of days of so many things."

Lorrey speaks of swamp kauri in reverential terms and uses the term "corpse," a personification that echoes Māori usage. Corpses bring obligations. In 2019, an energy company building a geothermal power station in Northland uncovered a bole, sixty-five feet long, and decided to donate it—three truck-sized segments—to the local family group (hapū). The community around Ngāwhā Springs proudly became its guardian (kaitiaki). Elders received the dismembered giant with the same ceremony performed for dead bodies, and gave the tree-body a blessing that combined Polynesian and Christian elements. The hapū plans to give away pieces of the tree for ceremonial uses. In the meantime, the twice-lived kauri lies in the parking lot by the communal space (marae) where people gather for birthdays and funerals.

In the form of data, the Ngāwhā swamp kauri has assumed a third life as the first scientifically studied tree that lived through the last excursion of the magnetic pole. Now and then, the North Pole wanders and, for years or centuries, makes off to Antarctica. Pole reversals bring increased solar radiation, because a wandering magnetic field is a weaker field. In 2021, Science published an article that refined the date range for the last excursion on the basis of radiocarbon from the tree's 1,700 rings. The onset occurred forty-one or forty-two millennia ago, in the Late Pleistocene.

What might have been a research note became a buzzy article. Without any proof except approximate coincidences, the authors conjectured that the lessening intensity of the geomagnetic field induced a "global environmental crisis," including climatic change, the mass extinction of Australia's megafauna, the end of the Neanderthals—and, to top it off, the birth of painting. Supposedly, hominins retreated to caves to avoid skin-damaging UV and began applying their red ochre sunscreen to rock walls in a burst of artistic inspiration. In short, the authors proposed that geomagnetic excursions could produce effects comparable to asteroid impacts and supervolcanic eruptions—a further extension of the intellectual return to catastrophism since the 1980s.[27]

To accompany their peer-reviewed speculation, the primary authors issued press releases with looser language. A "perfect cosmic storm" brought "apocalyptic conditions" like a "horror movie" or the "end of days." Because these imagined End Times began more or less 42,000 years BP, the authors hyped the transition as the "Adams Event," an unironic tribute to Douglas Adams, best-selling author of *The Hitchhiker's Guide to the Galaxy*. (The plot of the book, which begins with Earth's annihilation, turns on the number 42.) To promote the *Science* article, the academic employer of the second author produced a retroslick YouTube video with voiceover by comic actor Stephen Fry, who had previously narrated the audio version of Adams's novel. The proposed "Adams Event" smacked of citation-seeking, grandstanding, and reputation management, given that the lead author had recently lost his academic position (for the second time) as a result of misconduct.

Whether or not the hypothesized catastrophe was credible science, it was clickable news in a time of climate crisis. The *New York Times* and other media outlets obligingly provided coverage. In the attention economy of fast science, the discoverers of the kauri "Rosetta stone" demanded instant acclaim.

UNDEAD WOOD

Swamp kauri is an exemplar of what scientists call "subfossil wood"—old enough to be part of the geological record, but neither decomposed nor fossilized. It can persist on the surface if aridity remains low—as in the White Mountains, where bristlecone "driftwood" seems adrift in time. More typically, it occurs underground or underwater where local conditions inhibit decomposition.

Around the world, since ancient times, woodworkers have recovered subfossil wood, as exemplified by alerce in Chile. In the nineteenth century, such wood gathering turned commercial in a few locations, including the Pine Barrens of Cape May County, in southern New Jersey. The same Atlantic white cedar memorialized in Maya Lin's *Ghost Forest* was ghosted many times in the deep past. Over the Pliocene and Pleistocene epochs, repeatedly, the Atlantic rose and fell, ice sheets advanced and retreated, land subsided and rebounded, while moraine deposits and ocean sediments added to topographic change. At times, conditions shifted fast enough to submerge whole stands of conifers.

Cedar mining became an industry only after settlers logged most of the standing old-growth cedars. In a wetland near Dennisville, New Jersey, extractors found a large supply of timber six feet under—whole trees lying prostrate, one on top of the other, like a logjam. The "Great Cedar Swamp" became a big operation after the Civil War. The laborers tasked with getting out the material learned to read the underground forest. They drew a distinction between "windfalls" and "breakdowns." The former contained wood with the tight grain and "delicate flesh color" associated with old growth. Like gumdiggers in New Zealand, swamp rats probed the muck with a sharpened iron rod—a "progue"—six to eight feet long. After proguing a cedar, a shingle miner used a cutting spade to dislodge a chip that would float to the surface. If it smelled aromatic, he would excavate; then, right before the cavity refilled with water, he would axe the buoyant trunk from its intact roots. The trunk became bolts, and bolts became shingles. A proficient worker could split a thousand shingles per week. All this mucky work was done by hand. Although the last shingle miners died in the early twentieth century, their handiwork survives in Philadelphia, the main destination for Cape May's paleo-wood. When Independence Hall needed a new roof, renovators turned to this imperishable product.

The first capital of the United States has its own primeval trees beneath its colonial buildings, and below its African burial grounds, as exhumed during excavation of its subway. In 1931, two blocks from Independence Hall, workmen encountered a row of bald cypress stumps in upright position ten feet below sea level, thirty-eight feet below Locust Street. A massive specimen went on display at the Academy of Natural Sciences. In 1960, the University of Pennsylvania's radiocarbon lab (the same lab that dated many bristlecone samples) put a C-14 date on the "Subway Tree"—approximately 42,200 years old. In other words, it was an exact contemporary of the Ngāwhā kauri. Bald cypress didn't exist in Philadelphia when William Penn laid out the grid with streets named for trees, but now, in the twenty-first century, plantings of the species are common in landscaped settings because arborists expect it will do well in the city's increasingly subtropical climate. When I walk home from my workplace at Penn, I see bald cypresses and metasequoias by the postindustrial Schuylkill River.

Farther south, sediment from the Potomac River long ago buried a magnificent stand of bald cypress a few blocks from the current location of the White House. Workers found hundreds of big stumps in 1921 when excavating the foundation for the Mayflower Hotel, and hundreds more in 1961 when constructing a National Geographic Society building. These trees are radiocarbon dead, almost certainly older than 100,000 years—relics of the previous interglacial.

To imagine what might happen to the American Rome as Greenland and Antarctica continue to melt, a team of landscape designers submitted plans for a speculative memorial, "Climate Chronograph." This was the winning entry in "Memorials for the Future," a 2016 competition cosponsored by the NPS and the National Capital Planning Commission. The designers envisioned new plantings of Japanese cherries, row upon row, sloping downward to Hains Point, which juts into the Potomac. The trees would die, one row at a time, as the tidal river rose, and these "rampikes of lost cherries" would mark the "aftermath of the present." Invoking both ancient Egyptian Nilometers and current climate refugees, the designers called for a transitory monument to mark a change in time.[28]

Trump won the presidency two months after the competition, and soon the US government began deleting "climate change" from its websites. The existing Yoshino cherries of the Tidal Basin (two miles upstream from Hains Point) could not be censored. They had become their own chronograph. On a century-long scale, the average date for peak bloom had moved up one week. Every day, at high tide, water submerged the sidewalk, and tourists resorted to walking on roots. Individual cherries succumbed to flooding, with additional mortalities in the offing. By late 2020, the threat to the basin—including the Jefferson, FDR, and MLK memorials—was undeniable. A government-sponsored Tidal Basin Ideas Lab released five potential plans, each with a different matrix of adaptation, transformation, and relocation, all with an acceptance of the inevitability of change.

In some coastal regions, revelations of past change appear when beach-scouring storms coincide with low tides. "A very remarkable circumstance occurred," noted Gerald of Wales, the twelfth-century chronicler. "The sandy shores of South Wales, being laid bare by the extraordinary violence of a storm, the surface of the earth, which had been covered for many ages, re-appeared, and discovered the trunks of trees cut off, standing in the very sea itself, the strokes of the hatchet appearing as if made only yesterday." By this "wonderful revolution," a sandless beach with black soil appeared "like a grove cut down, perhaps, at the time of the deluge."[29] Such coastal reveals still happen. In Wales and Cornwall and North Yorkshire, stumps and roots from interglacial forests occasionally surface for a few wintry days, before disappearing again under sand and sea.

Ancient trees like these attracted the attention of British paleobotanist Clement Reid, author of *Submerged Forests* (1913). Reid—in collaboration with his wife, the paleobotanist Eleanor Mary Reid—anticipated the interdisciplinary work of climate historians. The Reids inferred that the southern North Sea used to be an alluvial plain, a land bridge connecting Britain and the Continent. By the 1990s, archaeologists had gained remote sensing tools to map this lost world, which they named Doggerland. The Mesolithic history of Britain became longer, wider, and noninsular. In the warming period of the early Holocene, Doggerlanders witnessed unprecedented change that was perceptible in a human lifetime, and narratable on a cultural timescale. We don't know the stories they told, or the emotions they felt, but they adapted and acclimated until at last they accepted the stillness of the open sea.

In its short evolutionary history, *Homo* has witnessed profound planetary changes. In Europe, before the geomagnetic reversal, Neanderthals and other hominins lived through the class 7 Campanian Ignimbrite volcanic eruption, an event large enough to cause multiyear global cooling. Contemporary Native peoples of the Pacific Northwest tell oral narratives that include specific tsunamis, volcanic eruptions, and megafloods thousands of years in the past—events that geologists have named and dated using different methods. Geological stories from Aboriginal Australia are even older.

One thing humankind has never experienced is a planet without ice. The timeline of our species overlaps that of current mountain glaciers. Given that fact, the stance of Big Oil in the late twentieth century represents the polar opposite of the precautionary principle. In 1977, an Exxon scientific adviser created a graph that projected a CO_2-induced "super-interglacial" that could potentially last centuries. He gave a presentation to upper management on the greenhouse effect and warned that "man has a time window of five to ten years" before hard decisions about energy had to be made.[30] Corporations like Exxon instead used short-term thinking to give fuel to denialism. Politicians chased the scent of money. As a result of business as usual, our planet has become an out-of-control experiment.

Decades after Exxon's "time window" closed, climate activists made cyclical declarations: humanity had only twelve years left, or eight years, or four, or one final year, to avert catastrophic warming. Ironically, short-term framing of long-term thinking did not produce rapid results. Those with greatest political and financial power felt the least urgency. It's one thing for boomers to speculate about life in 2100, another for teenagers to visualize their pre-tarnished golden years.

In the face of all those missed deadlines about future doomsdays, who cannot feel hopeless or blasé? It takes a pair of cultivated qualities—knowledge of history, faith in hope—to apprehend the time remaining for meaningful action and compassionate adaptation. The future has never been set in stone. Historical insight reveals that different politics and different economies are possible. Different planets, too. The only impossibility is returning to the most hospitable climate of the past.

Whatever comes next, it won't be unprecedented for the vegetal kingdom. Roughly 56 million years ago, the Earth underwent a rapid, intense warming event. Scientists call it the Paleocene-Eocene Thermal Max-imum (PETM). It probably resulted from a massive release of carbon into the atmosphere. The source is unknown; the effects are not. Oceans acidified, and precipitation intensified. This hyperthermal—the hottest planet since animals moved to land—lasted almost two hundred millen-nia. Throughout much of the Eocene epoch, 22 million years, the Earth was ice-free. Conifers retreated to refugia in midlatitudes, while in high latitudes they dominated. Beyond the Arctic Circle, trees covered the land—not shrubby growers living under adversity, but mixed deciduous forests with megaflora such as ginkgo and metasequoia.

The best place to see evidence of this past planet is the east coast of Axel Heiberg Island, the second most northerly Canadian isle at 80°N. Nobody lives here year-round, despite the presence of two science sta-tions, one for polar research, another for space research. In its dryness, cold, and exposure, the island resembles Mars. Every year since 1959, sci-entists have come here to measure the extent of White Glacier—now an official reference point for the UN Framework Convention on Climate Change. It's been retreating nearly continuously since the 1970s.

When government geologists surveyed Axel Heiberg for petroleum potential in 1985, they spotted from the air what looked like petrified trees lying on the permafrost. They landed their helicopter and found something more surprising: scores of logs and stumps jutting out from the barren slopes of the Geodetic Hills.

Paleobotanists identified the megaflora as *Metasequoia occiden-talis* from the Eocene. On close inspection, the wood wasn't petri-fied, nor coalified. Buried in silt, then frozen, the wood had been perfectly preserved—"mummified." If a wooden torso is the essential signifier of treehood, the trees of Axel Heiberg were indeed the oldest in the world. The lead geologist celebrated his eureka by kindling a fire with 45-million-year-old twigs and boiling water for tea time. "If people get to it, it's in trouble," he added. "Without people, it will be useful to scientists for a long time. Otherwise it becomes a curiosity, of little scientific value."[31]

Word got out immediately. Military officers from the spy station on nearby Ellesmere Island helicoptered in for visits, taking souvenirs. Soon enough, Arctic cruise ship operators added the site to their itineraries. Pocketful by pocketful, surface wood began to disappear from the edge of the far north. As of 1993, this ancestral forest officially belonged to the Grise Fiord Inuit community of Ellesmere Island—a result of Canada's historic Nunavut Land Claims Agreement—though none of the popular science articles called attention to that fact.

Eager to gather data before time ran out, a well-funded group of US scientists descended upon Axel Heiberg in 1999. Their chartered copters delivered Quonset huts, gasoline-powered generators, computers, and beer. The Canadian government approved: "The more we know of the [paleo] climate and vegetation," said a state geologist, "the better we'll be able to assess the oil and gas potential there."[32] Grad students who called themselves "diggers" cut trenches to expose whole trees. The activities of the Americans sparked a debate about Canadian pride and research ethics. The PI from the University of Pennsylvania felt aggrieved. "Some people think that this place is sacred, others want it for a picnic spot," said the professor. "We think it's a scientific resource."[33]

Out of place among the all-male team from Penn was a young paleobotanist from Johns Hopkins, Hope Jahren. Later, she would narrate her experience in a memoir, *Lab Girl*. In three summers, she obtained enough data to last a career. "We have no need to go back and disturb the site," she told an interviewer. "We need to worry about it lasting another 45 million years."[34]

Jahren's lab provided an answer to the obvious question: How did trees thrive at this extreme latitude, with no sunlight for months at a time, alternated by continuous weak light? (Unlike some islands and continents, Axel Heiberg has barely moved tectonically since the Eocene.) By studying the isotopic signatures of recovered wood, Jahren inferred a strange circulation pattern on the post-PETM planet. A south-to-north jet stream delivered humid equatorial air to the Arctic each summer, allowing for explosive 24-hour-a-day vegetative growth. The swampy polar forest enjoyed a full growing season in half the time, with temperatures ranging from mild to balmy. In a positive feedback loop, constant summer evapotranspiration from the trees helped to maintain a wet, ice-free climate.

Models for worst-case climate scenarios share similarities with reconstructed climates for PETM: increased intensity of droughts and storms in continental interiors; irregularity at midlatitudes; and dramatic shifts around the poles. Today, the Arctic warms faster than any part of Earth. In time, high latitudes should get wetter, too, for all that extra water vapor has to go somewhere. A burning planet is setting the scene for future northern forests. Seasonal darkness won't be enough to stop woody plants. Species from a who's who of conifer genera—*Abies, Ginkgo, Glyptostrobus, Larix, Picea, Pinus, Taxodium, Tsuga*—populated the Eocene Arctic alongside *Metasequoia*. Here stood one of the great forests of all time. The genomes of descendant species may retain enough phenotypic plasticity to allow for adaptation to a warm-again planet, even as new species evolve. Ancient DNA has advantages.

Currently, paleobotanists attempt to estimate CO_2 levels in the PETM through proxy data: the number of stomata, or gas-exchange pores, in fossilized *Ginkgo* foliage from the Rocky Mountains. The cuticles are so tough, they preserve the outlines of epidermal cells from the geologic past. Under the microscope, the most beautiful leaves in creation may yet allow scientists to pinpoint how many parts per million marks an atmospheric level of no return—a planet without ice. But that would be an academic exercise. Morally and politically, climate action requires no more climate research.

I think about the Eocene whenever I see ginkgoes, meaning almost every day. They grow from Philadelphia sidewalks that seem more inhospitable than Great Basin mountains. Contemplating all the lifetimes of these and other gymnosperms, I feel awe. The combination of lignin biosynthesis with photosynthesis, which goes back nearly half a billion years, remains a singular outcome of evolutionary history. Light-eating lignophytes are perfectly adapted for geotemporal existence. They stand as the ultimate terrestrials. They have a firmer grasp on earthly time than *Homo sapiens*.

In the longest term, woody plants and their microbial allies will reclaim this terrestrial kingdom. We're just passing through. But our passing can and should last indefinitely. There can be longevity under precarity. The future is long enough for worlds without number, if only we can see the fullness of time in the kindred we call trees. The idea is to share the Earth with them, living and dying in different ways at different speeds, as long as humanly possible.

One year at a time.

EPILOGUE: PROMETHEUS

THE OLDEST LIVING THING EVER KNOWN TO SCIENCE SUCCUMBED TO male knowledge seekers. Indeed, it was killed in the act of knowing, and thus did not become sacralized through dendrochronology or radiocarbon dating. Because this plant acquired its age honor posthumously, it belongs here in the book, after forty sections on lastingness.

In summer 1954, Dr. Edmund Schulman and technician Wes Ferguson went to White Pine County in eastern Nevada, bioprospecting for oldness. They found a 3,100-year-old bristlecone pine in the Schell Creek Range, but determined that tree rings from this part of the Great Basin were less sensitive than those in the White Mountains, where five-needled pines grew in the direct rain shadow of the High Sierra. The Tucsonans did not continue eastward to the Snake Range, because the habitat appeared too moist from a distance. Just as importantly, there was no road to the alpine zone below 13,064-foot Wheeler Peak, and tree-ring scientists of that era preferred to stay close to their vehicles.

The following year, a writer-naturalist named Darwin Lambert moved to Ely, population 3,500, seat of White Pine County. He was hired to direct the Chamber of Commerce and Mines; soon he took on additional roles as newspaper editor and legislator. Though educated at George Washington University, Lambert had grown up 150 miles to Ely's north, in a company town developed by the Southern Pacific Railroad and the Salt Lake-based Pacific Reclamation Company. Optimistically named Metropolis, Nevada, the town was meant to showcase the science of "dry farming," or the cultivation of nonirrigated crops on semiarid land. The venture attracted Latter-day Saints, including Lambert's parents. Raised Mormon, Lambert sang songs in church about making the desert blossom as the rose. By the time he left for DC at age eighteen to take a New Deal job, the desert had reclaimed Metropolis. In leafy Washington, Lambert majored in botany, stopped going to LDS services, and started to think of the great outdoors as his temple. He became the first employee of Shenandoah National Park.

In the late 1950s, there were enough Nevadans like Lambert—pro-government and pro-conservation—to prompt the state's congressional delegation (two senators and a single at-large representative) to request that the National Park Service study the Snake Range for potential designation as a national park. The Snakes seemingly had it all: a cave system (already managed by the NPS as a national monument), the second-highest peak in Nevada (already designated by the USFS as a scenic area), and bristlecone pine—an organic feature that swiftly gained notoriety after Schulman's 1958 article in *National Geographic*. When Ely hosted congressional hearings on the proposed park, the Nevada delegation heard clashing local viewpoints. The majority opposed the idea, characterizing Wheeler Peak as just another mountain, best used for grazing, mining, and hunting. Don't lock it up, they said; don't put it in a deep freeze. A minority, including Lambert, called it a crown jewel and a potential magnet for tourist dollars. Over time, the delegation moved toward a compromise proposal: a multiple-use national park or recreation area that would allow grazing and logging, much like the existing national forest, only with greater publicity.

Lambert moved on to Alaska, then back to Virginia, but remained active in Nevada politics, and followed the next round of hearings in DC in 1961. He and his allies promoted the tiny snowfield on Wheeler Peak as a "glacier," celebrated the mountain as a classic example of Humboldtian life zones, and shared photographs of pines reputed to be four millennia old. Adolph Murie, a major figure in US conservation, did double duty, writing assessments for the NPS and the Sierra Club. Murie asserted that these bristlecones had national significance and deserved park status by themselves. Each tree was uniquely fantastic, he wrote, like characters from Oz, with weird, hobgoblin shapes.[1]

Oblivious to the political currents swirling around the Snake Range, a doctoral candidate from the University of Kansas arrived for field research in 1963. His name was Donald Rusk Currey. Born and raised in California, trained in geology in Wyoming, Currey benefited from timing. That very year, the USFS had begun paving the rough road to the alpine zone—a preemptive move by the agency (based in the Department of Agriculture), which did not want to surrender the Wheeler Peak Scenic Area to the rival NPS (based in the Department of the Interior).

Currey's dissertation research, supported by the NSF, concerned global cooling in the previous three thousand years, with temperatures reaching their minima during the Little Ice Age. In 1963, when the CO_2 monitoring station on Mauna Loa was only five years old, geologists like Currey could still plausibly argue that the planet existed in the "Neoglacial" stade within the larger Holocene interglacial—a stade marked by alpine glaciers, not continental ice sheets. Currey traveled throughout the alpine American West looking for evidence of glaciation within the life spans of extant conifers. He used the presence or absence of trees on glacial deposits to corroborate his geomorphological dating. Having read Schulman's article, he was excited to see the large population of bristlecone on the moraines below Wheeler Peak.

The next summer, Currey, age thirty, now a predoctoral instructor in geography at the University of North Carolina, returned to the Snakes with a new grant to study the Little Ice Age. He determined that these bristlecones grew on "pre-Neoglacial" (that is, Pleistocene) deposits, meaning they were not entirely germane to his research. But he had brought a collection of increment borers, and he intended to use them. For centuries, male scientists had searched for fame and glory by age-dating trees, and he was next in line. Maybe Schulman had been wrong about Wheeler Peak. Maybe Currey would find a tree older than Methuselah; that would be a feather in his cap. With no training in forestry or dendrochronology, he bored his way through 113 trees, including several 3,000-year-olds, to which he gave alphanumeric identifiers starting with "WPN" for White Pine County. Using the workshop at Lehman

Caves National Monument as his temporary lab, he counted rings under a magnifier.[2]

The 114th tree comprised a massive slab that extended horizontally like a Chinese dragon. Mostly dead and sun-bleached, this sectorial organism still supported one healthy lead with bushy branches and purple cones dripping in sap. Hitting the pith of this thick, dense, resinous pine exceeded the coring dexterity Currey had acquired in the preceding weeks. He tried multiple oblique angles, with results indicating a tree beyond four millennia, but he broke all his borers in the process. Instead of letting it go, and letting the mystery be, Currey drove to Ely to talk to the district ranger. He asked for a wood cutting permit. It's all for science, he must have said. He knew from *National Geographic* that Schulman himself had felled at least one 4,000-year-old. It's just one more, he must have said. The district ranger agreed, and phoned his boss, the supervisor of Humboldt National Forest, who gave the okay.[3]

The Forest Service offered horses, a part-time employee, and a chainsaw. Five others came to watch: Currey, his research assistant, the national monument's chief naturalist, a seasonal cave guide, and the district ranger from Ely. The naturalist took a photo of Currey, wearing a cowboy hat, astride the bristlecone like a bull rider. Once again, the event wasn't meant to be. The chainsaw operator, Mike Drakulich, refused to fire up the engine. He was a tough guy—boxer, miner, bricklayer, rancher, Cat skinner, blackjack dealer—but, when he saw that tree, he felt uneasy. On the morrow, August 7, 1964, Currey and the district ranger returned with willing hands and a working saw.[4]

It would be a deathday to remember. Once horses, then trucks, had transported the weighty cross sections down the mountain, over slopes of debris, through groves of aspen; and once the sander had smoothed the jagged cut of the chainsaw, Currey started counting under magnification. It took days. He tallied many times, making sure he got his ring count right: 4,844. In his hands, he held the lifeless fragments of the newest oldest known—his omega to Schulman's Alpha.

When informing the Forest Service, Currey asked for a news embargo while he wrote an article for *Science*, the flagship journal for US scientists.

He wanted his "discovery" published quickly, with him as sole author. By early November, he'd heard back from the editors, who rejected the piece with the suggestion he try a specialty journal. Currey immediately rewrote the piece and submitted it to the organ of the Ecological Society of America. Given that the editorial turnaround for *Ecology* was six to eight months, Currey felt obligated to release the Forest Service from the requested pause. Go ahead and publicize the 4,900-year-old, if you like, he wrote in a letter.

Humboldt National Forest decided to wait because they had their own big plans. They removed and polished multiple cross sections—one for the local office, another for the regional office, and the prime piece for Ely's Hotel Nevada. They hired an artisan to inscribe upon it a timeline that would draw attention to the Wheeler Peak Scenic Area. Kennecott Copper Corporation, which owned a big open-pit mine nearby, helped pay for the display. Patterned after sequoia timelines, it ran from the START OF THE GREAT PYRAMIDS to the BIRTH OF THE SPACE AGE. The hotel unveiled the slab in its renamed Bristlecone Pine Room—a casino lobby—in July 1965, the same month that *Ecology* published "An Ancient Bristlecone Pine Stand in Eastern Nevada."[5]

The journal listed Currey's contribution as a report rather than a research article. An article would have required an argument. The closest thing to a takeaway was that Schulman's hypothesis of an east–west age gradient for the species may need to be revised. On the opening page, Currey used the present tense "is" when writing about the "oldest reported prior to 1965"; then, in a squirrely manner, he switched to the passive form of the past tense when implying its death: "A horizontal slab from the interval 18–30 inches above the ground and a smaller piece including the pith 76 inches above the ground were cut from the tree." When speaking to reporters in North Carolina, he seemed more comfortable presenting himself as the discoverer of the oldest, though he hastened to add that trees should never be cut down except to serve a purpose.[6]

Still feeling good about its authorization, Humboldt National Forest decided it could use additional chunks of the record holder, and sent a four-man crew to haul them out in autumn 1965. Their horses couldn't make it over the last mile of boulders, so the men used a stretcher. While maneuvering a three-hundred-pound slab over uneven ground at 10,000 feet, one of the workers, Fred Solace, collapsed of a heart attack. His anguished mates tried for two hours to resuscitate him. Solace, age 32, had a wife and infant son in Ely. His body descended the alpine zone in the stretcher intended for the bough. Obituarists couldn't make sense of what Solace was doing when he died, and they published contradictory accounts of "gathering" pine slabs, or "slips," or "slaps," or cones.

Darwin Lambert had fallen behind on Silver State news, having moved back to his adopted home in the Shenandoah Valley to become a freelance writer and full-time environmental activist. On New Year's Day, 1965, having read the report in *Ecology*, he wrote to the chief naturalist at Lehman Caves, desperate for information: Who is Currey? Was the tree actually cut down? He couldn't believe that the age champion of the world had been detected and destroyed simultaneously. What an indictment of USFS management, he thought, and the strongest possible argument for national park status. Within days, Lambert contacted the *New York Times* magazine, pitching a story. They turned him down, as did the *Christian Science Monitor* and *Reader's Digest*. Trying to get the inside scoop, he wrote an insincere letter to Currey, introducing himself as the former editor of the *Ely Daily Times*: "I congratulate you on the remarkable discovery and will be grateful if you would tell me anything which might be useful in establishment of the park." Meanwhile, Lambert's allies with the Nevada Outdoor Recreation Association accused the Forest Service of abetting an "eastern graduate student" in the "annihilation" of the oldest. They complained to Nevada's senators about that and the "feverish" road building on Wheeler Peak.

By March 1965, Lambert had decided it was his mission to turn the wronged pine into a political scandal. Currey's tree must become a "martyr," he wrote. He reached out to the leadership of the National Parks Association and the Sierra Club. Inspired by the club's coffee-table lament, *The Place No One Knew*—a book about the loss of Glen Canyon through damming—he referred to WPN-114 as the Tree No One Knew. He called the office of Nevada senator Alan Bible. The senator, deluged by complaints, formally questioned the chief of the USFS: How could a graduate student be allowed to do this? It was, responded the chief, a legitimate request from a scientist. Besides, it's a common kind of tree, not a vanishing species. There are lots of old pines like this. "The loss of this one tree seems to have stirred up a lot of sentiment," the chief concluded. "I think that is an understatement," replied Bible.[7]

Recriminations worked their way down the chain of command. When the Intermountain Regional Office in Ogden, Utah, conducted an internal investigation, the regional forester shook his head at his colleagues in Nevada. How could they not see the folly in cutting an ancient tree in a proposed national park, then seeking assistance from Kennecott, an opponent of said park?[8] To his embarrassment, the good ol' boys in Ely had turned pieces of WPN-114 into mementos, including a paperweight gifted to Nevada's congressman. The rangers pleaded their case, disingenuously. Currey needed to fell a tree to establish increment boring techniques, they said. We selected a pine that appeared in poor health; we had no reason to believe it was the oldest. Currey, for his part, went on the record in the passive voice: a cross section had been considered important as a "framework" around which to build a tree-ring chronology for "indirectly" dating glacial events.[9]

Dendrochronologists from Tucson disputed all of this. Val LaMarche and Wes Ferguson both visited White Pine County in 1965. LaMarche had no trouble coring the stump of the freak tree. He went to Ely to see the hotel's display case, which made him shudder. Ferguson examined another slab in storage and verified Currey's ring count. Shocked by the arboricide, the tree-ring expert addressed a letter to a high-ranking contact in the Forest Service's DC headquarters, only to receive this disappointing reply: We've been assured by our people on the ground that Currey believed the felling was necessary to secure scientific data. Whether that's true is not for us to say; perhaps you, a fellow scientist, should debate with Currey rather than us. We note that Dr. Schulman cut a bristlecone.[10]

Throughout 1966, more and more media outlets broadcast capsule items about the ex-oldest tree, while Lambert struggled to get long-form attention. Like Schulman with *National Geographic*, he needed help understanding his audience. Hoping to reapproach *Reader's Digest*, Lambert hired a big-time literary agent. From his office on Fifth Avenue, the agent typed a frank evaluation.[11] You have three available genres, he explained—opinion, which requires authority; information, which requires timeliness; and entertainment. Your article attempts all three at once, something to be devoutly avoided. You're not a botanical authority, and this tree tale isn't timely. To top it off, your writing is technical and pedantic instead of entertaining. Undaunted, Lambert tried again, and got a final rejection from *Reader's Digest*: although we share your basic outrage at the cutting of a 4,900-year-old tree, you just haven't put it over in a way we can use.

Lambert was enough of a writer to realize a victim needs a name. Back when he lived in Nevada, he'd given tours of the bristlecone forest to visiting environmentalists such as Adolph Murie and David Brower, and, like a good tourist guide, he'd thrown out fanciful monikers for various photogenic specimens: Buddha, Socrates, Storm King, Cliff-Clinger, the Witch, the Waif, and the Money Tree (because a photographer friend had sold so many prints of it). Lambert couldn't be sure, exactly, but he believed the one that Currey sectioned had been dubbed the Leaning Tower. In his role as obituarist, he decided to rename it Prometheus, after the fire-gifting Titan chained to a mountain for a countless duration. (The ancient dramatist Aeschylus played with different counts: thirteen generations, ten millennia, three myriads.)

Finally, in 1968, Lambert succeeded in publishing his interpretation of events in *Audubon*. The title was "Martyr for a Species," and the tagline read: EARTH'S OLDEST LIVING THING WAS CASUALLY KILLED (YES, MURDERED!) IN THE NAME OF SCIENCE. Neither masterful nor error-free, this secondhand version of events became canonical. The USFS and Currey lacked the moral authority to issue corrections or contradictions.[12]

As part of damage control, the Forest Service had hired the Laboratory of Tree-Ring Research to do a proper dendrochronological survey of the Snake Range population. The agency's none-too-secret hope was that Ferguson, the primary investigator, would find an older tree.

Ferguson came to an unfortunate conclusion in 1970: I regard WPN-114 as unique, he wrote, truly a singular age class. Only a few trees on the moraine surpassed 3,000 years. His colleague LaMarche had studied the death site and determined that drainage patterns had produced an anomalously dry microhabitat within a locale with relatively high precipitation. Here had occurred longevity under adversity on the smallest and longest scales.[13]

While the researchers from Tucson conducted their survey, Currey defended his degree and put his thesis in the icebox.[14] Beyond the *Ecology* report, no scientific publications resulted from his bristlecone fieldwork. He moved on to inanimate records of past climates, yielding WPN-114 to other interpreters. Lambert retold the martyrdom in a coffee-table book, *Timberline Ancients*, dedicated to Edmund Schulman. In his philosophical narration of tree-ring science, Lambert used Great Basin bristlecone pine as a conduit to planetary consciousness—what he and his coactivist wife, Eileen, called "earthmanship." To greater public exposure, the mountaineer and adventure photographer Galen Rowell published the first image of WPN-114 in 1974. In winter, on fiberglass skis, he had hunted for the stump for two days. The sight was most repugnant, he wrote.[15]

By 1985, the twentieth death anniversary of Prometheus, the story had become a chainsaw murder legend, and a reporter tracked down the key participants. Currey, now a professor at the University of Utah, took cover under his former lowly status. I was a graduate student at a university in another state; I had no power to order anything, he said. It's not easy being called a monster. Meanwhile, the former district ranger, who had escaped public shaming, felt no remorse. That tree was mostly dead, he said. It didn't look nice or distinguished. In fact, it looked common. And it wasn't the biggest. No one would have walked more than a hundred yards to see it. As a timeline, though, it educated a lot of people. I was thrilled to be a part of it, he said.[16]

Currey generally turned down interview requests, but he made two exceptions late in his career. In 1998, he asserted that the Forest Service had cut down as many as one hundred bristlecones in the cirque before he arrived. "I was amazed at the desecration," he claimed. He regretted his actions in retrospect, without second-guessing himself. "Out of 100,000 bristlecones or whatever, I'd only looked at 500. So, what were the odds I had the oldest?" In a subsequent film interview, he doubled down on his story of simple bad luck, and reverted to the passive collective voice, followed by an incredulous chuckle: "The tree that ended up being cut was literally the first old tree that we climbed to on the crest of the lateral moraine. Five minutes of looking is all that was involved."[17]

Around this time, Currey gifted his "cursed" relics (five-part slab, plus pith section) to a botanist at the University of Texas at Arlington, who subsequently donated them to the University of Arizona, where tree-ring scientists performed cross-dating.[18] They assigned the date 2936 BCE to the inner ring, meaning the tree had lived at least 4,900 years, probably five millennia.

In 2004, Don Currey died, age 70, never having been in control of his academic reputation. Great Basin National Park subsequently installed a cross section of WPN-114 in its visitor center and offered posthumous absolution: "Destruction Leads to Preservation: These ancient trees are now protected on federal lands, in part due to the public outcry over the loss of Prometheus. The researcher responsible for cutting down the tree later became one of the strongest advocates for the creation of Great Basin National Park." A friend of Currey's updated the late professor's Wikipedia entry in the form of a memorial biography, giving a more complex and sympathetic view of the career of the geomorphologist, who became an advocate for the conservation of Pleistocene geoheritage around Salt Lake City. Within forty-eight hours, a Wikipedia gatekeeper restored the "neutral point of view" about the grad student who felled the oldest living thing.

In an autobiography, published posthumously, Lambert issued his own numerical claim: his writings on Prometheus in magazines, books, and reprints added up to thirty-five million copies in a dozen languages.[19] The martyrdom story he had devised originally served a political purpose. It lost that purpose in 1986 when Congress created Great Basin National Park, at last placing Wheeler Peak under the jurisdiction of the NPS. But people kept telling versions of the narrative, abbreviated versions with embellishments and errors and fibs, because who could resist? Prometheus was perfect for campfires, podcasts, listicles: Top 5 Science Fails, Top 10 Biggest Fuck-Ups in History, 3 People Who Are Bigger Losers Than You, 5 High IQ Idiots Who Destroyed Irreplaceable Things.

Such glibness is a waste of a good story if not an insult to elderflora. In the terminal phase of the Holocene, the untimely demise of the epoch's eldest tree surely deserves a narrative more profound than "oops."

In my mind, Currey echoes Enkidu or Erysichthon—masculine violators who succeeded, not failed. It may seem ridiculous and unfair to assign mythic meaning to a minor twentieth-century Utahn, but the story should not be about Currey per se. In relation to the narrative, he's a type, and it's important to get the type right. His proto-professional identity as a graduate student is not the key fact. Many fully credentialed men of Currey's generation did wrongs in the name of science. In 1964, Currey was just another male researcher in pursuit of measurable data and the career validation that data provided. By the core principle of modern science, anything and everything could be datafied. Although quantitative information might initially lack for questions or applications—and monetizations—scientists had faith that quantification itself created positive good, and that more rows and columns of numbers led to more progress, as well as more citations.

In mythic terms, WPN-114 might have, in death, become another Tree of Knowledge, but only if the scientist had applied the rings within its trunk toward a meaningful question. Counterfactually, it could have stood as a Tree of Wisdom—if only the scientist and the ranger had chosen to withhold their Promethean power, and allowed the plant to live out its years. As events transpired, unwatched by gods and hamadryads, men tore down this Tree of Data because they could.

When I saw the remains, and touched the borehole stigmata, the scene was all gray—deadwood on quartzite under overcast sky. I hadn't anticipated what I would feel, and the depth of my mournfulness surprised me. I perceived, in a material and hyperlocal way, a moral problem at global scale. To understand how humankind is changing the planet, people must acquire long-term understanding of the Earth system, including tipping points, nonlinear dynamics, and feedback loops. Yet this acquisition entails ever more abstraction and datafication and despiritualization. We quantify everything under the sun, we store our quantifications in the cloud, and we continue to lose our planet. If all the redundant terabytes of Holocene climate data could be terrestrialized and concentrated into a single carbon-based form, I imagine they would resemble WPN-114—a lifeless tree at the threshold of an iceless cirque.

Every book of history, even one with a recapitulating epilogue, is linear, with a forced ending, and my omega is a benediction addressed to *Pinus longaeva*:

May there never be older living things in the world, precisely known.

May there always be timeful beings on Earth, with and without our knowing.

ACKNOWLEDGMENTS

This book would have been impossible for me to research without the assistance of three grant-giving institutions: the National Endowment for the Humanities (Public Scholars Program), the Alfred P. Sloan Foundation (Public Understanding of Science Program), and the Carnegie Corporation of New York (Andrew Carnegie Fellows Program).

Likewise, this book would have been impossible for me to write without the assistance of four scholarly institutions: the Rachel Carson Center at Ludwig-Maximilians-Universität München, the American Academy in Berlin, the College of Arts and Sciences at Stony Brook University, and the School of Arts and Sciences at the University of Pennsylvania.

At the beginning and the end of drafting, respectively, I worked in beautiful silence at the Michael King Writers Centre in Auckland, New Zealand, and the Ernest and Mary Hemingway House in Ketchum, Idaho.

At Basic Books, publisher Lara Heimert showed patience, editor in chief Brian Distelberg gave guidance, and associate editor Michael Kaler provided comments. Separately, Amyrose McCue Gill restored my faith in the craft of copyediting.

Fellow academics reviewed individual chapters: James Beattie, Anne Berg, Matthew Booker, Alex Chase-Levenson, Michael Cohen, Don Falk, Tom Lekan, Catherine McNeur, David Schoenbrun, Emily Wakild, Beth Wenger, Caroline Winterer.

Matt Ritter and Jenn Yost arranged Golden State interventions when I needed them most.

Many others assisted with expertise, encouragement, or hospitality. Here I list them, to the best of my memory, in alphabetical order:

ACKNOWLEDGMENTS

René Ahlborn / Craig Allen / Lisa Amati / Jenny Anderson / Aviva Arad / Ligia Arguilez / Chris Baisan / Tanya Bakhmetyeva / Barbara Bentz / Staffan Bergwik / Eunice Blavascunas / Gretel Boswijk / Peter Brown / Erika Bsumek / Andy Bunn / Tony Caprio / Octavia Carr / Chris Chetland / José Chueca / Valerie Cohen / Ed Cook / Peter Crane / Pearce Paul Creasman / Samantha D'Acunto / Sara Dant / Brian DeLay / Peter Del Tredici / Karen Elsbernd / Jenny Emery-Davidson / Don Falk / Clark Farmer / Zosia Farmer / Antonio Feros / Dan Flores / Tim Forsell / Anthony Fowler / David Frank / Eliza French / Cebron Fussell / Dan Gerstle / Larissa Glasser / Bob Goldberg / Paul Gootenberg / Wilko Graf von Hardenberg / Lisa Graumlich / Peggy Grove / Susan Grumet / Huia Hamon / Dave Hardin / Carter Hedberg / Arielle Helmick / Rodolfo Alfredo Hernández Rea / Tim Hills / Peter Holquist / Malcolm Hughes / Kuang-chi Hung / Cathy Hunter / Laura Hurtado / Anna Iwanik / Karl Jacoby / Beth James / Bogdan Jaroszewicz / Eva Jensen / Tom Kearns / Brandon Keim / Arthur Kiron / Tom Klubock / Sacha Kopp / Yolande Korb / Werner Krauß / Deborah Farmer Kris / Mike Kris / Ron Lanner / Doug Larson / Sara Lipton / Drew Lorrey / Gary Lowe / Magda Mączyńska / Christof Mauch / Don McGraw / Richard Menzies / Connie Millar / Juan Montes-Lara / Robert Nelson / Nick Okrent / Cindy Ott / Jonathan Palmer / Lisa Pearson / Neil Pederson / Alessandro Pezzati / Deborah Poole / Jenny Price / Mitch Provance / Megan Raby / Nishi Rajakaruna / Gregory Raml / Zander Rose / Amy Rule / Chris Sabella / Matt Salzer / Donna Sammis / Andy Sanders / Vince Santucci / Carol Scherer / Jack Schmidt / Katja Schmidtpott / Richard Schulman / Prerna Singh / David Stahle / Bill Stein / Nate Stephenson / Tania Stewart / Scotty Strachan / Ellen Stroud / Jane Sundberg / Jacque Sundstrand / Tom Swetnam / Ricky Tomczak / Jonathan Treat / Skip Vasquez / Denise Waterbury / Doron Weber / Jeff Weiss / Bede West / Richard White / Martha Williams / Richa Wilson / Scott Wing / Connie Woodhouse / Shang Yasuda.

I cannot adequately express my gratitude to these persons and institutions. Time and again, I was humbled by the kindness of strangers, the generosity of colleagues, and the goodness of friends. I pledge to give to others in turn.

BIBLIOGRAPHY

Elderflora is an edited abridgment of a thousand-page volume I drafted in my mind. The full-length mental version is an attempted synthesis of an interdisciplinary field that does not exist. From professional experience, I know my research was hopelessly incomplete and madly excessive. To catalog all my readings would be gratuitous. The following guide is therefore selective.

INTRODUCTION

On plants: Francis Hallé, *In Praise of Plants* (Portland, OR, 2002). On trees: Colin Tudge, *The Secret Life of Trees: How They Live, and Why They Matter* (London, 2005); Rémy J. Petit and Arndt Hampe, "Some Evolutionary Consequences of Being a Tree," *Annual Review of Ecology, Evolution, and Systematics* 37 (2006): 187–214; Suzanne Simard, *Finding the Mother Tree: Discovering the Wisdom of the Forest* (New York, 2021); and Shannon Mattern, "Tree Thinking," *Places Journal*, September 2021 [online only]. On conifers: Aljos Farjon, *A Natural History of Conifers* (Portland, OR, 2008); and the online Gymnosperm Database.

On forest fire: Stephen J. Pyne, *Fire: A Brief History*, 2nd ed. (Seattle, 2019). On deforestation: Michael Williams, *Deforesting the Earth: From Prehistory to Global Crisis, an Abridgment* (Chicago, 2006). On utilization: Joachim Radkau, *Wood: A History* (Cambridge, 2012). On state forestry: Brett Bennett, *Plantations and Protected Areas: A Global History of Forest Management* (Cambridge, MA, 2015). On Chinese forest use: Ian M. Miller, *Fir and Empire: The Transformation of Forests in Early Modern China* (Seattle, 2020).

On sacred groves: Jan Woudstra and Colin Roth, eds., *A History of Groves* (London, 2018). For ancient case studies: Darice Elizabeth Birge, "Sacred Groves in the Ancient Greek World" (PhD diss., UC Berkeley, 1982); and Alisa Hunt, *Reviving Roman Religion: Sacred Trees in the Roman World* (Cambridge, 2016). On Hebron's antediluvian oak: F. Nigel Hepper and Shimon Gibson, "Abraham's Oak of Mamre: The Story of a Venerable Tree," *Palestine Exploration Quarterly* 126, no. 2 (1994): 94–105.

On European arboreal traditions: Robert Pogue Harrison, *Forests: The Shadow of Civilization* (Chicago, 1992); Simon Schama, *Landscape and Memory* (New York, 1995); and Charles Watkins, *Trees, Woods and Forests: A Social and Cultural History*

(London, 2014). On sustainability in European thought: Paul Warde, *The Invention of Sustainability: Nature and Destiny, c. 1500–1870* (Cambridge, 2018).

On Frazer: Robert Ackerman, *J. G. Frazer: His Life and Work* (Cambridge, 1987). On his reception history: Mary Beard, "Frazer, Leach, and Virgil: The Popularity (and Unpopularity) of *The Golden Bough*," *Comparative Studies in Society and History* 34, no. 2 (April 1992): 203–224. On his literary legacy: Matthew Sterenberg, *Mythic Thinking in Twentieth-Century Britain: Meaning for Modernity* (London, 2013). On secularism: Charles Taylor, *A Secular Age* (Cambridge, MA, 2007).

On Western attitudes toward age: Pat Thane, ed., *The Long History of Old Age* (London, 2005); David Lowenthal, *The Past Is a Foreign Country—Revisited* (Cambridge, 2015), 206–301; and Nancy A. Pachana, *Ageing: A Very Short Introduction* (Oxford, 2016).

On arboreal longevity: Gianluca Piovesan and Franco Biondi, "On Tree Longevity," *New Phytologist* 231, no. 4 (August 2021): 1318–1337; Jonathan Silvertown, *The Long and the Short of It: The Science of Life Span and Aging* (Chicago, 2013); Edward Parker and Anna Lewington, *Ancient Trees: Trees That Live for a Thousand Years* (London, 2012); and articles by Sergi Munné-Bosch.

On age-dating: Doug Macdougall, *Nature's Clocks: How Scientists Measure the Age of Almost Everything* (Berkeley, 2008). On evidence: Michael Strevens, *The Knowledge Machine: How Irrationality Created Modern Science* (New York, 2020). On synchronization: Vanessa Ogle, *The Global Transformation of Time: 1870–1950* (Cambridge, MA, 2015).

On long nineteenth centuries: C. A. Bayly, *The Birth of the Modern World, 1780–1914* (Oxford, 2003); and Jürgen Osterhammel, *The Transformation of the World: A Global History of the Nineteenth Century* (Princeton, 2014 [2009]).

On multiscalar history: Deborah R. Coen, "Big Is a Thing of the Past: Climate Change and Methodology in the History of Ideas," *Journal of the History of Ideas* 77, no. 2 (April 2016): 305–321. On historians and temporalities: Penelope J. Corfield, *Time and the Shape of History* (New Haven, 2007). On the eco-philosophy of history: Dipesh Chakrabarty, *The Climate of History in a Planetary Age* (Chicago, 2021).

On "big history," see works by David Christian and Cynthia Stokes Brown. On "deep history," see works by Daniel Lord Smail; and David Graeber and David Wengrow. On the *longue durée*, see Jo Guldi and David Armitage, *The History Manifesto* (Cambridge, 2014), and the oeuvre of Fernand Braudel. On the "long now": Stewart Brand, *The Clock of the Long Now: Time and Responsibility* (New York, 1999). On "fast" and "slow" thinking, see works by Daniel Kahneman and Amos Tversky. On short time: Robert Pogue Harrison, *Juvenescence: A Cultural History of Our Age* (Chicago, 2014).

On deep pasts: Martin J. S. Rudwick, *Earth's Deep History: How It Was Discovered and Why It Matters* (Chicago, 2014); Pascal Richet, *A Natural History of Time* (Chicago, 2014); Peter Brannen, *The Ends of the World: Volcanic Apocalypses, Lethal Oceans, and Our Quest to Understand Earth's Past Mass Extinctions* (New York, 2017); Marcia Bjornerud, *Timefulness: How Thinking Like a Geologist Can Help Save the World* (Princeton, 2018); and Robert Macfarlane, *Underland: A Deep Time Journey* (New York, 2019).

On deep futures: Jan Zalasiewicz, *The Earth After Us: What Legacy Will Humans Leave in the Rocks?* (Oxford, 2008); David Archer, *The Long Thaw: How Humans Are*

Changing the Next 100,000 Years of Earth's Climate (Princeton, 2009); Curt Stager, *Deep Future: The Next 100,000 Years of Life on Earth* (New York, 2011); and David Farrier, *Footprints: In Search of Future Fossils* (New York, 2020).

On climate change and temporality, see works by Bronislaw Szerszynski; and collections edited by Bethany Wiggin et al., Kyrre Kverndokk et al., and Anders Ekström and Staffan Bergwik. On planetary thinking: Ursula Heise, *Sense of Place and Sense of Planet: The Environmental Imagination of the Global* (New York, 2008); Bruno Latour, *Down to Earth: Politics in the New Climatic Regime* (New York, 2018); and Thomas Nail, *Theory of the Earth* (Stanford, 2021).

On trees as geological agents: David Beerling, *The Emerald Planet: How Plants Changed Earth's History* (Oxford, 2007); and William J. Manning, *Trees and Global Warming: The Role of Forests in Cooling and Warming the Atmosphere* (Cambridge, 2019).

On humans as geological agents: J. R. McNeill and Peter Engelke, *The Great Acceleration: An Environmental History of the Anthropocene since 1945* (Cambridge, MA, 2014); Jeremy Davies, *The Birth of the Anthropocene* (Berkeley, 2016); Jan Zalasiewicz et al., eds., *The Anthropocene as a Geological Time Unit: A Guide to the Scientific Evidence and Current Debate* (Cambridge, 2019); Kathryn Yusoff, *A Billion Black Anthropocenes or None* (Minneapolis, 2019); and works by critics who use alternative names "Capitalocene" and "Plantationocene."

On religious and secular end times: John R. Hall, *Apocalypse: From Antiquity to the Empire of Modernity* (Malden, 2009). On finality and modernity: Frank Kermode, *The Sense of an Ending: Studies in the Theory of Fiction*, rev. ed. (Oxford, 2000); and Umberto Eco et al., *Conversations about the End of Time* (London, 1999). On environmentalist end times: Jacob Darwin Hamblin, *Arming Mother Nature: The Birth of Catastrophic Environmentalism* (Oxford, 2013); Frank Uekötter, ed., *Exploring Apocalyptica: Coming to Terms with Environmental Alarmism* (Pittsburgh, 2018); and David Sepkoski, *Catastrophic Thinking: Extinction and the Value of Diversity from Darwin to the Anthropocene* (Chicago, 2020).

On interspecies ethics: Donna J. Haraway, *Staying with the Trouble: Making Kin in the Chthulucene* (Durham, NC, 2016). On intergenerational ethics: Katrina Forrester, "The Problem of the Future in Postwar Anglo-American Political Philosophy," *Climatic Change* 151, no. 1 (2018): 55–66. On environmental storytelling: Anna Lowenhaupt Tsing, *The Mushroom at the End of the World: On the Possibility of Life in Capitalist Ruins* (Princeton, 2015); and Amitav Ghosh, *The Great Derangement: Climate Change and the Unthinkable* (Chicago, 2016).

CHAPTER 1. VENERABLE SPECIES

On the world's oldest text: A. R. George, *The Babylonian Gilgamesh Epic: Introduction, Critical Edition and Cuneiform Texts*, 2 vols. (Oxford, 2003). On its modern reception history: Vybarr Cregan-Reid, *Discovering Gilgamesh: Geology, Narrative and the Historical Sublime in Victorian Culture* (Manchester, 2015); and Theodore Ziolkowski, *Gilgamesh among Us: Modern Encounters with the Ancient Epic* (Ithaca, 2011).

On ancient utilization of cedar: Russell Meiggs, *Trees and Timber in the Ancient Mediterranean World* (Oxford, 1982); Marvin W. Mikesell, "The Deforestation of Mount Lebanon," *Geographical Review* 59, no. 1 (January 1969): 1–28; Nili Liphschitz and

BIBLIOGRAPHY

Gideon Biger, "Building in Israel Throughout the Ages—One Cause for the Destruction of the Cedar Forests of the Near East," *GeoJournal* 27, no. 4 (August 1992): 345–352; and Lara Hajar et al., "Environmental Changes in Lebanon during the Holocene: Man vs. Climate Impacts," *Journal of Arid Environments* 74, no. 7 (July 2010): 746–755. For historic descriptions of Mount Lebanon: [Edward James Ravenscroft], *The Pinetum Britannicum*, vol. 3 (Edinburgh, 1884), 247–298. For the earliest scientific account: J. D. Hooker, "On the Cedars of Lebanon, Taurus, Algeria, and India," *Natural History Review* 2, no. 5 (January 1862): 11–18.

On modern Lebanese conservation: E. W. Beals, "The Remnant Cedar Forests of Lebanon," *Journal of Ecology* 53, no. 3 (November 1965): 679–694; S. N. Talhouk et al., "Conservation of the Coniferous Forests of Lebanon: Past, Present and Future Prospects," *Oryx* 35, no. 3 (July 2001): 206–215; Myra Shackley, "Managing the Cedars of Lebanon: Botanical Gardens or Living Forests?" *Current Issues in Tourism* 7, no. 4/5 (2004): 417–425; Lara Hajar et al., "*Cedrus libani* (A. Rich) Distribution in Lebanon: Past, Present and Future," *Comptes Rendus Biologies* 333, no. 8 (August 2010): 622–630; and "Climate Change Is Killing the Cedars of Lebanon," *New York Times*, July 18, 2018.

On *Olea*: Fabrizia Lanza, *Olive: A Global History* (London, 2011); Angeliki Loumou and Christina Giourga, "Olive Groves: 'The Life and Identity of the Mediterranean,'" *Agriculture and Human Values* 20, no. 1 (March 2003): 87–95; and Concepcion M. Diez et al., "Olive Domestication and Diversification in the Mediterranean Basin," *New Phytologist* 206, no. 1 (April 2015): 436–447.

On Holy Land olives: Michal Bitton, "The Garden as Sacred Nature and the Garden as a Church: Transitions of Design and Function in the Garden of Gethsemane, 1800–1959," *Cathedra* 146 (December 2012): 27–66 [in Hebrew]; and Masha Halevi, "Contested Heritage: Multi-layered Politics and the Formation of the Sacred Space—the Church of Gethsemane as a Case Study," *Historical Journal* 58, no. 4 (December 2015): 1031–1058. On Palestinian olives: Irus Braverman, *Planted Flags: Trees, Land, and Law in Israel/Palestine* (Cambridge, 2009).

On *Ginkgo*: Terumitsu Hori et al., eds., *Ginkgo biloba—A Global Treasure* (Tokyo, 1997); Teris A. van Beek, ed., *Ginkgo biloba* (Amsterdam, 2000); Zhiyan Zhou and Shaolin Zheng, "The Missing Link in *Ginkgo* Evolution," *Nature* 423 (June 19, 2003): 821–822; Cindy Q. Tang et al., "Evidence for the Persistence of Wild *Ginkgo biloba* (Ginkgoaceae) Populations in the Dalou Mountains, Southwestern China," *American Journal of Botany* 99, no. 8 (August 2012): 1408–1414; Peter R. Crane, *Ginkgo: The Tree That Time Forgot* (New Haven, 2013); Peter R. Crane et al., "*Ginkgo biloba*: Connections with People and Art across a Thousand Years," *Curtis's Botanical Magazine* 30, no. 3 (October 2013): 239–260; Peter Del Tredici, "Wake Up and Smell the Ginkgoes," *Arnoldia* 66, no. 2 (2008): 11–21; Kuang-chi Hung, "Within the Lungs, the Stomach, and the Mind: Convergences and Divergences in the Medical and Natural Histories of *Ginkgo biloba*," in *Historical Epistemology and the Making of Modern Chinese Medicine*, ed. Howard Chiang (Manchester, 2015), 41–79; and Li Wang et al., "Multifeature Analyses of Vascular Cambial Cells Reveal Longevity Mechanisms in Old *Ginkgo biloba* Trees," *PNAS* 117, no. 4 (January 28, 2020): 2201–2210. On "A-bombed trees," see the website of Green Legacy Hiroshima.

BIBLIOGRAPHY

On *Ficus*: Mike Shanahan, *Gods, Wasps, and Stranglers: The Secret History and Redemptive Future of Fig Trees* (White River Junction, 2016).

For foundational texts of Lankan Buddhism: John S. Strong, *The Legend of King Aśoka: A Study and Translation of the* Aśokāvadāna (Princeton, 1983); and Douglas Bullis, *The* Mahavamsa: *The Great Chronicle of Sri Lanka* (Fremont, 1999).

On Anuradhapura: Elizabeth Nissan, "History in the Making: Anuradhapura and the Sinhala Buddhist Nation," *Social Analysis* 25 (September 1989): 64–77; H. S. S. Nissanka, ed., *Maha Bodhi Tree in Anuradhapura, Sri Lanka: The Oldest Historical Tree in the World* (New Delhi, 1994); K. M. I. Swarnasinghe, *World's Oldest Historical Sacred Bodhi Tree at Anuradhapura* (Erewwala, 2005); Karel R. van Kooij, "A Meaningful Tree: The Bo Tree at Anuradhapura, Sri Lanka," in *Site-Seeing: Place in Culture, Time and Space*, ed. Kitty Zijlmans (Leiden, 2006), 9–31; and Sujit Sivasundaram, *Islanded: Britain, Sri Lanka, and the Bounds of an Indian Ocean Colony* (Chicago, 2013).

On Bodh Gaya: Janice Leoshko, *Bodhgaya, the Site of Enlightenment* (Bombay, 1988); Steven Kemper, *Rescued from the Nation: Anagarika Dharmapala and the Buddhist World* (Chicago, 2015); David Geary, *The Rebirth of Bodh Gaya: Buddhism and the Making of a World Heritage Site* (Seattle, 2017); and K. T. S. Sarao, *The History of Mahabodhi Temple at Bodh Gaya* (Singapore, 2020). On pipals in contemporary Hinduism: David L. Haberman, *People Trees: Worship of Trees in Northern India* (Oxford, 2013).

On *Adansonia*: Thomas Pakenham, *The Remarkable Baobab* (New York, 2004); Rupert Watson, *African Baobab* (Cape Town, 2007); Gerald E. Wickens and Pat Lowe, eds., *The Baobabs: Pachycauls of Africa, Madagascar and Australia* (New York, 2008); Haripriya Rangan and Karen L. Bell, "Elusive Traces: Baobabs and the African Diaspora in South Asia," *Environment and History* 21, no. 1 (February 2015): 103–133; Witness Kozanayi et al., "Customary Governance of Baobab in Eastern Zimbabwe: Impacts of State-led Interventions," in *Governance for Justice and Environmental Sustainability: Lessons across Natural Resource Sectors in Sub-Saharan Africa*, ed. Merle Sowman and Rachel Wynberg (London, 2014), 242–262; and Adrian Patrut et al., "The Demise of the Largest and Oldest African Baobabs," *Nature Plants* 4 (July 2018): 423–426.

For autobiographies of tree planters in Africa: Richard St. Barbe Baker, *My Life, My Trees* (London, 1970); and Wangari Maathai, *Unbowed: A Memoir* (New York, 2006).

CHAPTER II. MEMENTO MORI

For a scientific overview of European yew: P. A. Thomas and A. Polwart, "*Taxus baccata* L.," *Journal of Ecology* 91, no. 3 (June 2003): 489–524.

For single-topic yew books: John Lowe, *The Yew-Trees of Great Britain and Ireland* (London, 1897); Vaughan Cornish, *The Churchyard Yew and Immortality* (London, 1946); E. W. Swanton, *The Yew Trees of England* (London, 1958); Richard Williamson, *The Great Yew Forest: The Natural History of Kingley Vale* (London, 1978); Hal Hartzell Jr., *The Yew Tree, a Thousand Whispers: Biography of a Species* (Eugene, 1996); Anand Chetan and Diana Brueton, *The Sacred Yew* (London, 1994), which should be read skeptically; Fred Hageneder, *Yew: A History* (Stroud, 2007); Robert Bevan-Jones, *The Ancient Yew: A History of* Taxus baccata (Oxford, 2017); and Tony Hall, *The Immortal Yew* (Richmond, 2018).

BIBLIOGRAPHY

On yew lore: John Brand and Henry Ellis, *Observations on the Popular Antiquities of Great Britain*, vol. 2 (London, 1854), 255–266; Walter Johnson, *Folk-Memory* (Oxford, 1908), 250–257; and idem, *Byways in British Archaeology* (Cambridge, 1912), 360–407. On mortuary practices: Thomas W. Laqueur, *The Work of the Dead: A Cultural History of Mortal Remains* (Princeton, 2015).

On yew wood weaponry: Robert Hardy, *Longbow: A Social and Military History*, 3rd ed. (London, 1992); and Matthew Strickland and Robert Hardy, *The Great Warbow: From Hastings to the Mary Rose* (Somerset, 2011).

On age-dating yews: Jeremy Harte, "How Old Is That Old Yew?" *At the Edge* 4 (1996): 1–9; Richard Mabey, *The Cabaret of Plants: Botany and the Imagination* (London, 2016), 47–63; and the website of the Ancient Yew Group.

On premodern British trees: Della Hooke, *Trees in Anglo-Saxon England: Literature, Lore and Landscape* (Woodbridge, 2013); and Michael D. J. Bintley, *Trees in the Religions of Early Medieval England* (Woodbridge, 2015). On the persistence of sacred landmarks: Alexandra Walsham, *The Reformation of the Landscape: Religion, Identity, and Memory in Early Modern Britain and Ireland* (Oxford, 2011). On pagans, ancient and modern: Philip C. Almond, "Druids, Patriarchs, and the Primordial Religion," *Journal of Contemporary Religion* 15, no. 3 (October 2000): 379–394; Ronald Hutton, *Blood and Mistletoe: A History of the Druids in Britain* (New Haven, 2009); and idem, *Pagan Britain* (New Haven, 2013).

On modern British memorialization of trees: Jacob George Strutt, *Sylva Britannica; or, Portraits of Forest Trees* (London, 1826); and Mary Roberts, *Ruins and Old Trees Associated with Memorable Events in English History* (London, 1843). On modern British tree planting: J. C. Loudon, *Arboretum et fruticetum Britannicum* (London, 1838); Paul A. Elliott et al., *The British Arboretum: Trees, Science and Culture in the Nineteenth Century* (Pittsburgh, 2011); and Paul A. Elliott, *British Urban Trees: A Social and Cultural History, c. 1800–1914* (Winwick, 2016).

CHAPTER III. MONUMENTS OF NATURE

On Humboldt: Andrea Wulf, *The Invention of Nature: Alexander von Humboldt's New World* (New York, 2015); and Laura Dassow Walls, *The Passage to Cosmos: Alexander von Humboldt and the Shaping of America* (Chicago, 2009).

On Canary Island dragon trees: Peter Mason, *Before Disenchantment: Images of Exotic Animals and Plants in the Early Modern World* (London, 2009); José Barrios García, "La imagen del drago de la Orotava (Tenerife) en la literatura y el arte: Apuntes para un catálogo cronológico (1770–1878)," *Coloquio de historia canario-americana* 19 (2014): 748–758; Manuel de Paz-Sánchez, "Un drago en *El Jardín de las Delicias*," *Flandes y canarias* 1 (2004): 13–109; and Alfredo Herrera Piqué, "El Árbol del Drago: Iconografía y referencias históricas," *Coloquio de historia canario-americana* 12 (1996): 163–183. On Venezuela's iconic tree: Elías Pino Iturrieta, *El divino Bolívar: ensayo sobre una religión republicana* (Caracas, 2003), 126–130; and Manuel Barroso Alfaro, *El samán de Güere* (Maracay, 2007).

On sacred oaks in Europe: John Walter Taylor, "Tree Worship," *Mankind Quarterly* 20, nos. 1–2 (1979): 79–141. On the forest of Fontainebleau: Caroline Ford, *Natural Interests: The Contest over Environment in Modern France* (Cambridge, MA,

2016). On the Polish forest in the age of partition: Tomasz Samojlik et al., eds., *Białowieża Primeval Forest: Nature and Culture in the Nineteenth Century* (Cham, 2020).

On Germans and their forests: Paul Warde, *Ecology, Economy, and State Formation in Early Modern Germany* (Cambridge, 2006); Jeffrey K. Wilson, *The German Forest: Nature, Identity, and the Contestation of a National Symbol, 1871–1914* (Toronto, 2012); Michael Imort, "A Sylvan People: Wilhelmine Forestry and the Forest as a Symbol of Germandom," in *Germany's Nature: Cultural Landscapes and Environmental History*, ed. Thomas Zeller and Thomas Lekan (New Brunswick, 2005), 55–80; and Carina Liersch and Peter Stegmaier, "Keeping the Forest above to Phase Out the Coal Below: The Discursive Politics and Contested Meaning of the Hambach Forest," *Energy Research & Social Science* 89 (July 2022): 102537 [online only].

On nature protection in prewar Germany: Thomas M. Lekan, *Imagining the Nation in Nature: Landscape Preservation and German Identity, 1885–1945* (Cambridge, MA, 2009); Friedemann Schmoll, *Erinnerung an die Natur: Die Geschichte des Naturschutzes im deutschen Kaiserreich* (Frankfurt, 2004); and William H. Rollins, *A Greener Vision of Home: Cultural Politics and Environmental Reform in the German* Heimatschutz *Movement, 1904–1918* (Ann Arbor, 1997).

On German natural monuments: Ernst-Rainer Hönes, "Über den Schutz von Naturdenk-mälern: Rund 100 Jahre Naturdenkmalpflege," *Die Gartenkunst* 16, no. 2 (2004): 193–232; Anette Lenzing, "Der Begriff des Naturdenkmals in Deutschland," *Die Gartenkunst* 15, no. 1 (2003): 4–27; and Hugo Conwentz, *The Care of Natural Monuments with Special Reference to Great Britain and Germany* (Cambridge, 1909).

On nature in the Third Reich: Frank Uekoetter, *The Green and the Brown: A History of Conservation in Nazi Germany* (Cambridge, 2006). On the concentration camp: David A. Hackett, ed., *The Buchenwald Report* (New York, 1997); and Paul Martin Neurath, *The Society of Terror: Inside the Dachau and Buchenwald Concentration Camps*, ed. Christian Fleck and Nico Stehr (Boulder, 2005). On the tree inside the camp: Gerhard Sauder, "Die Goethe-Eiche: Weimar und Buchenwald," in *Spuren, Signaturen, Spiegelungen: Zur Goethe-Rezeption in Europa*, ed. Bernhard Beutler and Anke Bosse (Köln, 2000), 473–499; and Magdalena Izabella Sacha, "Le chêne de Goethe ou la protection des monuments naturels dans le IIIe Reich," *Bulletin trimestriel de la Fondation Auschwitz* 92 (Juillet–Septembre 2006): 51–69.

On Hugo Conwentz and his milieu: Walther Schoenichen, *Naturschutz, Heimatschutz: Ihre Begründung durch Ernst Rudorff, Hugo Conwentz und ihre Vorläufer* (Stuttgart, 1954), which should be read skeptically; Jeffrey K. Wilson, "Imagining a Homeland: Constructing *Heimat* in the German East, 1871–1914," *National Identities* 9, no. 4 (December 2007): 331–349; and Lynn K. Nyhart, *Modern Nature: The Rise of the Biological Perspective in Germany* (Chicago, 2009).

On Alois Riegl, see books by Diana Reynolds Cordileone and Michael Gubser; and Margaret Olin, "The Cult of Monuments at a State Religion in Late 19th Century Austria," *Wiener Jahrbuch für Kunstgeschichte* 38 (1985): 177–198.

On European monuments: Miles Glendinning, *The Conservation Movement: A History of Architectural Preservation* (London, 2013); Astrid Swenson, *The Rise of Heritage: Preserving the Past in France, Germany and England, 1789–1914* (Cambridge, 2013);

BIBLIOGRAPHY

Françoise Choay, *The Invention of the Historic Monument* (Cambridge, 2001); and G. Baldwin Brown, *The Care of Ancient Monuments* (Cambridge, 1905).

On international conservation before World War II: Corey Ross, "Tropical Nature as Global *Patrimoine*: Imperialism and International Nature Protection in the Early Twentieth Century," *Past & Present*, suppl. 10 (2015): 214–239; Raf de Bont, "Borderless Nature: Experts and the Internationalization of Nature Protection, 1890–1940," in *Scientists' Expertise as Performance: Between State and Society, 1860–1960*, ed. Joris Vandendriessche et al. (London, 2015), 49–65; Joachim Radkau, *The Age of Ecology: A Global History* (Cambridge, 2014), 11–45; Bernhard Gissibl et al., eds., *Civilizing Nature: National Parks in Global Historical Perspective* (New York, 2012); and John Sheail, *Nature's Spectacle: The World's First National Parks and Protected Places* (London, 2010).

On US national monuments: Raymond Harris Thompson, ed., "'The Antiquities Act of 1906' by Ronald Freeman Lee," *Journal of the Southwest* 42, no. 2 (Summer 2000): 197–269; and Hal Rothman, *Preserving Different Pasts: The American National Monuments* (Urbana, 1997). On symbolic uses of US trees: Jared Farmer, "Taking Liberties with Historic Trees," *Journal of American History* 105, no. 4 (March 2019): 815–842.

On IUCN origins: Anna-Katharina Wöbse, "'The World After All Was One': The International Environmental Network of UNESCO and IUPN, 1945–1950," *Contemporary European History* 20, no. 3 (August 2011): 331–348; Stephen J. Macekura, *Of Limits and Growth: The Rise of Global Sustainable Development in the Twentieth Century* (Cambridge, 2015), 17–53; and Perrin Selcer, *The Postwar Origins of the Global Environment: How the United Nations Built Spaceship Earth* (New York, 2018). On the global problem of protected areas: E. O. Wilson, *Half-Earth: Our Planet's Fight for Life* (New York, 2016).

On world heritage origins: Melanie Hall, *Towards World Heritage: International Origins of the Preservation Movement 1870–1930* (London, 2011); and Lynn Meskell, *A Future in Ruins: UNESCO, World Heritage, and the Dream of Peace* (Oxford, 2018). On sacred natural sites: Robert Wild et al., eds., *Sacred Natural Sites: Guidelines for Protected Area Managers* (Gland, 2008); and Bas Verschuuren et al., eds., *Sacred Natural Sites: Conserving Nature and Culture* (London, 2010).

On ahuehuetes: Maximino Martínez, *Las pináceas mexicanas*, 3rd ed. (México, 1963), 161–212. On Aztec tree planting: Patrizia Granziera, "Concept of the Garden in Pre-Hispanic Mexico," *Garden History* 29, no. 2 (Winter 2001): 185–213; Susan Toby Evans, "Aztec Royal Pleasure Parks: Conspicuous Consumption and Elite Status Rivalry," *Studies in the History of Gardens and Designed Landscapes* 20, no. 3 (2000): 206–228; Paul Avilés, "Seven Ways of Looking at a Mountain: Tetzcotzingo and the Aztec Garden Tradition," *Landscape Journal* 25, no. 2 (January 2006): 143–157; and Barbara E. Mundy, *The Death of Aztec Tenochtitlan, the Life of Mexico City* (Austin, 2015), 52–71.

On arbonationalism in Mexico City: Manuel Rivera Cambas, *México pintoresco, artístico y monumental*, vol. 2 (México, 1882), 342–347; Enrique de Olavarría y Ferrari, *Crónica del undécimo congreso internacional de americanistas, primero reunido en México en octubre de 1895* (México, 1896), 67–75; and Enrique Plasencia de la Parra,

"Conmemoración de la hazaña épica de los niños héroes: Su origen, desarrollo y simbolismos," *Historia Mexicana* 45, no. 2 (October–December 1995): 241–279.

On El Bosque de Chapultepec: Emily Wakild, "Parables of Chapultepec: Urban Parks, National Landscapes, and Contradictory Conservation in Modern Mexico," in *A Land Between Waters: Environmental Histories of Modern Mexico*, ed. Christopher R. Boyer (Tucson, 2012), 192–217. On Mexican nature protection: Emily Wakild, *Revolutionary Parks: Conservation, Social Justice, and Mexico's National Parks, 1910–1940* (Tucson, 2011). On hydraulic Mexico City, see books by Vera S. Candiani, Casey Walsh, and Matthew Vitz; and Emily Wakild, "Naturalizing Modernity: Urban Parks, Public Gardens and Drainage Projects in Porfirian Mexico City," *Mexican Studies/Estudios Mexicanos* 23, no. 1 (Winter 2007): 101–123.

On dendroclimatology of Mexico: D. W. Stahle et al., "Major Mesoamerican Droughts of the Past Millennium," *Geophysical Research Letters* 38 (2011): L05703; and idem, "The Mexican Drought Atlas: Tree-Ring Reconstructions of the Soil Moisture Balance During the Late Pre-Hispanic, Colonial, and Modern Eras," *Quaternary Science Reviews* 149 (2016): 34–60.

On El Tule: Manuel Ortega Reyes, "El Sabino de Santa María del Tule," *La Naturaleza* 6 (1884): 110–115; Manuel Francisco Álvarez, *Las ruinas de Mitla y la arquitectura* (México, 1900), 1–38; Victor Jimenez, *El árbol de el Tule en la historia* (México, 1990); C. Conzatti, *Monograph on the Tree of Santa María del Tule* (México, 1934); John Skeaping, *The Big Tree of Mexico* (Bloomington, 1953), 72–84; "Grand Old Tree Has Nothing to Fear but Mexico," *New York Times*, July 29, 1995; Oscar Dorado et al., "The Árbol del Tule (*Taxodium mucronatum* Ten.) Is a Single Genetic Individual," *Madroño* 43, no. 4 (October–December 1996): 445–452; Zsolt Debreczy and István Rácz, "*El Arbol del Tule*: The Ancient Giant of Oaxaca," *Arnoldia* 57 (Winter 1997–1998): 3–11; Gerard Passola i Parcerissa, "Informe del estado general del Árbol de Santa María del Tule," unpublished report (Barcelona, 2011); and Ursula Thiemer-Sachse, "El Árbol de Tule: Un monument de importancia en el ideario de la gente indígena de Oaxaca," *Anthropos* 111 (2016): 99–112.

On syncretism: Judith Francis Zeitlin, "Contesting the Sacred Landscape in Colonial Mesoamerica," unpublished report (Los Angeles, 2008); and Patrizia Granziera, "The Worship of Mary in Mexico: Sacred Trees, Christian Crosses, and the Body of the Goddess," *Toronto Journal of Theology* 28, no. 1 (Spring 2012): 43–60.

On 1968: Eugenia Allier-Montaño, "Memory and History of Mexico '68," *European Review of Latin American and Caribbean Studies* 102 (October 2016): 7–25; and Luis Alberto Pérez-Amezcua, "Por 'su propia, peculiar, historia': 'La nueva mexicanidad' y la literatura," *Mitologías hoy* 16 (diciembre 2017): 93–105.

CHAPTER IV. PACIFIC FIRES

On biogeography of Zealandia: George Gibbs, *Ghosts of Gondwana: The History of Life in New Zealand* (Nelson, 2006); and Alan De Queiroz, *The Monkey's Voyage: How Improbable Journeys Shaped the History of Life* (New York, 2013).

For general histories of Aotearoa: James Belich, *Making Peoples: A History of the New Zealanders from Polynesian Settlement to the End of the Nineteenth Century* (Auckland, 1996); idem, *Paradise Reforged: A History of the New Zealanders from the 1880s to the*

Year 2000 (Honolulu, 2001); Michael King, *The Penguin History of New Zealand* (Auckland, 2003); Paul Moon, *Encounters: The Creation of New Zealand* (Auckland, 2013); and the government-run website, Te Ara: The Encyclopedia of New Zealand. On the peopling of Aotearoa: D. R. Simmons, *The Great New Zealand Myth: A Study of the Discovery and Origin Traditions of the Maori* (Wellington, 1976); and Richard Walter et al., "Mass Migration and the Polynesian Settlement of New Zealand," *Journal of World Prehistory* 30, no. 4 (December 2017): 351–376. On modern Māori canoe traditions: Michael King, *Te Puea: A Biography* (Auckland, 1977); and Anne Nelson, *Nga Waka Maori: Maori Canoes* (Wellington, 1991).

On fire in Aotearoa: George L. W. Perry et al., "Ecology and Long-Term History of Fire in New Zealand," *New Zealand Journal of Ecology* 38, no. 2 (2014): 157–176; and Stephen J. Pyne, *Vestal Fire: An Environmental History, Told Through Fire, of Europe and Europe's Encounter with the World* (Seattle, 2000). On environmental change in Aotearoa: Eric Pawson and Tom Brooking, eds., *Making a New Land: Environmental Histories of New Zealand* (Dunedin, 2013); Tom Brooking and Eric Pawson, eds., *Seeds of Empire: The Environmental Transformation of New Zealand* (London, 2011); and Herbert Guthrie-Smith, *Tutira: The Story of a New Zealand Sheep Station* (Seattle, 1999 [1921]).

On kauri biology: Gregory A. Steward and Anthony E. Beveridge, "A Review of New Zealand Kauri (*Agathis australis* [D.Don] Lindl.): Its Ecology, History, Growth and Potential for Management for Timber," *New Zealand Journal of Forestry Science* 40 (2010): 33–59.

On kauri history: A. H. Reed, *The New Story of the Kauri*, 3rd ed. (Wellington, 1964); E. V. Sale, *Quest for the Kauri* (Wellington, 1978); John Halkett and E. V. Sale, *The World of the Kauri* (Auckland, 1986); Gordon Ell, *King Kauri: Tales & Traditions of the Kauri Country of New Zealand* (Auckland, 1996); Joanna Orwin, *Kauri: Witness to a Nation's History* (Auckland, 2004); Keith Stewart, *Kauri* (North Shore, 2008); and Gretel Boswijk, "Remembering Kauri on the 'Kauri Coast,'" *New Zealand Geographer* 66, no. 2 (August 2010): 124–137.

On colonial warfare and the King Movement: Vincent O'Malley, *The Great War for New Zealand: Waikato, 1800–2000* (Wellington, 2016). On Māori history in the two greatest kauri forests: Mark Derby, " 'Fallen Plumage': A History of Puhipuhi, 1865–2015," Waitangi Tribunal report 1040-A61 (2016); and various reports on Waipoua filed in Te Roroa's tribunal claim (WAI 38): waitangitribunal.govt.nz.

On kauri gum: A. H. Reed, *The Gumdiggers: The Story of Kauri Gum* (Wellington, 1972); and Senka Božić-Vrbančić, *Tarara: Croats and Maori in New Zealand: Memory, Belonging, Identity* (Dunedin, 2008). For export statistics: *New Zealand Official Year-Book*; and the earlier *New Zealand Official Handbook*.

On forestry in New Zealand: Thomas E. Simpson, *Kauri to Radiata: Origin and Expansion of the Timber Industry of New Zealand* (Auckland, 1973); Michael Roche, *Forest Policy in New Zealand: An Historical Geography, 1840–1919* (Palmerston North, 1987); idem, *History of Forestry* (Auckland, 1990); Paul Star, "Native Forest and the Rise of Preservation in New Zealand (1903–1913)," *Environment and History* 8, no. 3 (August 2002): 275–294; James Beattie, "Environmental Anxiety in New Zealand, 1840–1941: Climate Change, Soil Erosion, Sand Drift, Flooding and Forest

Conservation," *Environment and History* 9 (November 2003): 379–392; and James Beattie and Paul Star, "Global Influences and Local Environments: Forestry and Forest Conservation in New Zealand, 1850s–1925," *British Scholar* 3, no. 2 (September 2010): 191–218.

On comparative settler colonialism: Thomas Dunlap, *Nature and the English Diaspora: Environment and History in the United States, Canada, Australia, and New Zealand* (Cambridge, 1999); Gregory A. Barton, *Empire Forestry and the Origins of Environmentalism* (Cambridge, 2002); and James Belich, *Replenishing the Earth: The Settler Revolution and the Rise of the Anglo-World, 1783–1939* (Oxford, 2011). On trans- and cis-Pacific connections: Ian Tyrrell, *True Gardens of the Gods: Californian-Australian Environmental Reform, 1860–1930* (Berkeley, 1999); and Edward Dallam Melillo, *Strangers on Familiar Soil: Rediscovering the Chile-California Connection* (New Haven, 2015).

On Monterey pine: Peter B. Lavery and Donald J. Mead, "*Pinus radiata*: A Narrow Endemic from North America Takes on the World," in *Ecology and Biogeography of Pinus*, ed. David M. Richardson (Cambridge, 1998), 432–449.

On alerce biogeography, biology, longevity, and ecology, respectively: Claire G. Williams, Victor Martinez, and Carlos Magni, "Ice-age Persistence of *Fitzroya cupressoides*, a Southern Hemisphere Conifer," *Japanese Journal of Historical Botany* 19, nos. 1–2 (April 2011): 101–107; Martin F. Gardner et al., eds., "*Fitzroya cupressoides*," *Curtis's Botanical Magazine* 16, no. 3 (August 1999): 229–240; Antonio Lara and Ricardo Villalba, "A 3,620-Year Temperature Record from *Fitzroya cupressoides* Tree Rings in Southern South America," *Science* 260, no. 5111 (May 21, 1993): 1104–1106; Thomas T. Veblen, "Temperate Forests of the Southern Andean Region," in *The Physical Geography of South America*, ed. Thomas T. Veblen et al. (Oxford, 2007), 217–231. On alerce clearance: Luis Otero Durán, *La huella del fuego: Historia de los bosques nativos* (Santiago, 2006); and Fernando Torrejón et al., "Consecuencias de la tala maderera colonial en los bosques de alerce de Chiloé, sur de Chile (Siglos XVI–XIX)," *Magallania* 39, no. 2 (2011): 75–95.

On Chilean natural historians and foresters: Sergio A. Castro et al., "Rodulfo Amando Philippi, el naturalista de mayor aporte al conocimiento taxonómico de la diversidad biológica de Chile," *Revista Chilena de Historia Natural* 79, no. 1 (2006): 133–143; Pablo Camus, *Ambiente, bosques y gestión forestal en Chile, 1541–2005* (Santiago, 2006); Fernando Ramírez Morales, "Los bosques nativos chilenos y la 'política forestal' en la primera mitad del Siglo XX," *Cuadernos de historia* 26 (Marzo 2007): 135–167; Alejandra Bluth Solari, *El aporte de la ingeniería forestal al desarrollo del país: Una reseña histórica de la profesión forestal en Chile* (Santiago, 2013); Patience A. Schell, *The Sociable Sciences: Darwin and His Contemporaries in Chile* (New York, 2013); Emily Wakild, "Protecting Patagonia: Science, Conservation and the Prehistory of the Nature State on a South American Frontier, 1903–1934," in *The Nature State: Rethinking the History of Conservation*, ed. Wilko Graf von Hardenberg et al., (London, 2017), 37–54; and Thomas Miller Klubock, "The Politics of Forests and Forestry on Chile's Southern Frontier, 1880s–1940s," *Hispanic American Historical Review* 86, no. 3 (August 2006): 535–570.

On Indigenous peoples in Araucanía: Joanna Crow, *The Mapuche in Modern Chile: A Cultural History* (Gainesville, 2013); and Pilar M. Herr, *Contested Nation: The*

Mapuche, Bandits, and State Formation in Nineteenth-Century Chile (Albuquerque, 2019).

On Chilean forest conservation in the twentieth century: Thomas Miller Klubock, *La Frontera: Forests and Ecological Conflict in Chile's Frontier Territory* (Durham, NC, 2014); Emily Wakild, "Purchasing Patagonia: The Contradictions of Conservation in Free Market Chile," in *Lost in the Long Transition: Struggles for Social Justice in Neoliberal Chile*, ed. William L. Alexander (Lanham, MD, 2009), 121–132; and articles with keywords "private protected areas" and "neoliberal conservation." On Rick Klein: Marc Cooper, "Alerce Dreams," *Sierra* 77, no. 1 (January–February 1992): 122–129; and Jimmy Langman, "The Untold Conservation Legacy of Rick Klein," *Patagon Journal* 18 (Spring 2018): 32–39. On Douglas Tompkins: Diana Saverin, "The Entrepreneur Who Wants to Save Paradise," *Atlantic*, September 15, 2014.

On California's tallest trees: Michael G. Barbour et al., *Coast Redwood: A Natural and Cultural History* (Los Olivos, 2001); and Reed F. Noss, ed., *The Redwood Forest: History, Ecology, and Conservation of the Coast Redwoods* (Washington, 2000). On paleobotany: Gary D. Lowe, *Geologic History of the Giant Sequoia and the Coast Redwood*, rev. ed. (Dublin, CA, 2014). On canopy science: Richard Preston, *The Wild Trees: A Story of Passion and Daring* (New York, 2007). On exploitation and preservation: Jared Farmer, *Trees in Paradise: A California History* (New York, 2013), 44–108; Susan R. Schrepfer, *The Fight to Save the Redwoods: A History of Environmental Reform, 1917–1978* (Madison, 1983); and Darren Frederick Speece, *Defending Giants: The Redwood Wars and the Transformation of American Environmental Politics* (Seattle, 2016).

On Yurok experiences: Lucy Thompson, *To the American Indian* (Eureka, 1916); T. T. Waterman, *Yurok Geography* (Berkeley, 1920); Lynn Huntsinger et al., "A Yurok Forest History," unpublished report (Berkeley, 1994); Tony Platt, *Grave Matters: Excavating California's Buried Past* (Berkeley, 2011); Benjamin Madley, *An American Genocide: The United States and the California Indian Catastrophe, 1846–1873* (New Haven, 2016); Thomas Buckley, *Standing Ground: Yurok Indian Spirituality, 1850–1990* (Berkeley, 2002); and Carolyn Kormann, "How Carbon Trading Became a Way of Life for California's Yurok Tribe," *New Yorker*, October 10, 2018.

On sugi: Aljos Farjon, "*Cryptomeria japonica*," *Curtis's Botanical Magazine* 16, no. 3 (August 1999): 212–228. On sugi longevity: Shigejiro Yoshida, "Information Collection of the Discs of Yaku-sugi, Old *Cryptomeria japonica* Trees More than 1,000 Years Old on Yakushima Island," *Journal of the Japanese Forestry Society* 99, no. 1 (2017): 46–49 [in Japanese]. On Formosan conifers: *Flora of Taiwan*, vol. 1, rev. ed. (Taipei, 1994).

On Japanese forestry: Conrad Totman, *The Green Archipelago: Forestry in Preindustrial Japan* (Berkeley, 1989); and John Knight, "From Timber to Tourism: Recommoditizing the Japanese Forest," *Development and Change* 31, no. 1 (January 2000): 341–359.

On Yakushima: Takahiro Iseki and Sachihiko Harashina, "A Study on Relationship Between Nature Oriented Tourism and Other Nature-Use-Activities: A Case Study of the Old Famous Cedar Tree, Jyoumonsugi in Yakushima Island," *Environmental Information Science* 35, no. 2 (2006): 43–52 [in Japanese]; Andrew Daniels, "Woodland Landscape in Edo Period Japan with Specific Reference to Yakushima in Satsuma Domain," *International Human Studies* 15 (March 2009): 31–50; idem, "Yakushima's

Kosugidani: Human Presence in an Okudake Woodland Landscape," *International Human Studies* 16 (March 2010): 1–19; Dajeong Song and Sueo Kuwahara, "Ecotourism and World Natural Heritage: Its Influence on Islands in Japan," *Journal of Marine and Island Cultures* 5, no. 1 (June 2016): 36–46; and Shigemitsu Shibasaki, "Yakushima Island: Landscape History, World Heritage Designation, and Conservation Status for Local Society," in *Natural Heritage of Japan: Geological, Geomorphological, and Ecological Aspects*, ed. Abhik Chakraborty et al. (Cham, 2018), 73–83.

On modern Shinto and trees: Wilbur M. Fridell, *Japanese Shrine Mergers, 1906–12: State Shinto Moves to the Grassroots* (Tokyo, 1973); Gaudenz Domenig, "Sacred Groves in Modern Japan: Notes on the Variety and History of Shinto Shrine Forests," *Asiatische Studien* 51 (1997): 91–121; and Aike P. Rots, *Shinto, Nature and Ideology in Contemporary Japan: Making Sacred Forests* (London, 2017). On eco-spirituality in postwar Japan: Shimazono Susumu and Tim Graf, "The Rise of the New Spirituality," in *Handbook of Contemporary Japanese Religions*, ed. Inken Prohl and John K. Nelson (Leiden, 2012), 459–485. On Jomon heritage: John Knight, "'Indigenous' Regionalism in Japan," in *Indigenous Environmental Knowledge and Its Transformations: Critical Anthropological Perspectives*, ed. Alan Bicker et al. (Amsterdam, 2000), 151–176; and Akio Mishima, *Jōmonsugi no keishō* (Tokyo, 1994).

On forestry in Taiwan: Kuo-Tung Ch'en, "Nonreclamation Deforestation in Taiwan, c. 1600–1976," in *Sediments of Time: Environment and Society in Chinese History*, ed. Mark Elvin and Liu Ts'ui-jung (Cambridge, 2009), 693–727; Tessa Morris-Suzuki, "The Nature of Empire: Forest Ecology, Colonialism and Survival Politics in Japan's Imperial Order," *Japanese Studies* 33, no. 3 (2013): 225–242; Kuang-Chi Hung, "When the Green Archipelago Encountered Formosa: The Making of Modern Forestry in Taiwan under Japan's Colonial Rule (1895–1945)," in *Environment and Society in the Japanese Islands: From Prehistory to the Present*, ed. Bruce L. Batten and Philip C. Brown (Corvallis, 2015), 174–193; Chao-Hsu Su, "[Shitarō Kawai and the Birth of the Alishan Forest Railway]," *Taiwan Forestry Journal* 38, no. 3 (June 2012): 74–81 [in Chinese]; and Hagino Toshio, *Chōsen, Manshū, Taiwan ringyō hattatsushiron* (Tokyo, 1965).

On Indigenous peoples in late colonial Taiwan: Paul D. Barclay, *Outcasts of Empire: Japan's Rule on Taiwan's "Savage Border," 1874–1945* (Oakland, 2018). On social complexities of postcolonial Taiwan: Melissa J. Brown, *Is Taiwan Chinese? The Impact of Culture, Power, and Migration on Changing Identities* (Berkeley, 2004). On Tayal communities and cypresses: Yih-ren Lin, "Politicizing Nature: The Maqaw National Park Controversy in Taiwan," *Capitalism Nature Socialism* 22, no. 2 (June 2011): 88–103; and David Reid, "Nation vs. Tradition: Indigenous Rights and Smangus," in *Taiwan Since Martial Law: Society, Culture, Politics, Economy*, ed. David Blundell (Taipei, 2012), 453–483.

CHAPTER V. CIRCLES AND LINES

On sequoia biology, ecology, biogeography, longevity, and size, respectively: Richard J. Hartesveldt et al., *The Giant Sequoia of the Sierra Nevada* (Washington, 1975); Nathan L. Stephenson, "Reference Conditions for Giant Sequoia Forest Restoration: Structure, Process, and Precision," *Ecological Applications* 9, no. 4 (November

1999): 1253–1265; Dwight Willard, *A Guide to the Sequoia Groves of California* (Yosemite National Park, 2000); Nathan L. Stephenson, "Estimated Ages of Some Large Giant Sequoias: General Sherman Keeps Getting Younger," *Madroño* 47, no. 1 (January–March 2000): 61–67; and Wendell D. Flint, *To Find the Biggest Tree* (Three Rivers, CA, 2002).

On sequoias in US history: Walter Fry and John R. White, *Big Trees*, rev. ed. (Stanford, 1938); Hank Johnston, *They Felled the Redwoods: A Saga of Flumes and Rails in the High Sierra* (Los Angeles, 1966); Dennis G. Kruska, *Sierra Nevada Big Trees: History of the Exhibitions, 1850–1903* (Los Angeles, 1985); Lori Vermaas, *Sequoia: The Heralded Tree in American Art and Culture* (Washington, 2003); Farmer, *Trees in Paradise*, 7–44; and William C. Tweed, *King Sequoia: The Tree That Inspired a Nation, Created Our National Park System, and Changed the Way We Think about Nature* (Berkeley, 2016).

On mammoths and mastodons in the US imagination, see books by Paul Semonin, Claudine Cohen, Keith Thomson, and Mark V. Barrow Jr. On paleobotany: Ralph W. Chaney, "A Revision of Fossil *Sequoia* and *Taxodium* in Western North America Based on the Recent Discovery of *Metasequoia*," *Transactions of the American Philosophical Society* 40, no. 3 (1950): 171–263; and Lowe, *Geologic History of the Giant Sequoia and the Coast Redwood.*

On John Muir, see biographies by Michael P. Cohen and Donald Worster; and Richard G. Beidleman, *California's Frontier Naturalists* (Berkeley, 2006). On Marxists in the Giant Forest: Daegan Miller, *This Radical Land: A Natural History of American Dissent* (Chicago, 2018), 161–212.

For introductions to tree-ring science: Valerie Trouet, *Tree Story: The History of the World Written in Rings* (Baltimore, 2020); and James H. Speer, *Fundamentals of Tree-Ring Research* (Tucson, 2010). On early cross-dating: R. A. Studhalter, "Tree Growth: Some Historical Chapters," *Botanical Review* 21 (January–March 1955): 1–72; and Rupert Wimmer, "Arthur Freiherr von Seckendorff-Gudent and the Early History of Tree-Ring Crossdating," *Dendrochronologia* 19, no. 1 (January 2001): 153–158. For US context: Christopher H. Briand et al., "Tree Rings and the Aging of Trees: A Controversy in 19th Century America," *Tree-Ring Research* 62, no. 2 (December 2006): 51–65.

On A. E. Douglass, see biographies by George Ernest Webb and Donald J. McGraw; and Stephen Edward Nash, *Time, Trees, and Prehistory: Tree-Ring Dating and the Development of North American Archaeology, 1914–1950* (Salt Lake City, 1999). On Ellsworth Huntington, see the biography by Geoffrey J. Martin; James Rodger Fleming, *Historical Perspectives on Climate Change* (Oxford, 1998), 95–106; and Daniel E. Bender, *American Abyss: Savagery and Civilization in the Age of Industry* (Ithaca, 2009), 40–68.

For national and international contexts for Douglass and Huntington: Susan Schulten, *The Geographical Imagination in America, 1880–1950* (Chicago, 2001); Jamie L. Pietruska, *Looking Forward: Prediction and Uncertainty in Modern America* (Chicago, 2017); David N. Livingstone, *The Geographical Tradition: Episodes in the History of a Contested Enterprise* (Oxford, 1992); and Neville Brown, *History and Climate Change: A Eurocentric Perspective* (London, 2001).

On temporality and science: Stephen Jay Gould, *Time's Arrow, Time's Cycle: Myth and Metaphor in the Discovery of Geological Time* (Cambridge, MA, 1987). On tree charts: Manuel Lima, *The Book of Trees: Visualizing Branches of Knowledge* (New York, 2014). On timelines: Daniel Rosenberg and Anthony Grafton, *Cartographies of Time: A History of the Timeline* (New York, 2010). For museum context on sequoia artifacts: Steven Conn, *Museums and American Intellectual Life, 1876–1926* (Chicago, 1998); and Teresa Barnett, *Sacred Relics: Pieces of the Past in Nineteenth-Century America* (Chicago, 2013).

On connections between eugenics, museums, and megafloral preservation: Alexandra Minna Stern, *Eugenic Nation: Faults and Frontiers of Better Breeding in Modern America*, 2nd. ed. (Berkeley, 2015); Jonathan Peter Spiro's biography of Madison Grant; and Brian Regal's biography of Henry Fairfield Osborn. On Anglo-Saxonism in the Golden State: Richard White, *California Exposures: Envisioning Myth and History* (New York, 2020).

On settler temporalities: John Demos, *Circles and Lines: The Shape of Life in Early America* (Cambridge, MA, 2004); Thomas M. Allen, *A Republic in Time: Temporality and Social Imagination in Nineteenth-Century America* (Chapel Hill, 2008); and Amy Kaplan, "Imperial Melancholy in America," *Raritan* 28, no. 3 (Winter 2009): 13–31.

On Native temporalities: Peter Nabokov, *A Forest of Time: American Indian Ways of History* (Cambridge, 2002); and Mark Rifkin, *Beyond Settler Time: Temporal Sovereignty and Indigenous Self-Determination* (Durham, NC, 2017). On Yosemite's Natives: Mark David Spence, *Dispossessing the Wilderness: Indian Removal and the Making of the National Parks* (New York, 1999). For statewide coverage: Damon B. Akins and William J. Bauer Jr., *We Are the Land: A History of Native California* (Oakland, 2021).

On fire policy in Sierran national parks: Alfred Runte, *Yosemite: The Embattled Wilderness* (Lincoln, 1990); Hal K. Rothman, *Blazing Heritage: A History of Wildland Fire in the National Parks* (New York, 2007); and Stephen Pyne, *California: A Fire Survey* (Tucson, 2016). On Sierran fire history: Thomas W. Swetnam and Christopher H. Baisan, "Tree-Ring Reconstructions of Fire and Climate History in the Sierra Nevada and Southwestern United States," in *Fire and Climatic Change in Temperate Ecosystems of the Western Americas*, ed. Thomas T. Veblen et al. (New York, 2003), 158–195; Jan W. Van Wagtendonk and Jo Ann Fites-Kaufman, "Sierra Nevada Bioregion," in *Fire in California's Ecosystems*, ed. Neil G. Sugihara et al. (Berkeley, 2006), 264–294; Thomas W. Swetnam et al., "Multi-Millennial Fire History of the Giant Forest, Sequoia National Park, California, USA," *Fire Ecology* 5, no. 3 (2009): 120–150; and Trouet, *Tree Story*, 181–197.

On past megadroughts: David W. Stahle and Jeffrey S. Dean, "North American Tree Rings, Climatic Extremes, and Social Disasters," in *Dendroclimatology: Progress and Prospects*, ed. Malcolm K. Hughes et al. (Dordrecht, 2011), 297–327; B. Lynn Ingram and Frances Malamud-Roam, *The West without Water: What Past Floods, Droughts, and Other Climate Clues Tell Us about Tomorrow* (Berkeley, 2013); and Harvey Weiss, ed., *Megadrought and Collapse: From Early Agriculture to Angkor* (Oxford, 2017).

On managing sequoias in crisis: Michelle Nijhuis, "How the Parks of Tomorrow Will Be Different," *National Geographic* 230, no. 6 (December 2016): 102–121; Madeline Ostrander, "For the National Parks, a Reckoning," *Undark*, September 13, 2017; Dahr

Jamail, *The End of Ice: Bearing Witness and Finding Meaning in the Path of Climate Disruption* (New York, 2019); Zach St. George, *The Journeys of Trees: A Story about Forests, People, and the Future* (New York, 2020); and the website of Sequoia & Kings Canyon National Park.

CHAPTER VI. OLDEST KNOWN

For Schulman's career biography: Donald J. McGraw, *Edmund Schulman and the "Living Ruins": Bristlecone Pines, Tree Rings and Radiocarbon Dating* (Bishop, CA, 2007). For lab context: Pearce Paul Creasman et al., "Reflections on the Foundation, Persistence, and Growth of the Laboratory of Tree-Ring Research, circa 1930–1960," *Tree-Ring Research* 68, no. 2 (2012): 81–89; Christine Hallman et al., "Status Report: Lost and Found: The Bristlecone Pine Collection," *Tree-Ring Research* 62, no. 1 (2006): 25–29; and Michael L. Morrison and Joseph M. Szewczak, "White Mountain Research Station, University of California," *Bulletin of the Ecological Society of America* 83, no. 1 (January 2002): 63–68.

On Schulman's milieux: Deborah Dash Moore et al., *Jewish New York: The Remarkable Story of a City and a People* (New York, 2007); Elliott Robert Barkan, *From All Points: America's Immigrant West, 1870s–1952* (Bloomington, 2007), esp. 307–310; Leonard Dinnerstein, "From Desert Oasis to the Desert Caucus: The Jews of Tucson," in *Jews of the American West*, ed. Moses Rischin and John Livingston (Detroit, 1991), 139–163; and Noah J. Efron, *A Chosen Calling: Jews in Science in the Twentieth Century* (Baltimore, 2014).

For Schulman's posthumous article: "Bristlecone Pine, Oldest Known Living Thing," *National Geographic* 113, no. 3 (March 1958): 354–372. On *National Geographic* in midcentury US culture, see books by Catherine A. Lutz and Jane L. Collins; Susan Schulten; Tamar Y. Rothenberg; and Stephanie L. Hawkins.

For Schulman's magnum opus: *Dendroclimatic Changes in Semiarid America* (Tucson, 1956). For accessible follow-ups: James Lawrence Powell, *Dead Pool: Lake Powell, Global Warming, and the Future of Water in the West* (Berkeley, 2008); William deBuys, *A Great Aridness: Climate Change and the Future of the American Southwest* (Oxford, 2011); and Ingram and Malamud-Roam, *West without Water*.

On Schulman's subdiscipline: H. C. Fritts, *Tree Rings and Climate* (London, 1976); and Malcolm K. Hughes et al., eds., *Dendroclimatology: Progress and Prospects* (Dordrecht, 2011). For profiles of the Laboratory of Tree-Ring Research: Scott Norris, "Reading Between the Lines," *BioScience* 50, no. 5 (May 2000): 389–394; and Michelle Nijhuis, "Written in the Rings," *High Country News*, January 24, 2005. For bristlecone research by LTRR scientists, see the online UA Campus Repository.

On climate reconstruction and modeling: Spencer R. Weart, *The Discovery of Global Warming*, rev. ed. (Cambridge, MA, 2008); Paul N. Edwards, *A Vast Machine: Computer Models, Climate Data, and the Politics of Global Warming* (Cambridge, MA, 2010); and Joshua P. Howe, *Behind the Curve: Science and the Politics of Global Warming* (Seattle, 2014). On cultural meanings of proxy data: Matthias Dörries, "Politics, Geological Past, and the Future of the Earth," *Historical Social Research* 40, no. 2 (2015): 22–36; and Alessandro Antonello and Mark Carey, "Ice Cores and the Temporalities of the Global Environment," *Environmental Humanities* 9, no. 2 (November 2017): 181–203. On the politicization of proxy data: Michael E. Mann, *The Hockey Stick and the Climate Wars: Dispatches from the Front Lines* (New York, 2012).

On taxonomy: D. K. Bailey, "Phytogeography and Taxonomy of *Pinus* Subsection *Balfourianae*," *Annals of the Missouri Botanical Garden* 57, no. 2 (1970): 210–249; and David M. Gates, "An Amateur Botanist's Great Discovery," *Missouri Botanical Garden Bulletin* 59, no. 3 (May–June 1971): 39–48. On *Pinus longaeva* vis-à-vis science studies and cultural studies: Michael P. Cohen, *A Garden of Bristlecones: Tales of Change in the Great Basin* (Reno, 1998). For meditations on the species: Gayle Brandow Samuels, *Enduring Roots: Encounters with Trees, History, and the American Landscape* (New Brunswick, 1999), 135–160; Ross Andersen, "The Vanishing Groves," *Aeon*, October 16, 2012; Valerie Mendenhall Cohen and Michael P. Cohen, *Tree Lines* (Reno, 2017); and Alex Ross, "The Past and the Future of the Earth's Oldest Trees," *New Yorker*, January 20, 2020.

For bioregional histories: Stephen Trimble, *The Sagebrush Ocean: A Natural History of the Great Basin* (Reno, 1989); Clarence A. Hall, ed., *Natural History of the White-Inyo Range, Eastern California* (Berkeley, 1991); and Donald K. Grayson, *The Great Basin: A Natural Prehistory*, rev. ed. (Berkeley, 2011).

On the ontology of things: Jane Bennett, *Vibrant Matter: A Political Ecology of Things* (Durham, NC, 2009); John Durham Peters, *The Marvelous Clouds: Toward a Philosophy of Elemental Media* (Chicago, 2015); and David Wood, *Thinking Plant Animal Human: Encounters with Communities of Difference* (Minneapolis, 2020).

CHAPTER VII. LATEST OLDEST

For updates to Schulman's "longevity under adversity": David W. Stahle, "Tree Rings and Ancient Forest Relics," *Arnoldia* 56, no. 4 (Winter 1996–1997): 2–10; D. W. Larson et al., "Evidence for the Widespread Occurrence of Ancient Forests on Cliffs," *Journal of Biogeography* 27, no. 2 (March 2000): 319–331; Neil Pederson, "External Characteristics of Old Trees in the Eastern Deciduous Forest," *Natural Areas Journal* 30, no. 4 (October 2010): 396–407; and Alfredo Di Filippo et al., "The Longevity of Broadleaf Deciduous Trees in Northern Hemisphere Temperate Forests: Insights from Tree-Ring Series," *Frontiers in Ecology and Evolution* 3, no. 46 (May 2015): 1–15.

On northern white cedar in Ontario: Peter E. Kelly and Douglas W. Larson, *The Last Stand: A Journey Through the Ancient Cliff-Face Forest of the Niagara Escarpment* (Toronto, 2007).

On the Cross Timbers: Richard V. Francaviglia, *The Cast Iron Forest: A Natural and Cultural History of the North American Cross Timbers* (Austin, 2000); and the website of the Ancient Cross Timbers Consortium. On bald cypress in North Carolina: David W. Stahle et al., "Longevity, Climate Sensitivity, and Conservation Status of Wetland Trees at Black River, North Carolina," *Environmental Research Communications* 1, no. 4 (2019): 1–8; and Ayurella Horn-Muller, "The Oldest Tree in Eastern US Survived Millennia—but Rising Seas Could Kill It," *Guardian*, August 1, 2021.

BIBLIOGRAPHY

On sweet chestnut: Jean-Robert Pitte, *Terres de Castanide: Hommes et paysages du Châtaignier de l'Antiquité à nos jours* (Paris, 1986); M. Conedera et al., "The Cultivation of *Castanea sativa* (Mill.) in Europe, from Its Origin to Its Diffusion on a Continental Scale," *Vegetation History and Archaeobotany* 13 (2004): 161–179; Genevieve Michon, "Revisiting the Resilience of Chestnut Forests in Corsica: From Social-Ecological Systems Theory to Political Ecology," *Ecology and Society* 16, no. 2 (June 2011): article 5 [online only]; and Paolo Squatriti, *Landscape and Change in Early Medieval Italy: Chestnuts, Economy, and Culture* (Cambridge, 2013).

On Brazil nut: Glenn H. Shepard Jr. and Henri Ramirez, "'Made in Brazil': Human Dispersal of the Brazil Nut (*Bertholletia excelsa*, Lecythidaceae) in Ancient Amazonia," *Economic Botany* 65, no. 1 (2011): 44–65; and Evert Thomas et al., "Uncovering Spatial Patterns in the Natural and Human History of Brazil Nut (*Bertholletia excelsa*) Across the Amazon Basin," *Journal of Biogeography* 42 (2015): 1367–1382.

On plant domestication in the Amazonian forest, see articles by lead authors Charles R. Clement and Carolina Levis. On tree species distribution, see articles by lead author Hans ter Steege. On visions of the Amazon, see books by Candace Slater; Susanna B. Hecht; and Eduardo Kohn. On the "pristine myth" and the "trouble with wilderness," see articles by William M. Denevan and William Cronon, respectively. For a demographic overview: Charles C. Mann, *1491: New Revelations of the Americas Before Columbus* (New York, 2005).

On *Metasequoia*: "Metasequoia After Fifty Years," special combined issue of *Arnoldia* 58/59, nos. 4/1 (1998–1999); Edmund H. Fulling, "Metasequoia—Fossil and Living—an Initial Thirty Year (1941–1970) Annotated and Indexed Bibliography with an Historical Introduction," *Botanical Review* 42, no. 3 (July–September 1976): 215–314; Jinshuang Ma, "The Chronology of the 'Living Fossil' *Metasequoia Glyptostroboides* (Taxodiaceae): A Review (1943–2003)," *Harvard Papers in Botany* 8, no. 1 (2003): 9–18; and Ben A. LePage et al., eds., *The Geobiology and Ecology of* Metasequoia (Dordrecht, 2005).

On *Wollemia* in New South Wales: James Woodford, *The Wollemi Pine: The Incredible Discovery of a Living Fossil from the Age of the Dinosaurs* (Melbourne, 2000); and John Pastoriza-Piñol, "*Wollemia nobilis*," *Curtis's Botanical Magazine* 24, no. 3 (August 2007): 155–161. On Huon pine in Tasmania: Garry Kerr and Harry McDermott, *The Huon Pine Story: A History of Harvest and Use of a Unique Timber*, 2nd ed. (Portland, Victoria, 2004); and articles by lead author Edward R. Cook.

On aspen clonality: Burton V. Barnes, "The Clonal Growth Habit of American Aspens," *Ecology* 47, no. 3 (May 1966): 439–447. On "Pando," see articles by lead author Paul C. Rogers. On aspen decline in the North American West, see the website of the Western Aspen Alliance. On the organism known as "Old Tjikko": G. L. Mackenthun, "The World's Oldest Living Tree Discovered in Sweden? A Critical Review," *New Journal of Botany* 5, no. 3 (2015): 200–204. On cycads: David L. Jones, *Cycads of the World: Ancient Plants in Today's Landscape*, 2nd ed. (Washington, 2002); and Loran M. Whitelock, *The Cycads* (Portland, OR, 2002). On box huckleberry: Rob Nicholson, "Little Big Plant, Box Huckleberry (*Gaylussacia brachycera*)," *Arnoldia* 68, no. 3 (2011): 11–18. On creosote: Frank C. Vasek, "Creosote Bush: Long-Lived

Clones in the Mojave Desert," *American Journal of Botany* 67, no. 2 (February 1980): 246–255. On other long-lived clonal organisms: Rachel Sussman, *The Oldest Living Things in the World* (Chicago, 2014). On largeness: Matthew D. LaPlante, *Superlative: The Biology of Extremes* (Dallas, 2019).

On plants and moral philosophy: Christopher D. Stone, *Should Trees Have Standing? Law, Morality, and the Environment*, 3rd ed. (New York, 2010); Matthew Hall, *Plants as Persons: A Philosophical Botany* (Albany, 2011); Michael Marder, *Plant-Thinking: A Philosophy of Vegetal Life* (New York, 2013); Robin Wall Kimmerer, *Braiding Sweetgrass: Indigenous Wisdom, Scientific Knowledge, and the Teachings of Plants* (Minneapolis, 2015); and Rob Nixon, "The Less Selfish Gene: Forest Altruism, Neoliberalism, and the Tree of Life," *Environmental Humanities* 13, no. 2 (November 2021): 348–371.

CHAPTER VIII. TIME TO MOURN

On the politics of acid rain and climate change: Naomi Oreskes and Erik M. Conway, *Merchants of Doubt: How a Handful of Scientists Obscured the Truth on Issues from Tobacco Smoke to Global Warming* (New York, 2010); Rachel Emma Rothschild, *Poisonous Skies: Acid Rain and the Globalization of Pollution* (Chicago, 2019); and Nathaniel Rich, *Losing Earth: A Recent History* (New York, 2019).

On the trope of the dying forest: Franz-Josef Brüggemeier, "*Waldsterben*: The Construction and Deconstruction of an Environmental Problem," in *Nature in German History*, ed. Christof Mauch (New York, 2004), 119–131; and Frank Uekötter and Kenneth Anders, "The Sum of All German Fears: Forest Death, Environmental Activism, and the Media in 1980s Germany," in *Exploring Apocalyptica: Coming to Terms with Environmental Alarmism*, ed. Frank Uekötter (Pittsburgh, 2018), 75–106.

On the contemporary Polish forest: Stuart Franklin, "Białowieża Forest, Poland: Representation, Myth, and the Politics of Dispossession," *Environment and Planning A* 34, no. 8 (2002): 1459–1485; Malgorzata Blicharska and Ann Van Herzele, "What['s] a Forest? Whose Forest? Struggles over Concepts and Meanings in the Debate about the Conservation of the Białowieża Forest in Poland," *Forest Policy and Economics* 57 (2015): 22–30; Eunice Blavascunas, *Foresters, Borders, and Bark Beetles: The Future of Europe's Last Primeval Forest* (Bloomington, 2020); and articles by Andrzej Bobiec.

On the US Northeast: Ellen Stroud, *Nature Next Door: Cities and Trees in the American Northeast* (Seattle, 2013); Charles D. Canham, *Forests Adrift: Currents Shaping the Future of Northeastern Trees* (New Haven, 2020); and Jonny Diamond, "The Old Man and the Tree," *Smithsonian* 52, no. 9 (January/February 2022): 32–43. On the coastal Southeast: Elizabeth A. Rush, *Rising: Dispatches from the New American Shore* (Minneapolis, 2018). On the Southwest: Cally Carswell, "The Tree Coroners," *High Country News*, December 16, 2013; and Craig D. Allen, "Forest Ecosystem Reorganization Underway in the Southwestern United States: A Preview of Widespread Forest Changes in the Anthropocene?" in *Forest Conservation and Management in the Anthropocene*, RMRS-P-71, ed. V. Alaric Sample and R. Patrick Bixler (Fort Collins, 2014), 103–123.

BIBLIOGRAPHY

On drivers of change in western North America: Andrew Nikiforuk, *Empire of the Beetle: How Human Folly and a Tiny Bug Are Killing North America's Great Forests* (Vancouver, 2011); Edward Struzik, *Firestorm: How Wildfire Will Shape Our Future* (Washington, 2017); Lauren Oakes, *In Search of the Canary Tree: The Story of a Scientist, a Cypress, and a Changing World* (New York, 2018); Daniel Mathews, *Trees in Trouble: Wildfires, Infestations, and Climate Change* (Berkeley, 2020); and articles on forests with keywords such as: disturbance, resilience, persistence, recovery, regeneration, reorganization, tree migration, range shifts, type conversion, and ecosystem change.

On interacting climatic stressors, search for high-citation articles with combinations of these keywords: climate-induced, drought-induced, heat-induced, climate threshold, canopy mortality, forest dieback, forest die-off, tree death, tree dieback, tree mortality, vulnerability, risk, and decline. For landscape-scale overviews: Constance I. Millar and Nathan L. Stephenson, "Temperate Forest Health in an Era of Emerging Megadisturbance," *Science* 349, no. 6250 (August 21, 2015): 823–826; James S. Clark et al., "The Impacts of Increasing Drought on Forest Dynamics, Structure, and Biodiversity in the United States," *Global Change Biology* 22, no. 7 (July 2016): 2329–2352; Rupert Seidl et al., "Forest Disturbances under Climate Change," *Nature Climate Change* 7 (June 2017): 395–402; Nate G. McDowell et al., "Pervasive Shifts in Forest Dynamics in a Changing World," *Science* 368, no. 6494 (May 29, 2020): eaaz9463 [online only]; articles by lead author Donald A. Falk; and the May 2022 special issue of *National Geographic*.

On bristlecone-limber dynamics: Constance I. Millar et al., "Recruitment Patterns and Growth of High-Elevation Pines in Response to Climatic Variability (1883–2013) in the Western Great Basin, USA," *Canadian Journal of Forest Research* 45, no. 10 (October 2015): 1299–1312. On bristlecone defenses, see articles by lead authors Barbara J. Bentz, Curtis A. Gray, and Justin B. Runyon. On evolution and ecology of *Pinus longaeva*: Constance I. Millar, "Impact of the Eocene on the Evolution of *Pinus* L.," *Annals of the Missouri Botanical Garden* 80, no. 2 (Spring 1993): 471–498; and Ronald M. Lanner, *The Bristlecone Book: A Natural History of the World's Oldest Trees* (Missoula, 2007). On biogeography: David Alan Charlet, *Nevada Mountains: Landforms, Trees, and Vegetation* (Salt Lake City, 2020). On Great Basin geology: Keith Heyer Meldahl, *Rough-Hewn Land: A Geologic Journey from California to the Rocky Mountains* (Berkeley, 2011).

On "fast" and "slow" science: Stuart Ritchie, *Science Fictions: How Fraud, Bias, Negligence, and Hype Undermine the Search for Truth* (New York, 2020); and Isabelle Stengers, *Another Science Is Possible: A Manifesto for Slow Science* (Cambridge, 2018). On Stewart Brand's worldview: books by Fred Turner; Andrew G. Kirk; and John Markoff. For a critique of the Clock of the Long Now: Evander L. Price, "Future Monumentality" (PhD diss., Harvard, 2019).

On possible futures: Yuval Noah Harari, *Homo Deus: A Brief History of Tomorrow* (London, 2016); Chris D. Thomas, *Inheritors of the Earth: How Nature Is Thriving in an Age of Extinction* (New York, 2017); and Rob Dunn, *A Natural History of the Future: What the Laws of Biology Tell Us about the Destiny of the Human Species* (New York, 2021).

BIBLIOGRAPHY

On the ecological importance of ancient trees: Charles H. Cannon et al., "Old and Ancient Trees are Life History Lottery Winners and Vital Evolutionary Resources for Long-Term Adaptive Capacity," *Nature Plants* 8 (February 2022): 136–145. On global decline of large old trees, see articles by lead author David B. Lindenmayer.

On shifting baselines: Peter S. Alagona et al., "Past Imperfect: Using Historical Ecology and Baseline Data for Conservation and Restoration Projects in North America," *Environmental Philosophy* 9, no. 1 (Spring 2012): 49–70; Robin Kundis Craig, "Perceiving Change and Knowing Nature: Shifting Baselines and Nature's Resiliency," *Environmental Law and Contrasting Ideas of Nature: A Constructivist Approach*, ed. Keith H. Hirokawa (Cambridge, 2014), 87–111; and Irus Braverman, "Shifting Baselines in Coral Conservation," *Nature and Space* 3, no. 1 (2020): 20–39.

On nostalgia: Renato Rosaldo, "Imperialist Nostalgia," *Representations* 26 (Spring 1989): 107–122; and Svetlana Boym, *The Future of Nostalgia* (New York, 2001). On climate anxiety and grief: Glenn A. Albrecht, *Earth Emotions: New Words for a New World* (Ithaca, 2019). On the youth climate movement: Daniel Sherrell, *Warmth: Coming of Age at the End of Our World* (New York, 2021).

On the bushman turned artist: *Rei Hamon: Artist of New Zealand—His Life and His Drawings* (Auckland, 1971). On the lost bird of the bush: Julianne Lutz Warren, "Huia Echoes," in *Future Remains: A Cabinet of Curiosities for the Anthropocene*, ed. Gregg Mitman et al. (Chicago, 2018), 71–80. On kauri dieback, search for articles on *Phytophthora agathidicida*, and consult the Kauri Dieback Programme: kauriprotection.co.nz.

On subfossil *Agathis*: Kate Evans, "Buried Treasure," *New Zealand Geographic* 142 (November–December 2016): 34–53; Andrew M. Lorrey and Gretel Boswijk, *Understanding the Scientific Value of Subfossil Bog (Swamp) Kauri* (Auckland, 2017); Andrew M. Lorrey et al., "The Scientific Value and Potential of New Zealand Swamp Kauri," *Quaternary Science Reviews* 183 (March 2018): 12–39; and Kate Evans, "Swamp Sentinels," *bioGraphic*, February 18, 2021. On subfossil *Taxodium*: David W. Stahle et al., "Tree-Ring Analysis of Ancient Baldcypress Trees and Subfossil Wood," *Quaternary Science Reviews* 34 (January 2018): 1–15. On subfossil *Metasequoia*: Jane E. Francis, "Polar Fossil Forests," *Geology Today* (May–June 1990): 92–95; A. Hope Jahren, "The Arctic Forest of the Middle Eocene," *Annual Review of Earth and Planetary Sciences* 35 (2007): 509–540; and Hong Yang and Qin Leng, "Old Molecules, New Climate: *Metasequoia*'s Secrets," *Arnoldia* 76, no. 2 (2018): 24–32.

On PETM proxy data: Melanie L. DeVore and Kathleen B. Pigg, "The Paleocene–Eocene Thermal Maximum: Plants as Paleothermometers, Rain Gauges, and Monitors," in *Nature Through Time: Virtual Field Trips Through the Nature of the Past*, ed. Edoardo Martinetto et al. (Cham, 2020); and articles by Scott L. Wing. On PETM and long-term thinking: Henrik H. Svensen et al., "The Past as a Mirror: Deep Time Climate Change Exemplarity in the Anthropocene," *Culture Unbound* 11, nos. 3–4 (2019): 330–352.

On Doggerland and rising sea levels, see books by Vincent L. Gaffney et al.; Brian M. Fagan; and Jim Leary. For a *longue durée* account of humans and climate: John L. Brooke, *Climate Change and the Course of Global History: A Rough Journey* (Cambridge, 2014).

EPILOGUE: PROMETHEUS

On conservation politics around Wheeler Peak: Darwin Lambert, *Great Basin Drama: The Story of a National Park* (Niwot, 1991). For other takes on WPN-114: Michael P. Cohen, *A Garden of Bristlecones: Tales of Change in the Great Basin* (Reno, 1998); Radiolab, "Oops," WNYC podcast, June 28, 2010; "The Ghost of Prometheus: A Long-Gone Tree and the Artist Who Resurrected Its Memory," *Los Angeles Times*, February 27, 2015; and "The World's Oldest Tree Might or Might Not Be Sitting in a Warehouse in Tucson," *Arizona Republic*, October 3, 2015.

NOTES

INTRODUCTION

1. Josephus, *Jewish Antiquities* 1.10.4; Josephus, *The Jewish War* 4.9.7.

2. Pliny, *Natural History* 16.85–89.

3. Evelyn, *Sylva*, 4th ed. (R. Scott, 1706), 3.3.

4. This famous phrase from 1812 comes (via French) from Georges Cuvier.

5. Joseph Conrad, *Youth and Two Other Stories* (New York, 1903), 105.

6. Edmund Schulman, "Longevity under Adversity in Conifers," *Science* 119 (March 26, 1954): 396–399.

7. Darwin, *On the Origin of Species* (London, 1859), 489.

8. Ecclesiastes 1:18.

9. In English, this legendary tree is known as the Cypress of Kashmar (also Keshmar, Kishmar). Its planting is described in the *Shahnameh*; its felling recounted in the *Dabistan*.

10. Ovid, *Metamorphoses* 8.738–878.

11. Evelyn, *Sylva*, 2.4.

12. This 1974 advertisement from the St. Regis Paper Company ran in *Forbes*, *Fortune*, and *Businessweek*, among other publications.

13. *The Complete Works of Zhuangzi*, trans. Burton Watson (New York, 2013), esp. chap. 4 ("In the World of Men") and chap. 20 ("The Mountain Tree").

14. Lucan, *Pharsalia* 3.399–452.

15. James George Frazer, *Aftermath: A Supplement to the Golden Bough* (London, 1936), vi.

16. Ecclesiastes 1:9.

17. William W. Kellogg and Margaret Mead, eds., *The Atmosphere: Endangered and Endangering* (Tunbridge Wells, 1980), xxi.

CHAPTER I. VENERABLE SPECIES

1. Title of a 2017 book by Robin Russell-Jones.

2. *The Pinetum Britannicum*, vol. 3 (Edinburgh, 1884), 282.

3. Alphonse de Lamartine, *A Pilgrimage to the Holy Land*, vol. 2 (Philadelphia, 1835), 157.

4. Richard F. Burton and Charles F. Tyrwhitt Drake, *Unexplored Syria*, vol. 1 (London, 1872), 99–107.

5. "The Cedar of Lebanon," *Sharpe's London Magazine* 1, no. 26 (April 25, 1846): 409–410.

6. Theophrastus, *Enquiry into Plants* 4.13.5.

7. Sophocles, *Oedipus at Colonus* 694–705.

8. Herodotus, *Histories* 8.55.

9. John 18:1.

10. "Excerpted Letters of Charles J. Langdon, 'Through the Holy Land,'" *Mark Twain Journal* 47, nos. 1/2 (Spring/Fall 2009): 72–83, quote on 79.

11. Josephus, *The Jewish War* 6.1.1.

12. "Olive Trees of Gethsemane among Oldest in World," *Reuters,* October 19, 2012.

13. *The Olive Tree in the Holy Land* (dir. Albert Knechtel, 2012). The tree is called al-Badawi.

14. "Olive Branch Ends Olympian Battle of Ancient Trees," *Telegraph*, August 28, 2004.

15. Alejandra Borunda, "Inside the Race to Save Italy's Olive Trees," *National Geographic*, August 10, 2018.

16. Darwin, *On the Origin of Species* (London, 1859), 319.

17. Joseph Needham, Christian Daniels, and Nicholas K. Menzies, *Science and Civilisation in China*, vol. 6, pt. 3 (Cambridge, 1996), 581.

18. "Gingo biloba," in *West-östlicher Divan* (Stuttgart, 1819), 125. Goethe dropped the *k*.

19. Marie C. Stopes, *A Journal from Japan* (London, 1910), 218.

20. A. C. Seward, *Links with the Past in the Plant World* (Cambridge, 1911), 121–133.

21. Interviewed by author, Boston, April 18, 2008.

22. Ernest Henry Wilson, *The Romance of Our Trees* (Garden City, NY, 1920), 61.

23. "An Easter Story from Japan," *Economist*, March 31, 2010.

24. Peter R. Crane, *Ginkgo: The Tree That Time Forgot* (New Haven, 2013), 15.

25. Charles S. Sargent, "Notes on Cultivated Conifers," *Garden and Forest* 10 (October 6, 1897): 391–392.

26. Title of a 2017 book by Richard O. Prum.

27. *Bhagavad Gita* 10.26.

28. Nineteenth-century writers distinguished between the "bodhi tree" or "Mahābodi Tree" in India and the "bo tree" in Ceylon; today, the latter conventionally goes by the triple honorific "Jaya Sri Maha Bodhi."

29. James Emerson Tennent, *Ceylon: An Account of the Island*, 3rd ed., vol. 2 (London, 1859), 613.

30. He used this phrase in his diary in 1931 upon hearing of her death; see Sangharakshita, *Anagarika Dharmapala: A Biographical Sketch and Other* Maha Bodhi *Writings* (Ledbury, 2013), 77.

31. Wilma Stockenström, *The Expedition to the Baobab Tree*, trans. J. M. Coetzee (Cape Town, 2008), 34.

32. Ibn Battuta, *Travels in Asia and Africa, 1325–1354*, trans. H. A. R. Gibb (London, 1929), 322.

33. "A Description of the *Baobab*," *Gentleman's Magazine* 33 (1763): 500–503.

34. Michel Adanson, *Familles des plantes* (Paris, 1763), ccxv–ccxxiii.

35. Humboldt, *Personal Narrative of Travels to the Equinoctial Regions of the New Continent during the Years 1799–1804*, trans. Helen Maria Williams, vol. 1 (London, 1814), 143; Gray, "Botany of the Southern States," *American Journal of Science and Arts* 20 (November 1855): 131–134; Lyell, *Principles of Geology*, vol. 3 (London, 1833), 99.

36. Quoted in P. N. Pearson and C. J. Nicholas, "'Marks of Extreme Violence': Charles Darwin's Geological Observations at St Jago (São Tiago), Cape Verde Islands," in *Four Centuries of Geological Travel: The Search for Knowledge on Foot, Bicycle, Sledge and Camel*, ed. Patrick N. Wyse Jackson (London, 2007), 239–253.

37. "Trees That Have Lived for Millennia Are Suddenly Dying," *Atlantic*, June 11, 2018; "Last March of the 'Wooden Elephants,'" *New York Times*, June 12, 2018.

38. "This Much I Know," *Guardian*, June 8, 2008.

CHAPTER II. MEMENTO MORI

1. *Macbeth* 4.1.26–27.

2. Thomas Wright, ed., *The Historical Works of Giraldus Cambrensis* (London, 1863), 125, 371.

3. A. W. Wade-Evans, ed., *Welsh Medieval Law* (Oxford, 1909), 248.

4. *Giraldus Cambrensis*, 109.

5. Archibald John Stephens, *The Statutes Relating to the Ecclesiastical and Eleemosynary Institutions of England, Wales, Ireland, India, and the Colonies*, vol. 1 (London, 1845), 31–32.

6. *The Faerie Queene* 1.1.9.

7. *Richard II* 3.2.117.

8. *Statutes at Large from the First Year of King Edward the Fourth to the End of the Reign of Queen Elizabeth*, vol. 2 (London, 1770), 30.

9. *Statut Warcki* (1423), sec. 25.

10. John Norden, *The Surveyor's Dialogue* (London, 1607), 213.

11. H. Bourde de la Rogerie, "Le Parlement de Bretagne, l'évêque de Rennes et les ifs plantés dans les cimetières, 1636–1637," *Bulletin et mémoires de la Société archéologique du Département d'Ille-et-Vilaine* 56 (1930): 99–108.

12. Evelyn, *Sylva*, 2nd ed. (London, 1670), 26.8–11. Evelyn's anti-coal pamphlet was *Fumifugium* (1661).

13. *Sylva*, 33.14.

14. "A Letter to Dr. William Watson, F.R.S. from the Hon. Daines Barrington, F.R.S. on the Trees which are supposed to be indigenous in Great Britain," *Philosophical Transactions* 59 (1769): 23–38.

15. Augustin Pyramus de Candolle, "Notice sur Longévité des Arbres, et les Moyens de la Constater," *Bibliothèque universelle* 47 (Mai 1831): 49–73; translated as "On the Longevity of Trees and the Means of Ascertaining It," *Edinburgh New Philosophical Journal* 15, no. 30 (October 1833): 330–348.

16. J. S. Henslow, *Principles of Descriptive and Physiological Botany* (London, 1835), 242–248.

17. J. E. Bowman, "On the Longevity of the Yew," *Magazine of Natural History* n.s. 1 (1837): 28–35.

18. *Parliament of Fowls*, line 180; *Palamon and Arcite*, line 961.

19. W. T. Bree, "Some Account of an Aged Yew Tree in Buckland Churchyard, Near Dover," *Magazine of Natural History* 6 (1833): 47–51.

20. Samuel Fergusson, *The Queen's Visit, and Other Poems* (Edinburgh, 1869).

21. Robert Christison, "The Exact Measurement of Trees (Part 3)—the Yew Tree," *Transactions and Proceedings of the Botanical Society* 13 (1879): 410–435.

22. "Yew-Trees," in *The Poetical Works of William Wordsworth*, vol. 2 (London, 1827), 53–54.

23. Henry John Elwes and Augustine Henry, *The Trees of Great Britain & Ireland*, vol. 1 (Edinburgh, 1906), vii–viii.

24. *Twelfth Night* 2.4.54.

25. "Transformations," in *The Poetical Works of Thomas Hardy*, vol. 1 (London, 1919), 443.

26. Robert Turner, *Botanologia: The Brittish Physician; or, the Nature and Vertues of English Plants* (London, 1687), 362–363.

27. "The Grave," in *The Poetical Works of Robert Blair* (London, 1802), 3.

28. George Collison, *Cemetery Interment* (London, 1840), 135.

29. Dickens, *Bleak House* (London, 1853), 106–107, 160.

30. Select Committee on Metropolitan Sewage Manure, Minutes of Evidence, June 26, 1846, in *Parliamentary Papers* 10 (London, 1846), 651.

31. Francis Darwin and A. C. Seward, eds., *More Letters of Charles Darwin*, vol. 2 (New York, 1903), 433.

32. This phrase comes from "Lament," in *Poetical Works of Thomas Hardy*, 443.

33. John Lowe Papers, Archives, Royal Botanic Gardens, Kew.

34. John Lowe, *The Yew-Trees of Great Britain and Ireland* (London, 1897), 35, 46, 2.

35. Walter Scott, *Rokeby* (Edinburgh, 1813), 118.

36. Godfrey Higgins, *Celtic Druids* (London, 1827), 25.

37. "Woods of Westermain," in George Meredith, *Poems and Lyrics of the Joy of Earth* (London, 1883), 1–27.

38. *Cobbett's Weekly Political Register* 14, no. 25 (December 17, 1808): 933.

39. J. B. Burke, *Historic Lands of England* (London, 1848), 132–133.

40. A. Conan Doyle, *White Company*, vol. 3 (London, 1891), 113.

41. Walter Johnson, *Byways in British Archaeology* (Cambridge, 1912), 360–407.

42. Marie Carmichael Stopes, *A New Gospel to All Peoples* (London, 1922), 6.

43. Vaughan Cornish, "The Geographical Aspect of Eugenics," *Public Health Journal* 16, no. 7 (July 1925): 317–319.

44. E. W. Swanton, *The Yew Trees of England* (London, 1958).

45. Toby Hindson, Andy Moir, and Peter Thomas, "Estimating the Ages of Yews—Challenging Constant Annual Increment as a Suitable Model," *Quarterly Journal of Forestry* 113, no. 3 (July 2019): 184–188; Moir quoted in Tim Pilgrim, "Scientists Poke a Hole in the Age of Trees," Phys.org, August 14, 2019.

46. Janis Fry with Allen Meredith, *The God Tree* (Taunton, 2012).

47. Eliot, "Little Gidding," lines 232–233.

48. Eliot, "Ash Wednesday," penultimate line.

49. Wordsworth, "Yew-Trees."

50. Leo H. Grindon, *The Trees of Old England* (London, 1870), 128.

51. Eleanor Hull, "The Hawk of Achill or the Legend of the Oldest Animals," *Folklore* 43, no. 4 (December 1932): 376–409.

CHAPTER III. MONUMENTS OF NATURE

1. Humboldt, *Cosmos: Sketch of a Physical Description of the Universe*, trans. Edward Sabine, vol. 2 (London, 1848 [1847]), 92.

2. J. Löwenberg, Robert Avé-Lallemant, and Alfred Wilhelm Dove, *Life of Alexander von Humboldt Compiled in Commemoration of the Centenary of His Birth*, ed. C. Bruhns, trans. Jane Lassell and Caroline Lassell, vol. 1 (London, 1873), 260–262.

3. Humboldt, *Views of Nature*, ed. Stephen T. Jackson and Laura Dassow Walls, trans. by Mark W. Person (Chicago, 2014 [1808]), 190–195.

4. Humboldt, *Personal Narrative of Travels to the Equinoctial Regions of the New Continent, During the Years 1799–1804*, trans. Helen Maria Williams, vol. 1 (London, 1814), 143.

5. "Le Dragonnier de l'Orotava," in *Vues de cordillères et monumens de peuples indigenes de l'Amérique* (Paris, 1810), pl. 69.

6. *Nuove Effemeridi Siciliane: Studi storici, letterari, bibliografici in appendice alla biblioteca storica e letteraria di Sicilia*, ser. 3, vol. 5, comps. V. Di Giovanni et al. (Palermo, 1877), 140–146.

7. Frederick Burkhardt and Sydney Smith, eds., *The Correspondence of Charles Darwin*, vol. 1 (Cambridge, 1985), 121–127.

8. R. D. Keynes, ed., *Charles Darwin's* Beagle *Diary* (Cambridge, 1988), 19.

9. E. O. Fenzi, "Destruction of the Famous Dragon Tree of Teneriffe," *Gardeners' Chronicle*, January 11, 1868, 30.

10. *Personal Narrative of Travels to the Equinoctial Regions of the New Continent*, vol. 3 (London, 1818), 524.

11. Here I borrow the title phrase from Mary Louise Pratt, *Imperial Eyes: Travel Writing and Transculturation* (London, 1992).

12. *Personal Narrative of Travels to the Equinoctial Regions of the New Continent*, vol. 4 (London, 1819), 113–117. Humboldt used the spelling "zamang del Guayre."

13. Georges Louis Leclerc, Comte de Buffon, *Natural History, General and Particular*, trans. William Smellie, vol. 9 (London, 1785), 273–276.

14. Augustin Pyramus de Candolle, "On the Longevity of Trees and the Means of Ascertaining It," *Edinburgh New Philosophical Journal* 15, no. 30 (October 1833 [1831]): 330–348.

15. Pál Rosti, *Uti Emlékezetek Amerikából* (Pest, 1861), 70–72.

16. C. Piazzi Smyth, *Teneriffe, an Astronomer's Experiment: or, Specialities of a Residence above the Clouds* (London, 1858), 418–427.

17. C. Piazzi Smyth, *Report on the Teneriffe Astronomical Experiment of 1856, Addressed to the Lords Commissioners of the Admiralty* (London, 1858), 566–567.

18. Juan de Torquemada, *Monarquía indiana* (Sevilla, 1615), quoted in William B. Taylor, *Theater of a Thousand Wonders: A History of Miraculous Images and Shrines in New Spain* (Cambridge, 2016), 52.

19. Details from James Lockhart, Susan Schroeder, and Doris Namala, eds. and trans., *Annals of His Time: Don Domingo de San Antón Muñón Chimalpahin Quauhtle-huanitzin* (Stanford, 2006), 119, 303.

20. Frances Calderón de la Barca, *Life in Mexico* (London, 1843), 57.

21. The adviser was José Yves Limantour y Márquez.

22. Joseph de Acosta, *Historia natural y moral de las Indias*, vol. 1 (Madrid, 1894 [1590]), 408.

23. Murphy D. Smith, "Of Philadelphia Philosophers and a Mexican Tree," *Manuscripts* 42, no. 4 (Fall 1990): 287–292.

24. Asa Gray, "The Longevity of Trees," in *Scientific Papers of Asa Gray*, vol. 2, ed. Charles Sprague Sargent (Boston, 1889 [1844]), 116.

25. Désiré Charnay and M. Viollet-le-Duc, *Cités et ruines Américaines* (Paris, 1863), 256–258.

26. Désiré Charnay, *The Ancient Cities of the New World: Being Travels and Explorations in Mexico and Central America from 1857–1882*, trans. J. Gonino and Helen S. Conant (London, 1887 [1885]), 259–261.

27. Eduard Mielck, *Die Riesen der Pflanzenwelt* (Leipzig, 1863), 122–123.

28. "The Oldest Living Thing on This Planet," *St. Louis Post-Dispatch Sunday Magazine*, August 13, 1922.

29. Cassiano Conzatti, *Monografía del árbol de Santa María del Tule* (México, 1921).

30. Lino Ramón Campos Ortega, *Boceto histórico sobre el ahuehuete de "El Tule"* (Oaxaca de Juárez, 1927). Campos Ortega borrowed source material from local historian Manuel Martínez Gracida.

31. Translated quotes from *Un beso a esta tierra* (dir. Daniel Goldberg Lerner, 1995).

32. Rosemary Lloyd, ed. and trans., *Selected Letters of Charles Baudelaire: The Conquest of Solitude* (Chicago, 1986), 59.

33. Quoted in Greg M. Thomas, *Art and Ecology in Nineteenth-Century France: The Landscapes of Théodore Rousseau* (Princeton, 2000), 217.

34. Théophile Thoré, "Par monts et par bois," pt. 1, *Le Constitutionnel* 531 (November 27, 1847): 1–2.

35. The German adjective *merkwürdige* suggests oddity, but "remarkable" is the equivalence historically used in English.

36. H. Conwentz, *Die Gefährdung der Naturdenkmäler und Vorschläge zu ihrer Erhaltung* (Berlin, 1904).

37. Its German name: *Staatliche Stelle für Naturdenkmalpflege in Preußen*.

38. Quoted in Friedemann Schmoll, *Erinnerung an die Natur: Die Geschichte des Naturschutzes im deutschen Kaiserreich* (Frankfurt, 2004), 93.

39. Hermann Löns, "Der Naturschutz und die Naturschutzphrase: Ein noch unbekannter Kampfruf von Hermann Löns," *Der Waldfreund* 5, no. 1 (February 1929): 3–13.

40. H. Conwentz, "On National and International Protection of Nature," *Journal of Ecology* 2, no. 2 (June 1914): 109–122.

41. Alois Riegl, *Der moderne Denkmalkultus: Sein Wesen und seine Entstehung* (Wien, 1903). A portion of this work was first translated into English by Kurt W. Forster as "The Modern Cult of Monuments," *Opposition* 25 (1982): 20–51.

42. Mariko Shinoda, "Scientists as Preservationists: Natural Monuments in Japan, 1906–1931," *Historia Scientiarum* 8, no. 2 (1998): 141–155, quote on 143.

43. The neologism by Jan Gwalbert Pawlikowski was *swojszczyzna*. Besides Pawlikowski, key figures in Polish nature protection included Marian Raciborski and Władysław Szafer.

44. Mickiewicz, *Pan Tadeusz* 4.36.

45. Here I borrow the title phrase from Celia Applegate, *A Nation of Provincials: The German Idea of Heimat* (Berkeley, 1990).

46. "Alocución a la Poesía (Fragmento de un poema inédito, titulado 'America,')" *La Biblioteca Americana*, vol. 2 (London, 1823), 12.

47. Fernando González, *Mi compadre* (Barcelona, 1934), 164.

48. Goethe, *Conversations with Eckermann*, trans. anon. (New York, 1901 [1836–1848]), 233–234.

49. *Charte des Fürstenthums Weimar* (1797) reprinted in Barbara Aehnlich und Eckhard Meineke, eds., *Namen und Kulturlandschaften* (Leipzig, 2015), 268.

50. Paul Martin Neurath, *The Society of Terror: Inside the Dachau and Buchenwald Concentration Camps* (Boulder, 2005 [1951]), 120.

51. "Die Eiche Goethes in Buchenwald," box 2, folder 20, Joseph Roth Papers, Leo Baeck Institute, New York City.

52. *Topographische Karte: Meßtischblatt 4933: Neumark in Thüringen* (Berlin, 1942).

53. Ernst Wiechert, *Forest of the Dead*, trans. Ursula Stechow (London, 1947), 78–79.

54. Edmund Polak, *Morituri* (Warszawa, 1968), 150–155; and idem, *Dziennik buchenwaldski* (Warszawa, 1983), 267–268.

55. "O dębie Goethego w obozie buchenwaldzkim," in *Ludwik Fleck: tradycje, inspiracje, interpretacje*, ed. Bożena Płonka-Syroka et al. (Wrocław, 2015), 189–191.

56. G. G. M. Pols-Harmsen, ed., *Een zondagskind in Buchenwald: Nico Pols, overleven om het te vertellen* (Zutphen, 2005), 72–74.

57. "Kunst im Widerstand: Gespräch mit Bruno Apitz," *Neue Deutsche Literatur* 24, no. 11 (November 1976): 18–26.

58. *Nationale Gedenkstätte Buchenwald auf dem Ettersberg bei Weimar* (Reichenbach, 1956), quoted in Volkhard Knigge et al., eds., *Versteinertes Gedenken: Das Buchenwalder Mahnmal von 1958*, vol. 1 (Spröda, 1988), 88.

59. Walther Schoenichen, *Unter den Bäumen einer alten Reichsstadt: Baumbuch der Stadt Goslar, ein Beitrag zur Pflege und Gestaltung des deutschen Heimatbildes* (Hanover, 1952).

60. Originally called IUPN, with *P* for "Protection," a relic of *Naturschutz*.

61. Confusingly, the US National Park Service has, since the 1960s, administered a voluntary registry of "national natural landmarks" comparable to Prussian natural monuments, minus the resonance.

62. Kendall R. Jones et al., "One-Third of Global Protected Land Is under Intense Human Pressure," *Science* 360 (May 18, 2018): 788–791.

63. "Atentado historico en Mexico: Manos criminales incendiaron el Árbol de la Noche Triste," *La Opinion*, September 11, 1981, 8.

64. The organizer was Juan I. Bustamante.

65. Jaime Larumbe Mendoza compiled a local catalog of notable trees, a project that overlapped with a national-level project: Fernando Vargas Márquez, *Compendio de árboles históricos y notables de México* (México, 1996). In Oaxaca, tree-registering culminated in *Nuestras raíces: Catálogo de Árboles Notables y Emblemáticos del Estado de Oaxaca* (Oaxaca de Juárez, 2020).

66. The activist was Richard Torres.

67. "Big Tree of Tule Re-examined," *Science News-Letter* 24, no. 639 (July 8, 1933), 21.

68. Edmund Schulman, "Dendrochronology in Mexico, I," *Tree-Ring Bulletin* 10, no. 3 (1944): 18–24.

69. The artist was Francisco Verástegui.

70. The founder was Jorge Augusto Velasco.

71. Antonio Velasco Piña, *Regina: 2 de octubre no se olvida* (México, 1987).

CHAPTER IV. PACIFIC FIRES

1. The Māori name for the plant has also been rendered as courie, cowdi, cowdie, cowdy, cowri, cowrie, cowry, kaudi, kawdi, koudi, kouri, kowde, kowdie, kowri.

2. R. D. Keynes, ed., *Charles Darwin's* Beagle *Diary* (Cambridge, 1988), 391.

3. Richard Taylor, *Te Iki a Maui, or New Zealand and Its Inhabitants* (London, 1855), 136.

4. John Logan Campbell, *Poenamo: Sketches of the Early Days in New Zealand* (London, 1881), 79–82.

5. J. B. Marsden, ed., *Memoirs of the Life and Labors of the Rev. Samuel Marsden* (London, 1838), 136–137.

6. The resister was Hōne Heke (Ngāpuhi).

7. The king was Pōtatau Te Wherowhero (Waikato).

8. Quotation and details from Florence Keene, *Under Northland Skies: Forty Women of Northland* (Whangarei, 1984), 69–74.

9. Ferdinand von Hochstetter, *New Zealand: Its Physical Geography, Geology and Natural History*, trans. Edward Sauter (Stuttgart, 1867), 149.

10. Entries for February 12–13, 1904, John Muir Journal no. 69 ("January–May 1904, World Tour, Part V"), Holt-Atherton Special Collections, University of the Pacific, Stockton, California.

11. *New Zealand Parliamentary Debates* 16 (Wellington, 1874), 360.

12. Ibid., 351.

13. The forester was Inches Campbell-Walker, who came by invitation of premier Julius Vogel.

14. T. Kirk, "Report on Native Forests and the State of the Timber-Trade," *Appendix to the Journals of the House of Representatives*, 1886 Session I, C-3.

15. "Annual Report on Crown Lands Department," *Appendix to the Journals of the House of Representatives*, 1889 Session I, C-1.

16. George S. Perrin, "Report Upon the Conservation of New Zealand Forests," *Appendix to the Journals of the House of Representatives of New Zealand*, 1897 Session II, C-8.

17. W. P. Reeves, *New Zealand and Other Poems* (London, 1898), 4–8. In its revised, canonized form, the poem bears this subtitle: "A Lament for the Children of Tāne."

18. *New Zealand Illustrated Magazine* 4 (July 1901): 752.

19. Also known as alerce chileno, alerce patagónico, Patagonian cypress, red cypress, red cedar, redwood of the Andes, Andean redwood, South American redwood, redwood of the south. *Alerce* is commonly (mis)translated into English as "larch." The Mapudungun word for alerce has been rendered as laguán, lahuál, lahuán, lahuén.

20. "Mapuche" is a catch-all for Mapudungun-speaking peoples of Chile's southern zone—peoples who would not necessarily have thought of themselves collectively before the nineteenth century.

21. Keynes, *Charles Darwin's* Beagle *Diary*, 285.

22. C. Martin, "Pflanzengeographisches aus Llanquihue und Chiloé," *Verhandlungen des Deutschen wissenschaftlichen Vereins zu Santiago de Chile* 3 (1893–1898): 507–522.

23. Rodulfo Amando Philippi, *Elementos de historia natural* (Santiago, 1866), 186.

24. Federico Albert, *Los bosques en el país* (Santiago, 1903), 45, 128.

25. Bailey Willis, *Northern Patagonia: Character and Resources*, vol. 1 (New York, 1914), 366–373.

26. "Reminiscences of Mendocino," *Hutchings' Illustrated California Magazine* 3 (October 1858): 146–160, 177–181, quotes on 154 and 157.

27. *History of Humboldt County, California, with Illustrations* (San Francisco, 1881), 140.

28. *First Biennial Report of the California State Board of Forestry* (Sacramento, 1886), 137–157.

29. Willis Linn Jepson, *The Trees of California*, 2nd ed. (Berkeley, 1923), 15–16.

30. Ernest Henry Wilson, *The Conifers and Taxads of Japan* (Cambridge, 1916), 67.

31. The naturalists included Manabu Miyoshi and Kumagusu Minakata.

32. Kuang-Chi Hung, "When the Green Archipelago Encountered Formosa: The Making of Modern Forestry in Taiwan under Japan's Colonial Rule (1895–1945)," in *Environment and Society in the Japanese Islands: From Prehistory to the Present*, ed. Bruce L. Batten and Philip C. Brown (Corvallis, 2015), 174–193, quote on 174.

33. Alishan has also been rendered as A-li-shan, Ari-san, Arisan, Arizan, Mt. Ari, Ari Mountain.

34. Also called red hinoki, red cypress, Taiwan cypress, Formosan cypress, Chinese cypress, Yunnan cypress—and *hongkuai* or *hong gui* from Mandarin.

35. "Letter of E. H. Wilson," *Journal of the International Garden Club* 2, no. 2 (June 1918): 237–238; E. H. Wilson, "A Phytogeographical Sketch of the Ligneous Flora of Formosa," *Journal of the Arnold Arboretum* 2, no. 1 (July 1920): 25–41.

36. I have rendered his name the European way (personal name, family name), following his own publications in German.

37. "Describes Million Dollar Forestry Project," *American Lumberman* (April 15, 1911): 48.

38. Ibid.

39. Poultney Bigelow, "Colonial Japan" [part five], *Japan* 11, no. 9 (June 1922): 11–14, based on a talk Bigelow gave in Taipei, as reported in *Yale Alumni Weekly*.

40. The figure 7,200 comes from an unreliable 1968 estimate based on size. In current popular usage, the term *yakusugi* suggests 1,000+ years, while *jōmonsugi* suggests

3,000+ years. Recent tree-ring studies suggest that 2,000 years is the effective limit of sugi longevity.

41. The leadership was Edgar Wayburn and David Brower.

42. Elwood Maunder with Charles Buchwalter, *Four Generations of Management: The Simpson-Reed Story: An Interview with William G. Reed* (Santa Cruz, 1977), 85–86, 117–127.

43. Rafael Elizalde Mac-Clure, *La sobrevivencia de Chile: La conservación de sus recursos naturales renovables*, 2nd ed. (Santiago, 1970), esp. 308–311.

44. Ministerio de agricultura, "Declara monumento natural a la especie forestal alerce," Decreto 490, October 1, 1976.

45. In the colonial era, Spanish authorities generically used the word *Huilliche* (southern people) to refer to a geographical subset of Mapudungun-speaking "Araucanians" who lived between Río Toltén and the Chiloé Archipelago. In contemporary Chile, the state classifies self-identified Huilliches as Mapuches.

46. The NGO was Comité nacional pro defensa de la fauna y flora (CODEFF).

47. The organization was Environmental Protection Information Center (EPIC).

48. Notably Arne Næss, George Sessions, and Bill Devall.

49. "Clear-Cut Disaster," *Chicago Tribune*, December 12, 1994.

50. Notably La fundación lahuén and Defensores del Bosque Chileno.

51. "From Thousand-Year-Old Sentinel to Traffickers' Booty," *New York Times*, June 2, 2005.

52. See "Report of the Royal Commission on Forestry," *Appendix to the Journals of the House of Representatives*, 1913 Session I, C-12; and the many additional reports of commissioner Leonard Cockayne.

53. D. E. Hutchins, *A Discussion of Australian Forestry* (Perth, 1916), 389–396; and idem, *New Zealand Forestry* (Wellington, 1919).

54. W. R. McGregor, *The Waipoua Forest: The Last Virgin Kauri Forest of New Zealand* (Auckland, 1948).

55. "Annual Report of the Director of Forestry for the Year Ended 31st March, 1948," *Appendix to the Journals of the House of Representatives*, 1948 Session I, C-3.

56. Waitangi Tribunal, *The Te Roroa Report* (Wellington, 1992), 1–3.

57. Te Roroa Manawhenua Trust, "Effects Assessment, Waipoua Kauri National Park" (October 2016), accessed through tertoroa.iwi.nz.

58. Muir, *My First Summer in the Sierra* (Boston, 1911), 211.

CHAPTER V. CIRCLES AND LINES

1. Bonnie Johanna Gisel, ed., *Kindred & Related Spirits: The Letters of John Muir and Jeanne C. Carr* (Salt Lake City, 2001), 119.

2. *Description of the Big Black Walnut Tree* (Philadelphia, 1827); *A Description of the Large Black Walnut Tree* (London, 1828).

3. *Gardeners' Chronicle*, February 3, 1855, 70.

4. The fullest documentation of exhibitions is found within the Gary D. and Myrna R. Lowe Collection Relating to the Big Tree of California (M2147), Department of Special Collections and University Archives, Stanford University.

5. I obtained transcripts of the complete extant correspondence of Ephraim Cutting through the generosity of Gary D. Lowe.

6. "Trees in California," *The Friend* 30, no. 48 (August 8, 1857): 380.

7. Samuel Bowles, *Across the Continent: A Summer's Journey to the Rocky Mountains, the Mormons, and the Pacific States, with Speaker Colfax* (Springfield, MA, 1865), 237.

8. *A Pen Picture of the Kaweah Co-Operative Colony Co.* (San Francisco, 1889).

9. "Boole's Daughter Declares Fresnan Named Famous Tree," *Fresno Bee*, January 2, 1951.

10. "The Big Tree," *Santa Cruz Sentinel*, September 20, 1902.

11. "Asks to Be Buried as 'Smith,'" *San Francisco Call*, October 25, 1903.

12. "Find Body of Engineer in the Bay," *San Francisco Call*, December 30, 1904.

13. Handbills quoted in Charles Coleman Sellers, *Mr. Peale's Museum: Charles Willson Peale and the First Popular Museum of Natural Science and Art* (New York, 1980), 142.

14. Genesis 6:4. Emerson's statement of 1871 recollected in John Muir, *Our National Parks* (Boston, 1901), 134.

15. J. G. Lemmon, "Big Trees," *Pacific Rural Press* 29 (April 18, 1885): 374.

16. Asa Gray, "Sequoia and Its History," in *Darwiniana: Essays and Reviews Pertaining to Darwinism* (New York, 1876), 205–235.

17. Lyell, *Principles of Geology*, vol. 2, 2nd ed. (London, 1833), 147.

18. Summarized and quoted from John Muir, "The Royal Sequoia" (1875), in *John Muir Summering in the Sierra*, ed. Robert Engberg (Madison, 1984), 121–137; idem, "On the Post-Glacial History of Sequoia Gigantea," *Proceedings of the American Association for the Advancement of Science* (1877): 242–253; and idem, "The New Sequoia Forests of California," *Harper's* 57 (November 1878): 813–827.

19. King, *Mountaineering in the Sierra Nevada*, 4th ed. (Boston, 1874), 43.

20. John Gill Lemmon, *Third Biennial Report of the California State Board of Forestry for the Years 1889–1890* (Sacramento, 1890), 166.

21. Mrs. Frank Leslie, *California: A Pleasure Trip from Gotham to the Golden Gate, April, May, June, 1877* (New York, 1877), 259.

22. B. E. Fernow, "Ring-Growth in Trees," *Nation* 46 (January 12, 1888): 29.

23. Quoted in Berthold Seemann, "On the Mammoth-Tree of Upper California," *Annals and Magazine of Natural History* 3, no. 15 (March 1859): 161–175.

24. J. D. Whitney, *The Yosemite Guide-Book* (Cambridge, MA, 1869), 154; J. D. Whitney for the Geological Survey of California, *The Yosemite Book* (New York, 1868), 116.

25. C. B. Bradley, "A New Study of Some Problems Relating to the Giant Trees," *Overland Monthly & Out West Magazine* 7 (March 1886): 305–316.

26. William Russel Dudley, "The Vitality of the *Sequoia gigantea*," in *Dudley Memorial Volume* (Stanford, 1913), 33–42.

27. Quotes from S. D. Dill to Morris K. Jesup, October 30, 1891; and "Report of S. D. Dill Relative to His Journey on the Pacific Coast in Collecting Wood Specimens, Autumn 1891," box 4, folder 4A, Forestry Hall Papers (DR 091). Both are located in Central Archives, American Museum of Natural History (AMNH), New York City.

28. "A Forest Monarch," *Wood-Worker* 11, no. 11 (January 1893): 22.

29. "A Tree That Teaches History," *Strand Magazine* 26 (December 1903): 791–793.

30. George H. Sherwood, "The Sequoia: A Historical Review of Biological Science," *Supplement to American Museum Journal* 2, no. 8 (November 1902); Hermon C. Bumpus, "Extension of Education to Adults: How Adult Education Is Being Furthered by the Work of the American Museum of Natural History," *Journal of Social Science* 42 (September 1904): 144–151, quote on 147.

31. Unbeknownst to Douglass, the principle of cross-dating had been independently established by multiple German-language botanists prior to him. His contemporary, the Swedish geologist Gerard De Geer, founded "geochronology" using varves (annual clay deposits) in a technique analogous to tree-ring science.

32. Quotes from A. E. Douglass, "Survey of Sequoia Studies, II," *Tree Ring Bulletin* 12, no. 2 (October 1945): 10–16; and idem, *Climatic Cycles and Tree Growth, Volume I: A Study of the Annual Rings of Trees in Relation to Climate and Solar Activity* (Washington, 1919).

33. Ellsworth Huntington, *The Climatic Factor as Illustrated in Arid America* (Washington, 1914), 139.

34. Ellsworth Huntington, "The Secret of the Big Trees," *Harper's* 125 (July 1912): 292–302.

35. Ellsworth Huntington, "Tree Growth and Climatic Interpretations," in *Quaternary Climates*, Carnegie Institution Publication 352 (Washington, 1925), 182.

36. Harry H. Laughlin, ed., *The Second International Exhibition of Eugenics* (Baltimore, 1923). I viewed photographs of Huntington's chart in the Central Archives of AMNH and read relevant correspondence in the Ellsworth Huntington Papers, group 1, series 3, box 43, folder 1205, Special Collections, Sterling Library, Yale University.

37. A. E. Douglass, *Climatic Cycles and Tree Growth, Volume III: A Study of Cycles* (Washington, 1936).

38. "Tree Ring Research Uncovers Mysteries of Past Centuries," *Arizona Daily Star*, February 24, 1939.

39. Harry Milton Riseley, "1,341 Years Old When It Died," *National Magazine* 21 (October 1904): 186–189, quote on 189.

40. Barry Ahearn, ed., *Pound/Cummings: The Correspondence of Ezra Pound and E. E. Cummings* (Ann Arbor, 1996), 39–40.

41. Transcript of speech by E. T. Scoyen, box 197, folder 11, subject file 701-04.6, Sequoia & Kings Canyon National Park Archives, Three Rivers, California.

42. Mary McCarthy, *Birds of America* (New York, 1965), 335–336.

43. This report by Emilio Meinecke can be found as an appendix to Richard John Hartesveldt, "The Effects of Human Impact upon *Sequoia gigantea* and Its Environment in the Mariposa Grove, Yosemite National Park, California (PhD diss., University of Michigan, 1962), 286–302.

44. "Fallen Sequoias: Loved to Death?" *Philadelphia Inquirer*, April 17, 2003.

45. John R. White, "A Living Memorial," *American Forestry* 29, no. 358 (October 1923): 588.

46. George L. Mauger, "Felling the Leaning Sequoia Tree," box 257, folder 6, Sequoia & Kings Canyon National Park Archives.

47. "Californian Giants," *Chambers's Journal* 6, no. 155 (December 20, 1856): 398–399.

48. F. E. Olmsted, "Fire and the Forest—the Theory of 'Light Burning,'" *Sierra Club Bulletin* 8 (January 1911): 43–47.

49. Marsden Manson, "Preserving the Forests by Fire," in *Should the Forests Be Preserved?* (San Francisco, 1903), 38.

50. Aldo Leopold, "'Piute Forestry' vs. Forest Fire Prevention," *Southwestern Magazine* 2, no. 3 (March 1920): 12–13.

51. A. S. Leopold et al., *Wildlife Management in the National Parks* (Washington, 1963), 3–5.

52. Peter Steinhart, "Tree-Climbin' Man," *New West* 6, no. 11 (November 1981): 108–109, 149.

53. "New Age Estimated for Sequoia," *San Francisco Chronicle*, December 8, 2000.

54. A. Park Williams et al., "Large Contribution from Anthropogenic Warming to an Emerging North American Megadrought," *Science* 368, no. 6488 (April 17, 2020): 314–318; and a follow-up [*Science*] article by the same lead author in 2022.

55. A. J. Wells, "Helping the Sierra Sequoias," *Sunset* 16, no. 3 (January 1906): 280–284.

56. California wildfire emission estimates from the website of the California Air Resources Board.

57. Christy Brigham quoted in "They're Among the World's Oldest Living Things: The Climate Crisis Is Killing Them," *New York Times*, December 9, 2020; and "Hundreds of Towering Giant Sequoias Killed by the Castle Fire—a Stunning Loss," *Los Angeles Times*, November 16, 2020.

58. George W. Bush, "Remarks at Sequoia National Park," May 30, 2001, as reprinted in *Weekly Compilation of Presidential Documents*, June 4, 2001.

CHAPTER VI. OLDEST KNOWN

1. The diary runs from July 19, 1927, to June 1, 1932, with a hiatus from August 27, 1930, to January 30, 1932. Edmund's surviving nephew, Richard Schulman, kindly allowed me to read it.

2. Rebekah Kohut, *As I Know Them: Some Jews and a Few Gentiles* (Garden City, NY, 1929).

3. The Public Records Office at the University of Arizona gave me access to personnel files of Edmund and Alsie Schulman (formerly Alsie Raffman, née French). I also consulted Ancestry.com and Tucson's Jewish newspaper, the *Arizona Post*.

4. The Andrew Ellicott Douglass Papers (AZ 072) at the University of Arizona Libraries, Special Collections, contains ninety linear feet of material, including restricted documents related to the Laboratory of Tree-Ring Research. David Frank, current lab director, permitted access to those documents. In addition, Frank allowed me to peruse the lab's internal archive, including Schulman's professional correspondence, research materials created by other faculty members over the decades, and administrative files. This sizable archive has been roughly organized, but not cataloged like a regular collection in a library. For multiple reasons, then, I cannot provide box and folder numbers for most of the primary sources from which the middle part of this chapter derives.

5. Richard Schulman gave me copies of family photographs and personal letters, including some written by Edmund's wife, Alsie.

6. Edmund Schulman, "Runoff Histories in Tree Rings of the Pacific Slope," *Geographical Review* 35, no. 1 (January 1945): 59–73.

7. Edmund Schulman, "Tree-Ring Hydrology of the Colorado River Basin," *Laboratory of Tree-Ring Research Bulletin* 2 (Tucson, 1946).

8. *Arizona Post*, October 15, 1946.

9. A. E. Douglass, "The Significance of Honor Societies," *Phi Kappa Phi Journal* 6, no. 1 (October 1926): 3–6.

10. See, for example, "The Recognition of Israel," *Arizona Daily Star*, May 16, 1948, 10.

11. Nello Pace and S. F. Cook, "California's Laboratory above the Clouds," *Research Reviews* 5 (March 1952): 1–7.

12. Edmund Schulman, *Dendroclimatic Changes in Semiarid America* (Tucson, 1956), 34.

13. Hans E. Suess, "Radiocarbon Concentration in Modern Wood," *Science* 122, no. 3166 (September 2, 1955): 415–417.

14. Roger Revelle and Hans Suess, "Carbon Dioxide Exchange Between Atmosphere and Ocean and the Question of an Increase of Atmospheric CO_2 During the Past Decades," *Tellus* 9, no. 1 (1957): 18–27.

15. Edmund Schulman, "Longevity under Adversity in Conifers," *Science* 119, no. 3091 (March 26, 1954): 396–399.

16. "'Oldest Living Thing' a Pine," *San Francisco Chronicle*, October 7, 1956.

17. "The Oldest Thing Alive," *Life* 41, no. 21 (November 19, 1956): 69–70; "Scientists Find Oldest Object," *Tucson Daily Citizen*, October 1, 1956.

18. "Dr. Schulman Unlocks Living Secrets," *Arizona Post*, October 12, 1956.

19. Cathy Hunter of the National Geographic Society's Archives and Special Collections gave me access to Schulman's editorial file, preserved on microfiche.

20. The late David Hardin, Interpretive Ranger of Inyo National Forest, gave me access to Schulman's political correspondence, plus hundreds of pages of US Forest Service documents relating to the establishment, management, and interpretation of the Ancient Bristlecone Pine Forest. At the time of my perusal, in 2018, these unsorted papers were located in the attic of the visitor's center at Schulman Memorial Grove.

21. Douglas R. Powell quoted in *Methuselah Tree* (dir. Ian Duncan), *Nova*, season 28, episode 11, PBS, December 11, 2001.

22. Richard Schulman, interviewed by author, San Diego, January 19, 2017.

23. Edmund Schulman, "Bristlecone Pine, Oldest Known Living Thing," *National Geographic* 113, no. 3 (March 1958): 354–372.

24. "Dr. Edmund Schulman of Tree-Ring Fame Dies," *Tucson Daily Citizen*, January 9, 1958.

25. C. W. Ferguson, "Bristlecone Pine: Science and Esthetics," *Science* 159, no. 3817 (February 23, 1968): 839–846.

26. LeRoy C. Johnson and Jean Johnson, "Methuselah: Fertile Senior Citizen," *American Forests* 84 (September 1978): 29–31, 43.

27. *Restoring the Quality of Our Environment: Report of the Environmental Pollution Panel, President's Science Advisory Committee* (Washington, 1965), 126.

28. E. K. Ralph, H. N. Michael, and M. C. Han, "Radiocarbon Dates and Reality," *MASCA Newsletter* 9, no. 1 (1973): 1–20.

29. Quote on p. 81.

30. Details from "Scientist Tracks Old Logs for Prehistoric Clues," *New York Times*, December 27, 1975; Tom Gidwitz, "Telling Time," *Archaeology* 54, no. 2 (March–April 2001): 36–41; and Henry N. Michael Papers (PU-Mu. 0069), Penn Museum Archives, Philadelphia.

31. Harold C. Fritts, *Bristlecone Pine in the White Mountains of California: Growth and Ring-width Characteristics* (Tucson, 1969).

32. Valmore C. LaMarche Jr. and Katherine K. Hirschboeck, "Frost Rings in Trees as Records of Major Volcanic Eruptions," *Nature* 307 (January 12, 1984): 121–126.

33. Valmore C. LaMarche Jr. et al., "Increasing Atmospheric Carbon Dioxide: Tree Ring Evidence for Growth Enhancement in Natural Vegetation," *Science* 225, no. 4666 (September 7, 1984): 1019–1021.

34. Interviewed by author by video, November 16, 2017.

35. Michael E. Mann, Raymond S. Bradley, and Malcolm K. Hughes, "Global-Scale Temperature Patterns and Climate Forcing over the Past Six Centuries," *Nature* 392 (April 23, 1998): 779–787; Michael E. Mann and Raymond S. Bradley, "Northern Hemisphere Temperatures During the Past Millennium: Inferences, Uncertainties, and Limitations," *Geophysical Research Letters* 26, no. 6 (March 15, 1999): 759–762.

36. *Congressional Record* 149 (July 28, 2003), S19943.

37. Stephen McIntyre and Ross McKitrick, "The M&M Critique of the MBH98 Northern Hemisphere Climate Index: Update and Implications," *Energy & Environment* 16, no. 1 (2005): 69–100.

38. National Research Council, *Surface Temperature Reconstructions for the Last 2,000 Years* (Washington, 2006).

39. "Scientists, Volunteers Seek Tree Rings to Lengthen Ancient Bristlecone Record," *UArizona News*, August 27, 2002. Schulman died of cerebral hemorrhage, not heart attack.

40. Pearce Paul Creasman, interviewed by author, Tucson, September 19, 2017.

41. Bryant Bannister, Robert E. Hastings Jr., and Jeff Banister, "Remembering A. E. Douglass," *Journal of the Southwest* 40, no. 3 (Autumn 1998): 307–318.

42. Fusa Miyake et al., "Large ^{14}C Excursion in 5480 BC Indicates an Abnormal Sun in the Mid-Holocene," *PNAS* 114, no. 5 (January 31, 2017): 881–884. Colloquially, these excursions are called "Miyake events," after the lead author.

43. Interviewed by author, September 20, 2017 (in Tucson), and May 3, 2021 (by phone).

44. Alex Ross, "The Past and the Future of the Earth's Oldest Trees," *New Yorker*, January 20, 2020. Supposed victims of the "curse": Edmund Schulman (1909–1958); Fred V. Solace (1933–1965); Charles Wesley Ferguson (1922–1986); Valmore C. La-Marche Jr. (1937–1988); Donald Alan Graybill (1942–1993); Donald Rusk Currey (1934–2004); and F. Craig Brunstein (1951–2008).

45. *Methuselah Tree* (dir. Ian Duncan).

CHAPTER VII. LATEST OLDEST

1. Interviewed by the author by video, December 12, 2017.

2. Kathleen Wong, "Drought Strikes Centuries-Old California Oaks," Phys.org, December 14, 2016.

3. Especially *Juniperus occidentalis* (western juniper), *J. osteosperma* (Utah juniper), and *J. scopulorum* (Rocky Mountain juniper). In the Southwest, *J. deppeana* (alligator juniper) probably lives just as long, but old specimens generally hollow out, making exact age determinations impossible.

4. Washington Irving, *A Tour on the Prairies* (London, 1835), 186.

5. Interviewed by the author by phone, May 4, 2018.

6. Notably Bob Leverett and Mary Byrd Davis.

7. The forester was Zhan Wang (1911–2000). Moudao, located in Hubei Province, has also been spelled Modaoqi, Modaoxi, Motaochi.

8. E. D. Merrill to Henry Hicks, January 27, 1947, E. D. Merrill Papers, series 1, folder 1, Arnold Arboretum Horticultural Library. The name of Hsen-Hsu Hu (1894–1968) has also been rendered as HU Xiansu.

9. E. D. Merrill, "A Living *Metasequoia* in China," *Science* 107 (February 6, 1948): 140; idem, "Metasequoia, Another 'Living Fossil,'" *Arnoldia* 8, no. 1 (March 5, 1948): 1–8. "Johnny Appleseed" detail from Richard A. Howard to University of California Radio Office, April 26, 1969, Merrill Papers, series 1, folder 10, Arnold Arboretum.

10. Chaney to Merrill, January 27, 1948, Elmer Drew Merrill Papers, series 2, box 1, Mertz Library, New York Botanical Garden; Milton Silverman, "Science Makes a Spectacular Discovery," *San Francisco Chronicle*, March 25, 1948, 1–2; Milton Silverman, *Search for the Dawn Redwoods* (self-published, 1990), copy in Helen Crocker Russell Library of Horticulture, San Francisco Botanical Garden.

11. Silverman's front-page stories appeared in the following issues of the *San Francisco Chronicle* in 1948: March 25, 26, 28, 29, 30, and April 5.

12. J. Linsley Gressitt, "The California Academy–Lingnan Dawn-Redwood Expedition," *Proceedings of the California Academy of Sciences* 28, no. 2 (July 15, 1953): 25–58.

13. Silverman, *Search for the Dawn Redwoods*, 130–131.

14. Quotes from E. D. Merrill to John E. Gribble, April 2, 1952, Merrill Papers, series 2, box 1, New York Botanical Garden; David C. D. Rogers, "Professors Squabble over Seeds from China's Living Fossil Trees," *Harvard Crimson*, October 9, 1952; E. D. Merrill to Aubrey Drury, June 8, 1954, Merrill Papers, series 1, folder 6, Arnold Arboretum; and Richard Evans Schultes, "Elmer Drew Merrill: An Appreciation," *Taxon* 6, no. 4 (May 1957): 89–101.

15. "Incredible, Secret Firefighting Mission Saves Famous 'Dinosaur Trees,'" *Sydney Morning Herald*, January 15, 2020.

16. Interviewed by the author, Palisades, New York, January 23, 2018.

17. Michael C. Grant, Jeffry B. Mitton, and Yan B. Linhart, "Even Larger Organisms," *Nature* 360 (November 19, 1992): 216.

18. "The World's Largest Known Organism Is in Utah—and It's Dying," *City Weekly* (Salt Lake City), November 20, 2013.

19. Michelle Nijhuis, "The Quaking Giant," February 16, 2017, lastwordonnothing .com.

20. Karen Mock, interviewed for "The World's Largest Known Organism in Trouble," *Living on Earth* (syndicated radio program), February 1, 2013.

21. Russ Beck, "Pando: The World's Largest Discovered Organism," January 13, 2017, wildaboututah.org.

22. Charles J. Chamberlain, "A Round-the-World Botanical Excursion," *Popular Science Monthly* 81 (November 1912): 417–433.

23. "Giant Aristocrat Destroyed," *Rockhampton Evening News*, March 16, 1936.

24. "Famous Giant Tree Resurrected," *Rockhampton Evening News*, August 13, 1936.

25. In the clippings file for *L. peroffskyana* at Mertz Library, New York Botanical Garden.

26. "Redwood Tree Still Oldest Thing on Earth," *Santa Rosa Press Democrat*, June 20, 1937.

27. This syndicated cartoon ran the week of October 17, 1938.

28. Now protected as Hoverter and Sholl Box Huckleberry Natural Area.

29. Frederick V. Coville, "The Threatened Extinction of the Box Huckleberry, *Gaylussacia brachycera*," *Science* 50, no. 1280 (July 11, 1919): 30–34.

30. Hon. Richard M. Simpson, June 5, 1952, 82nd Cong., 2nd sess., *Congressional Record*, Appendix, 3478–3479.

31. Quoted in a letter from Bryant Bannister to *National Geographic*, March 21, 1958, located in the internal archive of the Laboratory of Tree-Ring Research, University of Arizona.

32. "Earth's Oldest Living Bushes Endangered," *Los Angeles Times*, March 7, 1983. My description also draws on coverage and correspondence in Frank C. Vasek papers at Special Collections & University Archives, UC Riverside Library.

33. Richard W. Stoffle, Michael Evans, and David Halmo, *Native American Plant Resources in the Yucca Mountain Area, Nevada* (Las Vegas, 1989), 6–13.

34. "Open to Interpretation," *Nature* 453, no. 7197 (June 12, 2008): 824.

35. Augustin Pyramus de Candolle, "On the Longevity of Trees and the Means of Ascertaining It," *Edinburgh New Philosophical Journal* 15, no. 30 (October 1833 [1831]): 330–348.

CHAPTER VIII. TIME TO MOURN

1. Here I borrow the title phrase from Naomi Oreskes and Erik M. Conway, *Merchants of Doubt: How a Handful of Scientists Obscured the Truth on Issues from Tobacco Smoke to Global Warming* (New York, 2010). The best-known German scientists were Bernhard Ulrich and Peter Schütt.

2. Rio Declaration on Environment and Development, Principle 15.

3. Kossak's fear: "Śmierć Puszczy."

4. Interviewed by the author, Harvard Forest, April 20, 2018.

5. Interviewed by the author, Santa Fe, New Mexico, July 16, 2018.

6. Chris A. Boulton, Timothy M. Lenton, and Niklas Boers, "Pronounced Loss of Amazon Rainforest Resilience Since the Early 2000s," *Nature Climate Change* 12, no. 3 (March 2022): 271–278.

7. Nate G. McDowell quoted in Craig Welch, "The Grand Old Trees of the World Are Dying, Leaving Forests Younger and Shorter," *National Geographic*, May 28, 2020 [online only].

8. Steven G. McNulty, Johnny L. Boggs, and Ge Sun, "The Rise of the Mediocre Forest: Why Chronically Stressed Trees May Better Survive Extreme Episodic Climate Variability," *New Forests* 45, no. 3 (May 2014): 403–415.

9. Brian V. Smithers et al., "Leap Frog in Slow Motion: Divergent Responses of Tree Species and Life Stages to Climatic Warming in Great Basin Subalpine Forests," *Global Change Biology* 42, no. 2 (February 2018): 1–16.

10. This AP story, dated September 13, 2017, appeared widely and prompted this reply: Jared Farmer, "Slow Trees and Climate Change: Why Bristlecone Pine Will Still Outlive You," *Los Angeles Times*, October 13, 2017.

11. Interviewed by the author, Albany, California, January 17, 2018, with follow-up video call on May 26, 2021.

12. Interviewed by the author multiple times, remotely in October–November 2017, and in person in the Snake Range in June 2018.

13. Stewart Brand, *The Clock of the Long Now: Time and Responsibility* (New York, 1999), 3.

14. Details and quote from Executive Director Alexander (Zander) Rose, interviewed by the author, San Francisco, January 18, 2018.

15. The artist is Jonathon Keats.

16. Charles H. Cannon, Gianluca Piovesan, and Sergi Munné-Bosch, "Old and Ancient Trees Are Life History Lottery Winners and Nurture Vital Evolutionary Resources for Long-Term Adaptive Capacity," *Nature Plants* 8, no. 2 (February 2022): 136–145.

17. A coinage of William R. Moomaw.

18. "'I Call Them My Gentle Giants': Why Artist Maya Lin Planted 49 Towering Cedar Trees in the Middle of New York City," *ArtNet*, May 12, 2021.

19. James H. Wandersee and Elisabeth E. Schussler, "Preventing Plant Blindness," *American Biology Teacher* 61, no. 2 (February 1999): 82–86.

20. Interviewed by the author, June Lake, California, September 10, 2017.

21. I borrow this phrase from Renato Rosaldo.

22. A coinage of Glenn Albrecht.

23. I saw this film, and other family mementos I quote from, thanks to Huia Hamon, granddaughter of Rei, who died in 2008.

24. "Disabling Accident Led to New Career as Landscape Artist," *Church News*, August 28, 1993, 5. Thanks to curator Laura Hurtado, I saw the untitled drawing in the basement storage room of the Church History Museum, Salt Lake City, Utah.

25. Don Enders, "The Sacred Grove," February 20, 2019, at history.churchofjesus christ.org.

26. Interviewed by the author, Auckland, February 20, 2018.

27. Alan Cooper et al., "A Global Environmental Crisis 42,000 Years Ago," *Science* 371, no. 6531 (February 19, 2021): 811–818.

28. The design team, Azimuth Land Craft, was composed of Erik Jensen and Rebecca Sunter: future.ncpc.gov.

29. Thomas Wright, ed., *The Historical Works of Giraldus Cambrensis* (London, 1863), 413.

30. J. F. Black to F. G. Turpin, June 6, 1978, summarizing a presentation from July 1977, available at climatefiles.com.

31. "Forestry Frozen in Time," *Maclean's*, September 8, 1986.

32. "Unearthing a Frozen Forest," *Time*, June 24, 2001.

33. "Arctic Fossil Forest Sparks U.S.–Canada Research War," *Nunatsiaq News*, July 23, 1999.

34. "Scientist's Notebook: Working in the Arctic," *Science News for Students*, December 16, 2011.

EPILOGUE: PROMETHEUS

1. National Park Service, Region Four Office, *Results of Field Investigations for Proposed National Park in the Snake Range of Eastern Nevada* (San Francisco, 1959).

2. The following details come from newspapers published in Ely and Reno as well as two sets of primary sources. A smaller set, collected by Wes Ferguson, exists in scattered form within the internal archive of the Laboratory of Tree-Ring Research at the University of Arizona. A larger set (including contemporaneous letters, memoranda, news clippings, manuscript drafts, and interview notes) was assembled by Darwin Lambert, who donated them to Great Basin National Park. Because neither collection is professionally archived, nor open to the general public, I cannot provide box and folder numbers.

3. The district ranger was Donald E. Cox; his supervisor was Wilford L. "Slim" Hansen.

4. His given name was Milan Joseph Drakulich. The identity of other employee(s), the one(s) who did the deed, was not recorded. Cox may have helped. Currey's research assistant, Jeffrey Ward, avoided opprobrium. The chief naturalist was Keith A. Trexler.

5. *Ecology* 46, no. 4 (July 1965): 564–566.

6. "Tree Found That's 4,900 Years Old," *Charlotte Observer*, August 12, 1965.

7. US Senate, 89th Cong., 2nd. sess., *Hearings Before a Subcommittee of the Committee on Appropriations on H.R. 14215*, pt. 2 (1966), 1157–1165. The chief was Edward P. Cliff.

8. The regional forester, Floyd Iverson, named Robert A. Rowen as new supervisor of Humboldt National Forest, replacing Slim Hansen, who retired in December 1965.

9. Currey to Assistant Regional Forester John Mattoon, March 21, 1966.

10. The author of this letter was Frederick W. Grover, Director of the Division of Land Classification.

11. The agent was Scott Meredith.

12. Darwin Lambert, "Martyr for a Species," *Audubon* 70 (May–June 1968): 50–55.

13. C. W. Ferguson, *Dendrochronology of Bristlecone Pine in East-Central Nevada* (Tucson, 1970); Valmore C. LaMarche Jr., "Environment in Relation to Age of Bristlecone Pines," *Ecology* 50, no. 1 (January 1969): 53–59.

14. Donald R. Currey, "Neoglaciation in the Mountains of the Southwestern United States" (PhD diss., Univ. of Kansas, 1969).

15. David Muench and Darwin Lambert, *Timberline Ancients* (Portland, OR, 1972); Galen Rowell, "The Rings of Life," *Sierra Club Bulletin* 59 (September 1974): 5–7, 36–37.

16. "Legend of Killing World's Oldest Living Thing Won't Die," *Reno Gazette-Journal*, October 6, 1985.

17. "High in California's White Mountains Grows the Oldest Living Creature Ever Found," *San Francisco Chronicle*, August 23, 1998; *Methuselah Tree* (dir. Ian Duncan), *Nova*, season 28, episode 11, Public Broadcasting Service, December 11, 2001.

18. The botanist was Howard J. Arnott; the tree-ring scientists were Matt Salzer and Chris Baisan.

19. Darwin Lambert, *Earth Sweet Earth: My Life Inside Nature* (Spokane, 2014), 350.

TAXONOMIC INDEX

Note: Plants have shifting names in multiple languages. Common names can lead us astray: most "cedars" fall outside *Cedrus*; certain "pines" are far removed from *Pinus*. Even binomial nomenclature can be unstable. "Species" itself is a contested concept, particularly since the rise of genomics. Scientific names are useful for taxonomy and for communicating science—and have been written into environmental law—but they should never be taken as perfect descriptions of fixed realities.

MAIN INDEX

Note: Individual trees can be found under species headings as well as the main heading "named woody plants." Entries in boldface appear in the taxonomic index.

acid rain, 128, 306
Adanson, Michel, 59–60, 62, 73
afforestation, 42, 71, 78, 171, 324; tree-planting initiatives and programs, 12, 63, 79, 158, 324
Africa, 6, 58–64, 73, 118, 158, 322; East, 59, 61–62, 63–64, 97–100; North, 66, 98; Southern, 62–63, 294; Sub-Saharan, 59; West, 58, 60–61, 98, 101. *See also* Canary Islands; Cape Verde; slavery
age-dating, 10, 60, 84, 88, 351. *See also* dendrochronology; radiocarbon dating
agroforestry, 12, 274–277, 279
ahuehuete: El Árbol de la Noche Triste, 105–107, 129, 133; El Árbol del Tule (El Tule), 97, 108–111, 129–132, 243; El Sargento (El Árbol de Moctezuma; El Centinela), 107, 128, 134
Albert, Federico, 147–148
alerce: La Silla del Presidente, 147
Allen, Craig, 311–313
Alps, 45, 274, 338

Amazonia, 144, 267, 273, 277–280, 313. *See also* Brazil
American Association for the Advancement of Science (AAAS), 189, 190, 282
American Museum of Natural History, 195–197, 202, 206, 226
American Revolution, 138, 204, 205, 229
Ancient Yew Group, 90–93, 94
Anglicanism (Church of England), 87, 88, 90, 93, 127; burial practices, 82–83; decline of, 65, 83; ownership of churchyards, 92. *See also* Christianity
Anthropocene epoch, 24–26, 322, 331
antiquities, trees as, 54, 101, 201, 293, 309, 335; and the ancient past, 3, 5, 73, 112, 176; Augustin Pyramus de Candolle and, 73, 103; and biblical chronologies, 6, 35; giant sequoia, 180, 183, 188; remarkable trees, 7, 14, 116, 117; thousand-year mark for, 91, 113; yew, 72, 76, 77, 91.

Credit: Annette Hornischer, courtesy American Academy in Berlin

Jared Farmer is the Walter H. Annenberg Professor of History at the University of Pennsylvania. A former Andrew Carnegie Fellow, he is the author of several books, including *On Zion's Mount: Mormons, Indians, and the American Landscape*, which won the Francis Parkman Prize. He lives in Philadelphia.